D1674227

Kenneth A. Jackson

Kinetic Processes

Related Titles

Capper, P., Rudolph, P. (eds.)

Crystal Growth Technology

Semiconductors and Dielectrics

2010

ISBN: 978-3-527-32593-1

Dubois, J.-M., Belin-Ferré, E. (eds.)

Complex Metallic Alloys

Fundamentals and Applications

2010

ISBN: 978-3-527-32523-8

Kainer, K. U. (ed.)

Magnesium

8th International Conference on Magnesium Alloys and their Applications

2010

ISBN: 978-3-527-32732-4

Kumar, C. S. S. R. (ed.)

Mixed Metal Nanomaterials

2009

ISBN: 978-3-527-32153-7

Kumar, C. S. S. R. (ed.)

Nanostructured Oxides

2009

ISBN: 978-3-527-32152-0

Kumar, C. S. S. R. (ed.)

Metallic Nanomaterials

2009

ISBN: 978-3-527-32151-3

Levitin, V., Loskutov, S.

Strained Metallic Surfaces

Theory, Nanostructuring and Fatigue Strength

2009

ISBN: 978-3-527-32344-9

Herlach, D. M. (ed.)

Phase Transformations in Multicomponent Melts

2009

ISBN: 978-3-527-31994-7

Scheel, H. J., Capper, P. (eds.)

Crystal Growth Technology

From Fundamentals and Simulation to Large-scale Production

2008

ISBN: 978-3-527-31762-2

Köhler, M., Fritzsche, W.

Nanotechnology

An Introduction to Nanostructuring Techniques

2007

ISBN: 978-3-527-31871-1

Pfeiler, W. (ed.)

Alloy Physics

A Comprehensive Reference

2007

ISBN: 978-3-527-31321-1

Kenneth A. Jackson

Kinetic Processes

Crystal Growth, Diffusion, and Phase Transitions in Materials

Second, Completely Revised and Enlarged Edition

WILEY-VCH Verlag GmbH & Co. KGaA

The Authors

Prof. Kenneth A. Jackson
600 WindSpirit Circle
Prescott AZ 86303
USA

Library of Congress Card No.: applied for

British Library Cataloguing-in-Publication Data
A catalogue record for this book is available from the British Library.

Bibliographic information published by the Deutsche Nationalbibliothek
The Deutsche Nationalbibliothek lists this publication in the Deutsche Nationalbibliografie; detailed bibliographic data are available on the Internet at http://dnb.d-nb.de.

Typesetting Thomson Digital, Noida, India
Printing and Binding betz-druck GmbH, Darmstadt
Cover Design Grafik-Design Schulz, Fußgönheim

Printed in the Federal Republic of Germany
Printed on acid-free paper

ISBN: 978-3-527-32736-2

Contents

Kinetic Processes: Crystal Growth, Diffusion, and Phase Transitions in Materials. Kenneth A. Jackson

ISBN: 978-3-527-32736-2

Preface to the First Edition

This book is based on a course on Kinetic Processes which I taught for several years in the Materials Science Department at the University of Arizona. This is a required course for first year graduate students, although some of the material contained in the book would be suitable for a lower level course. The course initially derived from a series of lectures which I gave at Bell Labs, augmented by course notes from a similarly titled course at MIT. The content of the book has a highly personal flavor, emphasizing those areas to which I have made scientific contributions. I have concentrated on developing an understanding of kinetic processes, especially those involved in crystal growth, which is, perhaps, the simplest form of a first order transformation. The book assumes a basic understanding of thermodynamics, which underlies all kinetic processes and can be used to predict transformation kinetics for simple cases. The understanding of the complexities of crystal growth has developed significantly over the past several decades, but it is a wonderfully complex process, with still much to be learned. I have tried to present a coherent account of these processes, based on my view of the subject, which is available at present only in a dispersed form in the published literature, but it has not been assembled and coordinated as I have attempted to do here.

The book concentrates on atomic level processes and on how these processes translate into the microscopic and macroscopic descriptions of kinetic processes. It is aimed at a level appropriate for practitioners of materials processing. I have kept the mathematics at the minimum level necessary to expose the underlying physics. Many of my mathematically inclined friends will cringe at the simplified treatments which I present, but nevertheless, I suspect that non-mathematically inclined students will struggle with them. My colleagues in the crystal growth community, on the other hand, will cringe at my over-simplified descriptions of how single crystals and thin films, the basic materials for high-tech devices, are produced.

There are two streams of context in this book. One concentrates on basic kinetic processes, and the other on modern applications, where these kinetic processes are of critical importance. These two streams are interleaved. The book starts with an introduction to the basis of classical kinetics, the Boltzmann distribution. The following four chapters deal with diffusion processes in fluids, in amorphous

Kinetic Processes: Crystal Growth, Diffusion, and Phase Transitions in Materials. Kenneth A. Jackson

ISBN: 978-3-527-32736-2

materials, in simple crystals, and in semiconductors. This is followed by Chapter 6, on ion implantation, the important method for doping semiconductors, and includes a discussion of Rutherford backscattering. The next chapter introduces the diffusion equation, and some standard solutions. Chapter 8 deals with Stefan problems which are moving boundary problems encountered in phase transformations. Chapter 9 contains a general description of the kinetic processes involved in phase transformations, and is followed by Chapter 10 which contains a brief description of the methods used for growing single crystals. This is not intended to teach anyone how to grow crystals: there are individual books on several of these methods. Chapter 11 describes segregation at a moving interface, and is followed by Chapter 12, on the interface instabilities which can result from this segregation. These instabilities are described by non-linear equations which have been studied extensively, but are far beyond the scope of this book. Chapter 13 outlines some aspects of chemical kinetic theory, and is followed Chapter 14 on the formal aspects of phase transformations. Chapter 15 treats the initial formation of a new phase by a nucleation process. The next few chapters are on atomic processes at surfaces. Chapter 16 outlines adsorption, surface nucleation and epitaxial growth. This treatment only scratches the surface of the knowledge which has been accumulated by surface scientists. Chapter 17 discusses methods for the deposition of thin films, and Chapter 18 is on plasmas, which are used for both deposition and etching. Chapter 19 discusses rapid thermal processing, which is used to control and fine tune thermal annealing. The next few chapters return to fundamental considerations. Chapter 20 discusses the kinetics of first order phase transformations, and the following chapter discusses the important role of the surface roughening transition in these processes. The final chapters are on kinetic processes in alloys. Chapter 22 is on equilibrium in alloys and on growth processes in alloys near equilibrium. It is followed by a discussion, in Chapter 23, of phase separation, also known as spinodal decomposition. Chapter 24 is on rapid phase transformations, where kinetic processes modify the usual equilibrium segregation; where the rate of motion of the interface is comparable the rate of diffusive motion of the atoms. Chapter 25 contains a brief account of coarsening, sintering and grain growth, which applies not only to alloys. Again, much more is known about these processes than could be included here. Chapter 26 presents a discussion of dendritic growth, including a simple mathematical model. This growth mode is an extreme version of interfacial instabilities as discussed in Chapter 12, and has been the focus of extensive mathematical modeling, including the development of the phase field method. Chapter 27 discusses the formation of a two phase solid from a single phase liquid. The final Chapter 28, discusses an important aspect of the formation of the grain structure in metal castings. It is by no means an introduction to the computer models of segregation and fluid flow which are used to design castings today.

Most of what I know about this subject I have learned from my colleagues over the years. I would like to take this opportunity express the enjoyment I have experienced working with them, especially John Hunt and George Gilmer, without whose contributions this book would be a lot thinner. Colleagues at Bell Labs, including Harry Leamy, Kim Kimerling, John Weeks, Rudy Voorhoeve, Ho-Sou Chen, Bill

Pfann, Richard Wagner, Bob Batterman, Jim Patel, John Hegarty, Kurt Nassau, Chuck Kurkjian, Ben Greene, Ray Wolfe, Ken Benson, Dennis Maher, David Joy, Helen Farrell, George Peterson, Walter Brown, Charlie Miller, Reggie Farrow, . . ., (the list is endless) have all been involved in my education. I began my interest in this area under the tutelage of Bruce Chalmers, and I owe much to fellow students Bill Tiller, Dick Davis, Don Uhlmann, Jacques Hauser, Bob Fliescher, and Jim Livingston, as well as to discussions with leaders in the field, David Turnbull and Charles Frank. My other great source of inspiration has been my colleagues in the field of crystal growth, Bob Sekerka, Bob Laudise, Franz Rosenberger, John Wilkes, Alex Chernov, Dave Brandle, Vince Fratello, Joe Wenkus, Don Hurle, Brian Mullen, . . . (another endless list). I would also like to acknowledge the direct and indirect contributions of my graduate students at the University of Arizona, Kirk Beatty, Don Hilliard, Katherine Gudgel, Mollie Minke, and Dan Bentz.

I would like to express my appreciation of the extensive and important contributions in the areas of crystal growth, phase transformations and materials processing, including some of my efforts, which have been made possible by the sponsorship of NASA.

I would like to thank Russell Linney for a critical reading of the manuscript, and Harry Sarkas, Franz Rosenberger, and Joe Simmons, as well as many of those mentioned above, for encouragement on this project.

This book would not have been possible without the continued support and understanding of a very special person, Gina Kritchevsky.

Prescott, Arizona
June 26, 2004

Kenneth A. Jackson

Preface to the Second Edition

Over the past few years I have been developing a new method for describing diffusion during phase separation and ordering. This is introduced in Chapter 23, but it fits in with the basic fabric of this book, which focuses on how atom motions determine the kinetic properties of materials. In this new method, compositions are defined only on lattice sites, and interactions between neighboring sites are introduced. These interactions are similar to those used in Monte Carlo simulations. The free energy of a site is determined from its composition and the compositions of the neighboring sites. The chemical potentials for each species are calculated from the free energy, and these give the activities, which feed into the diffusion equations. The resulting diffusion equations are no more difficult to deal with than standard diffusion equations in a numerical scheme, but they can describe the processes of phase separation and ordering.

The usual treatment of spinodal decomposition includes a "gradient energy" term. This term was introduced to limit the rapid growth of very short wavelength perturbations. When the analysis is done on a lattice, the shortest possible wavelength is automatically limited. The gradient energy term is replaced with a second difference term that derives naturally from the presence of the lattice. Concentrations are no longer defined on volume elements that are small compared to the atomic volume.

This new method provides a significantly improved overall picture of phase separation. The diffusion equations describe the motion of atoms when they cluster or order, and indicate that these motions are simple and to be expected. I believe that this scheme has excellent pedagogical potential, because of its conceptual simplicity, and because students can now study the details of various behavior such as precipitation, coarsening, ordering, the propagation of instabilities, and so on, as readily as they can model the behavior of compositional changes due to standard diffusion. I am pleased to introduce this method into the new edition of my book.

In displaying and analyzing the results of this new scheme, I made extensive use of the ImageJ software made available by W. S. Rasband of the U.S. National Institutes of Health, Bethesda, Maryland, USA http://rsb.info.nih.gov/ij/.

Kinetic Processes: Crystal Growth, Diffusion, and Phase Transitions in Materials. Kenneth A. Jackson

ISBN: 978-3-527-32736-2

There have also been significant changes to Chapter 18 on plasmas, due to suggestions made by Don Hilliard.

I am still grateful to all of my former colleagues at Bell Labs, from whom I learned so much. Franz Rosenberger and Joe Simmons have encouraged me and supported my efforts to revise the story about phase separation. Above all I wish to express my thanks and appreciation to my wife, Gina Kritchevsky, who has encouraged me and endured hours and days with me at my computer while I explored details of this new method. It has been a very exciting time for me, and she has shared my enthusiasm.

Prescott, Arizona
November 9, 2009

Kenneth A. Jackson

1
Introduction

The aim of this book is to provide an understanding of the basic processes, at the atomic or molecular level, which are responsible for kinetic processes at the microscopic and macroscopic levels.

Many of the rate processes dealt with in this book are classical rate processes described by Boltzmann statistics. That is, the rate at which a process occurs is given by an expression of the form:

$$R = R_0 \exp(-Q/kT) \tag{1.1}$$

The exponential term is known as the Boltzmann factor, k is Boltzmann's constant and Q is called the activation energy. The Boltzmann factor gives the fraction of atoms or molecules in the system which have an energy greater than Q at the temperature T. So the rate at which the process occurs depends on a prefactor, R_0, which depends on geometric details of the path, the atom density, and so on, times the number of atoms which have enough energy to traverse the path.

1.1
Arrhenius Plot

Taking the logarithm of both sides of Equation 1.1:

$$\ln R = \ln R_0 - Q/kT \tag{1.2}$$

So that plotting $\ln R$ vs $1/T$ gives a straight line with slope $-Q/k$, as illustrated in Figure 1.1. This kind of plot is known as an Arrhenius plot.

If the rate process has a single activation energy, Q, over the range of the measurement, this suggests strongly that the mechanism controlling the rate is the same over that range. If the slope changes, or if the curve is discontinuous, the mechanism controlling the rate has changed.

Kinetic Processes: Crystal Growth, Diffusion, and Phase Transitions in Materials. Kenneth A. Jackson

ISBN: 978-3-527-32736-2

Figure 1.1 Arrhenius Plot.

1.2 The Relationship between Kinetics and Thermodynamics

There is a simple relationship between the rate equation above and thermodynamics, which can be written:

$$G = -kT \ln R = Q - kT \ln R_0 \tag{1.3}$$

This implies that the rate at which atoms leave a state depends on their properties in that state. Comparing Equation 1.3 with $G = H - TS$, it is evident that Q is related to the enthalpy, and the entropy, S, is $k 1n R_0$. If two states or phases are in equilibrium, their free energies are equal, which is equivalent to the statement that rates of transition back and forth between the two states are the same.

The relationship between thermodynamics and kinetics will be a recurring theme in this book. The thermodynamics formalism was developed during the last century, based on the understanding of steam engines. If it were being developed today by materials scientists, it would be done in terms of rate equations, which are formally equivalent, but much more amenable to physical interpretation.

The origin of the Boltzmann factor will be outlined below, but first, we will attempt to answer the question:

1.2.1 What is Temperature?

We all know what temperature is: it is something that we measure with a thermometer. The temperature scales which we use are defined based on fixed temperatures, such as the melting point of ice and the boiling point of water. We use the thermal expansion of some material to interpolate between these fixed points. But what is the physical meaning of temperature? What is being measured with this empirical system?

The simplest thermometer to understand is based on an ideal gas, where we can relate the temperature to the pressure of a gas in a container of fixed volume. The pressure on a wall or a piston derives from the force exerted on it by atoms or molecules striking it. The pressure is due to the change of momentum of the atoms or molecules which hit the piston. The force, F, on the piston is the change in momentum per second of the atoms or molecules striking it.

So we can write:

$$F = (\text{change in momentum per molecule}) \times (\text{number of molecules per second striking the piston})$$

If v_x is the component of the velocity of a molecule in the x direction, then the change in momentum when the molecule makes an elastic collision (an elastic collision is one in which the molecule does not lose any energy) with the piston is $2mv_x$, since the momentum of the incident molecule is reversed during the collision.

As illustrated in Figure 1.2, only the molecules within a distance $v_x t$ of the wall will strike it during time t, so the number of molecules hitting the wall per second is given by the number of molecules per unit volume, n, times the area of the piston, A, times $v_x/2$, since half of the molecules are going the other way. The force on the piston is thus:

$$F = (2mv_x)\left(\frac{1}{2}nv_x A\right) \tag{1.4}$$

And the pressure on the piston is:

$$P = \frac{F}{A} = nmv_x^2 \tag{1.5}$$

Writing $n = N/V$, and using the ideal gas law, we can write:

$$PV = Nmv_x^2 = NkT \tag{1.6}$$

This indicates that the thermal energy of an atom, kT, is just the kinetic energy of the atom. A more refined analysis relating the pressure to the motion of atoms in three dimensions gives:

$$kT = \frac{1}{3}mv_{\text{rms}}^2 = \frac{1}{2}m\bar{v}^2 \tag{1.7}$$

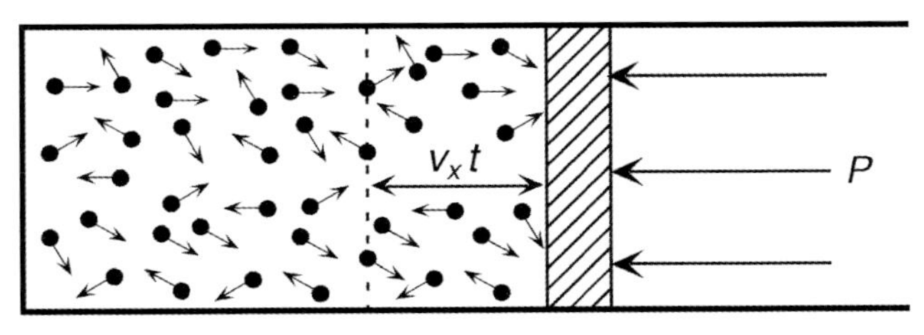

Figure 1.2 The change in momentum of the molecules striking a piston creates pressure on the piston.

where v_{rms} is the root mean square average velocity of the atoms, and $\bar{v}$ is the most probable velocity.

A familiar form of the ideal gas law uses the gas constant R, rather than Boltzmann's constant, k. The two are related by $R = N_0 k$, where N_0 is Avogadro's number, which is the number of molecules in a mol. RT is the average thermal energy of a mol of atoms, kT is the average thermal energy of one atom.

$$\begin{aligned}\text{Gas constant:}\quad R &= 1.98\ \text{cal mol}^{-1}\ \text{K}^{-1}\\ &= 0.00198\ \text{kcal mol}^{-1}\ \text{K}^{-1}\\ &= 8.31\ \text{J mol}^{-1}\ \text{K}^{-1}\\ &= 8.31 \times 10^{7}\ \text{erg mol}^{-1}\ \text{K}^{-1}\end{aligned}$$

$$\begin{aligned}\text{Boltzmann's constant:}\quad k &= 8.621 \times 10^{-5}\ \text{eV atom}^{-1}\ \text{K}^{-1}\\ &= 1.38 \times 10^{-16}\ \text{erg atom}^{-1}\ \text{K}^{-1}\\ &= 1.38 \times 10^{-23}\ \text{J atom}^{-1}\ \text{K}^{-1}\end{aligned}$$

$$1\ \text{eV atom}^{-1} = 23\ \text{kcal mol}^{-1}$$

$$\frac{R}{k} = \frac{8.31 \times 10^{7}}{1.38 \times 10^{-16}} = 6.023 \times 10^{23}\ \text{molecules mol}^{-1} = \text{Avogadro's Number}$$

The important point from the above discussion is that the temperature is a measure of the average *kinetic* energy of the molecules, not the total energy, which includes potential energy and rotational energy, just the kinetic energy. Temperature is a measure of how fast the atoms or molecules are moving, on average.

1.3 The Boltzmann Distribution

Where does $N = N_0 \exp(-E/kT)$come from?

We saw above that kT is a measure of the average kinetic energy of the atoms or molecules in an assembly. The Boltzmann function gives the fraction of atoms which have enough energy to surmount an energy barrier of height E, in an assembly of atoms with an average energy kT.

Let us look at an atom in a potential field. Gravity is a simple one.

The force of gravity on an atom is:

$$F = mg \tag{1.8}$$

The associated gravitational potential is:

$$\int_0^h F dx = mgh \tag{1.9}$$

which is the energy required to lift an atom from height 0 to height h in gravity. mgh is the potential energy of an atom at height h.

For N atoms in a volume V at temperature T, the ideal gas law states:

$$PV = NRT \tag{1.10}$$

or

$$P = nkT \tag{1.11}$$

where $n = N/V$, the number of atoms per unit volume.

The change in pressure, $\mathrm{d}P$, which will result from a change in atom density, $\mathrm{d}n$, is

$$\mathrm{d}P = kT\mathrm{d}n \tag{1.12}$$

In a gravitational field, the number of atoms per unit volume decreases with height, and $n\mathrm{d}h$ is the number of atoms between heights h and $h + \mathrm{d}h$.

The change in pressure between h and $h + \mathrm{d}h$ due to the weight of the atoms in $\mathrm{d}h$ is:

$$\mathrm{d}P = P_{h+\mathrm{d}h} - P_h = -mgn\mathrm{d}h \tag{1.13}$$

Combining Equations 1.12 and 1.13 gives:

$$\frac{\mathrm{d}n}{n} = -\frac{mg}{kT}\mathrm{d}h \tag{1.14}$$

or:

$$n = n_0 \exp\left(-\frac{mgh}{kT}\right) \tag{1.15}$$

which is the variation of atom density with height in the atmosphere due to the earth's gravitational field.

mgh is the potential energy of an atom at height h. This analysis is similar for a generalized force field, and the distribution of the atoms in the force field has the same form:

$$n = n_0 \exp\left(-\frac{E_\mathrm{p}}{kT}\right) \tag{1.16}$$

where E_p is the potential energy of the atoms in the force field.

The kinetic energy distribution of the atoms can be derived by examining the kinetic energy that an atom at height zero needs to reach a height h (ignoring scattering).

In order to reach a height h, an atom must have an upwards kinetic energy equal to, or greater than, the potential energy at height h. It must have an upwards velocity greater than u, given by:

$$\frac{1}{2}mu^2 \geq mgh \tag{1.17}$$

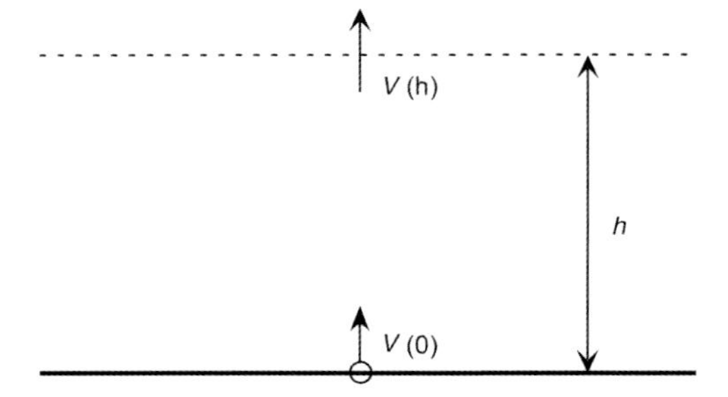

Figure 1.3 An atom with enough kinetic energy at height 0 can reach height h.

With a velocity less than u, the atom will not get to height h. In Figure 1.3, the number of atoms passing through the plane at h is the number moving upwards at height zero with velocity greater than u, so

$$n_{v>o}(h) = n_{v>u}(0) \tag{1.18}$$

Assuming that the temperature is the same at zero and at h, the distribution of velocities will be the same, and the only difference will be the density of atoms, so:

$$\frac{n_{v>u}(0)}{n_{v>0}(0)} = \exp\left(-\frac{mgh}{kT}\right) = \exp\left(-\frac{mu^2}{2kT}s\right) \tag{1.19}$$

or

$$n_{v>u} = n_0 \exp\left(-\frac{E_k}{kT}\right) \tag{1.20}$$

where E_k is the kinetic energy of the atom. This is the Boltzmann distribution of velocities. It says that the fraction of atoms with velocity (or kinetic energy) greater than some value is given by a Boltzmann factor with the average kinetic energy of the atoms in the denominator of the exponent.

Recalling that kT is the average kinetic energy of the atoms, the exponent is just the ratio of two kinetic energies. The Boltzmann factor describes the spread in the distribution of energies of the atoms. Given an average kinetic energy kT per atom, the Boltzmann factor tells how many of the atoms have an energy greater than a specific value.

In general, if there is a potential energy barrier of height Q, the Boltzmann factor says that a fraction of the atoms given by $\exp(-Q/kT)$ will have enough kinetic energy to get over the barrier. This is independent of how the potential varies along the path. The atom just has to have enough kinetic energy at the start to surpass the barrier. So, in general, if there is a potential barrier along a path, the atoms which are going in the right direction take a run at it, and those atoms which are going fast enough will make it over the barrier. And the fraction of the atoms which will make it over the barrier is given by the Boltzmann factor.

The original derivation of Boltzmann was concerned with atomic collisions between gas atoms, and the distribution of velocities and the spread in energy which the collisions produce. That is a very complex problem, but the result is remarkably simple. We have derived it crudely from Newton's laws and our practical definition of temperature which is based on the ideal gas law.

1.4 Kinetic Theory of Gases

From statistics based on Newton's laws, Boltzmann derived that the probability $P(v)$ that an atom of mass m will have a velocity v at a temperature T is given by:

$$P(v) = 4\pi\left(\frac{m}{2\pi kT}\right)^{\frac{3}{2}} v^2 \exp\left(-\frac{mv^2}{2kT}\right) \tag{1.21}$$

where k is Boltzmann's constant. This distribution is illustrated in Figure 1.4.

The average velocity is given by:

$$\bar{v} = \int_0^\infty vP(v)\,\mathrm{d}v = \sqrt{\frac{8kT}{\pi m}} \tag{1.22}$$

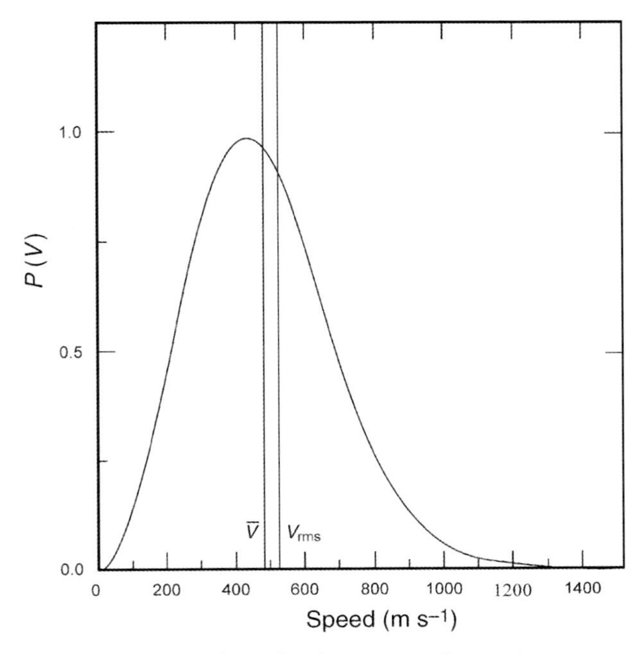

Figure 1.4 Typical velocity distribution, $P(v)$, of atoms in a gas at room temperature.

On average, the velocity of an atom in any one direction is $\overline{v_x} = \overline{v_y} = \overline{v_z} = \sqrt{\frac{2\,kT}{\pi m}}$. The root mean square velocity is slightly larger than the average velocity, and is given by $v_{\text{rms}} = \sqrt{\frac{3\,kT}{m}}$. kT is a measure of the average *kinetic* energy of the atoms in the gas. This is also true in liquids and solids. At room temperature, $\bar{v}_x$ is about 480 m s^{-1}, or about 1060 mph.

1.5 Collisions

Two atoms of diameter d will collide if their centers pass within d of each other. We can imagine a cylinder with a diameter $2d$, twice the diameter of an atom. The length of the cylinder, l, is such that the volume of the cylinder is the average volume per atom in the gas, $1/n$, where n is the number of atoms per unit volume in the gas. We then have: $l\pi d^2 = 1/n$. When an atom traverses a distance l, it is likely to collide with another atom. The average distance between collisions is thus given approximately by l, which is equal to $1/\pi d^2 n$. A rigorous treatment for the average distance between collisions in a gas, which is known as the mean free path, λ, gives:

$$\lambda = \frac{1}{\sqrt{2}\pi d^2 n} \tag{1.23}$$

which is smaller by $\sqrt{2}$ than our crude estimate. For an ideal gas, $n = P/kT$, where P is the pressure, so that the mean free path can be written:

$$\lambda = \frac{kT}{\sqrt{2}\pi d^2 P} \tag{1.24}$$

The diffusion coefficient (which will discussed in more detail in the next chapter) is given by:

$$D = \frac{\bar{v}\lambda}{3} \tag{1.25}$$

The viscosity of a gas is given by:

$$\eta = \frac{mn\bar{v}\lambda}{3} \tag{1.26}$$

A surprising result is obtained by inserting the value of the mean free path Equation 1.23, into Equation 1.26 for the viscosity: the viscosity is independent of the pressure on the gas at a given temperature.

The thermal conductivity of a gas is given by the specific heat times the diffusion coefficient:

$$K = \frac{C_V\bar{v}\lambda}{3} \tag{1.27}$$

where C_V is the specific heat of the gas at constant volume. The flux of atoms through unit area in unit time is given by:

$$J = \frac{n\overline{v_x}}{2} = n\sqrt{\frac{kT}{2\pi m}} = \frac{P}{\sqrt{2\pi mkT}} \tag{1.28}$$

which is an expression we will use later for the flux of gas atoms to a surface. For one atmosphere pressure of nitrogen gas at room temperature, the flux is about 6×10^{23} atoms $cm^{-2} s^{-1}$.

Further Reading

The definition of temperature and the derivation of the Boltzmann function are taken from:

Feynman, R.P., Leighton, R.B., and Sands, M. (1963) Ch. 40, in *The Feynman Lectures on Physics*, vol. **1**, Addison-Wesley Publ. Co., Reading MA.

Huang, K. (1963) *Statistical Mechanics*, John Wiley & Sons, Inc., New York, NY.

Lupis, C.H.P. (1983) *Chemical Thermodynamics of Materials*, North-Holland, New York, NY.

Problems

1.1. The specific heat of a monatomic gas is 3 Nk. What are the thermal conductivities of helium and argon at room temperature?

1.2. How many atoms of argon at a pressure of one atmosphere are incident on a square centimeter of surface at room temperature in one microsecond?

2
Diffusion in Fluids

2.1
Diffusion in a Gas

Diffusion describes the motion of atoms, so let us examine the net flux of atoms, J, across a plane in space.

In Figure 2.1, the density of gas atoms to the left of the plane is n^-, and the density to the right is n^+. The average velocity of the atoms crossing a plane perpendicular to the x-axis is $\bar{v}_x$. Denoting the average distance which an atom travels between collisions, known as the mean free path, by λ, and denoting the average time between collisions, called the mean free time, by τ, then the average velocity is given by:

$$\bar{v}_x = \lambda/\tau \tag{2.1}$$

The number of atoms crossing the plane from left to right in time Δt is one half of the number within a distance $\bar{v}_x\ \Delta t$ of the boundary, and similarly from right to left.

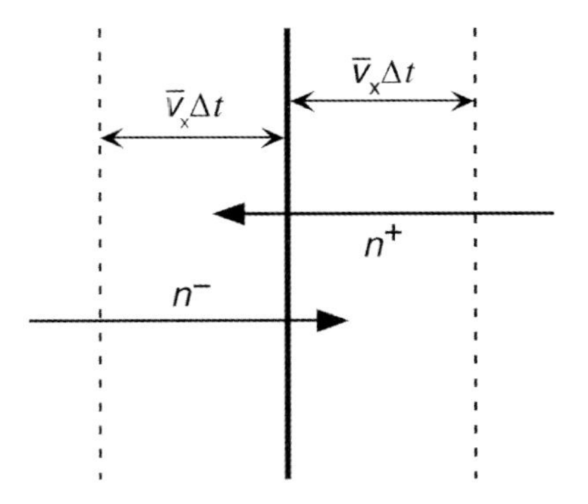

Figure 2.1 Flux of atoms across a plane.

Kinetic Processes: Crystal Growth, Diffusion, and Phase Transitions in Materials. Kenneth A. Jackson

ISBN: 978-3-527-32736-2

And so the net flux is:

$$J_x = \frac{\frac{1}{2}n^- \bar{v}_x \Delta t - \frac{1}{2}n^+ \bar{v}_x \Delta t}{\Delta t} = \left(\frac{n^- - n^+}{2}\right)\bar{v}_x \tag{2.2}$$

The densities of atoms on the two sides of the plane, n^- and n^+, should be taken as the densities within a mean free path of the plane, and the difference between these two densities can be written in terms of the gradient in the density of atoms:

$$n^+ - n^- = \frac{dn}{dx}\Delta x = \frac{dn}{dx}2\lambda \tag{2.3}$$

where the densities are taken to be within a distance λ of each side of the plane.

The net flux can then be written:

$$J = -\bar{v}_x \lambda \frac{dn}{dx} = -\frac{1}{3}\bar{v}\lambda \frac{dn}{dx} \tag{2.4}$$

Where $\bar{v}$ is the average velocity in three dimensions. Comparing this equation with Fick's first law, which is an empirical equation describing the diffusion flux in a concentration gradient in terms of the diffusion coefficient, D:

$$J = -D\frac{dn}{dx} \tag{2.5}$$

results in a diffusion coefficient given by:

$$D = \frac{1}{3}\lambda \bar{v} \tag{2.6}$$

as in Equation 1.25. The expression for the average velocity of the atoms in a gas, Equation 1.22 and for the mean free path, λ, Equation 1.23 can be substituted into this equation to express the diffusion coefficient in terms of the gas pressure, atom density, and so on. Equation 1.22 indicates that the average velocity of an atom is proportional to $\sqrt{1/m}$, so Equation 2.6 says that light atoms will diffuse much more rapidly than heavy ones. Since heat is carried by the diffusive motion of the atoms in a gas, the thermal conductivity of a gas is also much larger for gases which are composed of light atoms.

It is worth noting that gases can move very quickly by flow or convection, and this is usually much faster than diffusion.

2.2 Diffusion in Liquids

Diffusion in a liquid is similar to that in a gas, but different. The atoms are also randomly distributed, but they are much closer together. As a result, the collisions are much more complicated many-body interactions, rather than simple two-body collisions. These complex collisions are amenable to computer simulation, but not to mathematical analysis. The atoms or molecules in a liquid are surrounded by other atoms or molecules, which form a cage. An atom or molecule moves within its cage, at thermal velocities, and many of the properties of the liquid can be described by the atom or molecule bumping off its neighbors in its cage. This process is dominated by the repulsive part of the atomic potential. For diffusion, however, the cage must move. And this motion will depend on the differences in momentum and velocity between an atom and the other atoms or molecules with which it collides. The net motion produced by each collision, which is a diffusion step, Λ, depends on the spread of velocities. So the diffusion coefficient in the liquid has an Arrhenius dependence on temperature, whereas the mean free path is relatively independent of temperature.

If we estimate the distance in a liquid which an atom can move before it bumps into another atom, its mean free path, λ, to be about 3×10^{-11} m, which is about 1/10 of an atomic diameter, and assume that the atom is moving at the thermal velocity, $\bar{\nu} \approx 1000\ \mathrm{m\,s^{-1}}$, then the time between collisions, $\tau = \frac{\lambda}{\bar{\nu}} \approx 3 \times 10^{-14}$ s.

The atom position moves an average distance $\Lambda \approx 3 \times 10^{-11}$ m with each collision, so the diffusion coefficient, D, which is given by the square of this distance divided by the time between collisions:

$$D = \frac{1}{6}\frac{\Lambda^2}{\tau} = \frac{1}{6}\frac{\Lambda^2 \bar{\nu}}{\lambda} \approx 5 \times 10^{-9}\ \mathrm{m^2\,s^{-1}} \tag{2.7}$$

There is a factor of 1/6 because the diffusion step can be in any direction. This value of the diffusion coefficient is about the right value for many liquids. As pointed out above, the distance Λ depends on the spread in velocities of the atoms, and so it increases with temperature.

2.2.1 Diffusion Distances

The distance over which diffusion will occur in time t is given approximately by $\sqrt{Dt}$.

For a liquid diffusion coefficient of $10^{-8}\ \mathrm{m^2\,s^{-1}}$, diffusion will occur over a distance of 1 μm in about 200 μs. Diffusion over a distance of 10 cm will take approximately 2 million seconds, or 23 days. That is why you should stir your coffee, rather than wait for the sugar to diffuse throughout the cup.

Liquids can move rapidly by convection and flow.

2.2.2 Molecular Dynamics Simulations of Diffusion in Liquids

Modeling the properties of a liquid is much more complex than modeling the properties of a gas. In a gas, the atoms or molecules are separated in space. The collisions which occur are predominantly between two atoms or molecules. Other atoms or molecules in the gas have little effect on the collision, and so each collision can be treated separately, using Newton's laws, and then averages can be taken to obtain the velocity distributions on which the properties depend. This is a fairly complex mathematical problem, which was solved by Boltzmann.

In a liquid, the atoms or molecules are so close to each other that the potential field from one atom or molecule can be felt by many atoms in its vicinity. These many-body collisions are difficult to treat analytically, and this makes it very difficult to derive velocity distributions for liquid atoms. So the properties of liquids which depend on these velocity distributions cannot be determined analytically. However, they are amenable to computer simulation; they can be determined readily using molecular dynamics (MD) computer simulations. The development of our understanding of the properties of liquids has relied heavily on MD computer simulations.

Molecular dynamics computer simulations are very simple, in principle. The simulations start with a collection of atoms in the computer, each with a specific position and velocity. The subsequent position and velocity of each of the atoms is calculated in a series of small time steps, taking into account all the interactions with other atoms. The atoms do not move very far between the time steps.

The interactions between the atoms are calculated as follows. The atoms exert forces on each other, which are obtained from an assumed interatomic potential. The simplest of these is the Lennard-Jones potential [1]:

$$V = \frac{a}{r^{12}} - \frac{b}{r^{6}} \tag{2.8}$$

Here a and b are constants, and r is the distance between the two atoms. The first term is a repulsive term which increases the potential energy rapidly if the atoms get too close together. The second term is an attractive term which decreases the potential energy when the atoms approach each other from far away. The potential V has a minimum at some distance, r_0, which will be approximately the equilibrium separation between the atoms. The depth of the minimum is the energy gained when the atoms come together, that is, the binding energy between the atoms, E_0. The constants a and b can be readily expressed in terms of r_0 and E_0. The force between any two atoms is given by the derivative of the potential energy with respect to r. This gives the force between any two atoms at any given separation, r. The force is repulsive for $r < r_0$, and attractive for $r > r_0$. For the Lennard-Jones potential, the magnitude of the force depends on the constants a and b, or equivalently, on r_0 and E_0.

The initial position of each atom in an MD simulation is known, and so the total net force exerted on any one atom by its all its neighbors can be calculated from the positions. From the total net force, the acceleration of the atom is calculated using Newton's second law. This is then used to calculate the new velocity of the atom a time step later. The new position of the atom a time step later is calculated from its old positions and its velocity.

The MD simulation consists of doing this over and over for each of the atoms in the system. The total time of the simulation and the number of atoms in the simulation are limited by the size of the computer and the patience of the simulator, usually tens of thousands of atoms for times of less than a nanosecond.

So the simulation is very simple, in principle. It just depends on a having a force law for the interactions between the atoms, and then solving Newton's equation over and over. This type of simulation reproduces the behavior of materials very well. For example, the Lennard-Jones potential describes the behavior of argon very well, and also does a reasonable job for most metals. At high temperatures, the atoms behave like gas atoms. At lower temperatures, they condense to a liquid configuration in the simulations, and at a still lower temperature they crystallize to a face centered cubic structure. Many of the other properties, such as the specific heat, sound velocity, diffusion coefficient, and so on, of the liquid and solid are faithfully reproduced. This demonstrates that the behavior of the atoms in the real world depends on simple interactions between the atoms.

The simulations for silicon are done using the Stillinger–Weber potential [2], which is a three-body potential. That is, the interactions between the atoms depend on both the distance between the atoms and the angle between the neighboring atoms. With this potential the atoms crystallize in the diamond cubic structure at low temperatures. The Stillinger–Weber potential was devised to simulate the properties of silicon.

Figure 2.2 is a plot of the logarithm of the diffusion coefficient in liquid argon and in liquid silicon, plotted against T_M/T, the melting point divided by the temperature. These data were obtained from molecular dynamics simulations, using the Lennard-Jones potential for argon, and the Stillinger–Weber potential for silicon. The figure suggests that the liquid behavior is similar for both potentials. Indeed, the structure and properties of liquids depend much more on the repulsive part of the potential than the attractive part. The liquid atoms tend to jostle around with more or less the thermal velocity until they bump into one another.

The diffusion coefficients in Figure 2.2 can be described by the equation:

$$D = D_0 \exp(-Q/kT) \tag{2.9}$$

Where the temperature dependence derives from the temperature dependence of the mean displacement distance, Λ, in Equation 2.7. The slope of the line on the plot

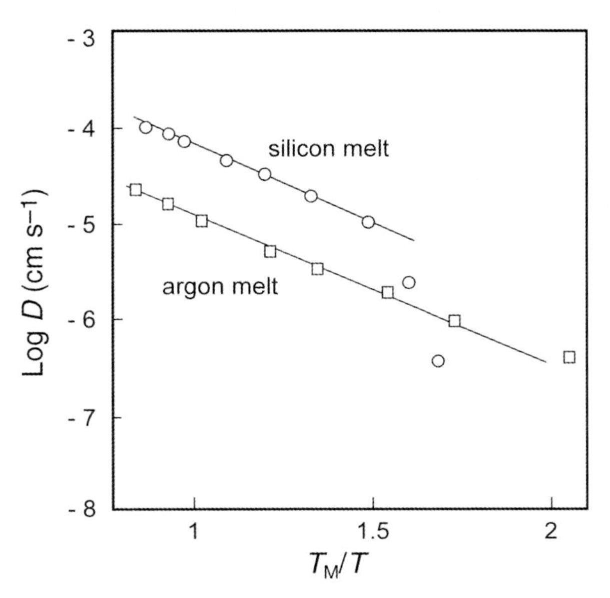

Figure 2.2 Diffusion coefficients from MD simulations for liquid argon using the Lennard-Jones potential, and for liquid silicon using the Stillinger–Weber potential.

gives the activation energy, Q. The two diffusion coefficients have the same slope when plotted against T_M/T, which indicates that the activation energy is proportional to the melting point. Both sets of data were taken in the vicinity of the melting points of the respective materials. The structures of the two liquids are similar and this result suggests that the spread in thermal velocities, which gives rise to the temperature dependence of Λ, depends similarly on temperature for the two materials.

In Equation 2.7 the diffusion coefficient is proportional to the average thermal velocity of the atoms, $\bar{v}$, which is proportional to $\sqrt{kT}$. The data in the figure superimpose if the diffusion coefficients are divided by $\sqrt{T_M}$.

These molecular dynamics data suggest that the simple relationships outlined in the previous section for the relationship between diffusion coefficients and atom velocities provide a reasonable approximation.

2.2.3
Measurement of Diffusion Coefficients in Liquids

Experimental measurements of the diffusion coefficient, D, are made by preparing a sample in an initial configuration for which a solution to the diffusion equation is known, permitting diffusion to occur for some time, determining the final composition distribution by some chemical analysis method, and then fitting the measured data with the formal solution, using D as the fitting parameter. Usually the measurements are made for a variety of times to ensure that a single value of D will fit all the data. Measurements are made for a variety of temperatures in order to obtain the temperature dependence of D.

A major problem with measuring the diffusion coefficient in liquids is convective mixing, which can be much faster than diffusion, especially over long distances. Measurements have been made in fine bore capillary tubes, where convection should be minimal.

The effects of convection can be minimized by keeping the more dense liquid at the bottom of the container.

A clever experimental set-up for measuring liquid diffusion coefficients is illustrated in Figure 2.3. The apparatus consists of a stack of discs which can be rotated

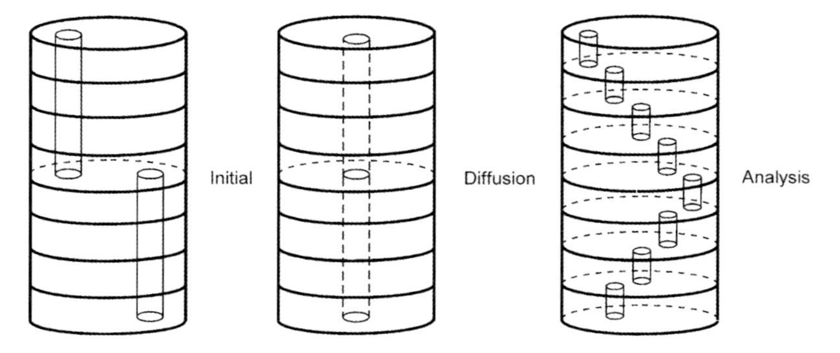

Figure 2.3 Apparatus for measuring liquid diffusion.

about a common axis.The discs have off-axis holes which can be aligned. The holes in the top half of the discs and the bottom half of the discs are aligned as shown in the initial configuration, and the starting material is inserted. The holes are then aligned to permit interdiffusion. The discs are then rotated as shown to provide separate samples for compositional analysis.

References

1 Lennard-Jones, J.E. (1924) *Proc. Roy. Soc.*, **A106**, 463.

2 Stillinger, F.H. (1985) Weber, T., *Phys. Rev. B*, **31**, 5262.

Further Reading

Kittel, C. (1980) Kroemer, H., *Thermal Physics*, W.H. Freeman, San Francisco, CA.

Chandler, D. (1987) *Introduction to Statistical Mechanics*, Oxford University Press, New York, NY.

Ragone, D.V. (1995) *Thermodynamics of Materials*, John Wiley & Sons, Inc., New York, NY.

Dzugutov, M. (1996) *Nature*, **381**, 137.

Hoyt, J.J., Asta, M., and Sadigh, B. (2000) *Phys. Rev. Lett.*, **85**, 594.

Problems

2.1. Write the equation for the Lennard-Jones potential in terms of r_0 and E_0.

2.2. Plot the Lennard-Jones potential for $r_0 = 3 \times 10^{-10}$ m and $E_0 = 2$ eV.

3
Diffusion in Amorphous Materials

3.1
Amorphous Materials

There are different classes of amorphous materials, and the diffusion process differs in each.

Amorphous materials which are formed by quenching from a liquid are called glasses. These are materials which can be readily cooled into a glassy state from the liquid because their molecular motion is so sluggish. They usually have a relatively high viscosity, and, on cooling, it is possible to reach the glass transition temperature before crystallization occurs. This is easier if the material is also difficult to crystallize. A complex liquid structure, which increases the viscosity of the liquid, or a low melting point, which delays crystallization, also assists in accessing the glassy state. Very rapid quenching is more likely to get a glass-forming material into a glassy state.

There are network glass-formers such as silica, and there are also molecular glass formers, such as Salol and glycerol. There are also metallic glasses, which form from liquid alloys on rapid cooling. These are alloys containing metals and metalloids. Some of these have a strongly depressed melting point because of the alloying, but they all have complex crystalline structures, so that crystallization is slow. It is believed that most materials can be prepared as glasses by sufficiently rapid quenching. However, there is a notable exception: no pure metal has been prepared in an amorphous state.

Amorphous materials can also be formed by deposition onto cold substrates, or by ion implantation damage.

Amorphous materials are usually not the lowest free energy configuration, but their random arrangement persists because the atoms do not have enough mobility to rearrange into a lower free energy crystalline array. For all practical purposes, these materials are stable solids below their glass transition temperatures.

Most solid polymers are glasses. Only very regular polymers, for example polyethylene, crystallize, and then the crystal size is usually extremely small. The glass transition temperature of a polymeric material is a very important parameter in many applications.

Kinetic Processes: Crystal Growth, Diffusion, and Phase Transitions in Materials. Kenneth A. Jackson

ISBN: 978-3-527-32736-2

3.2
Network Glass Formers

3.2.1
Silica

Silica (SiO_2) and silicates are the most common materials in the earth's crust. Silica is a network glass-former. In amorphous silica, each silicon atom is surrounded by four oxygen atoms. Each oxygen atom is between two silicon atoms. The arrangement of the atoms is not regular as it is in quartz, which is a crystalline form of silica. This arrangement satisfies the bonding requirements of the atoms involved, and the bonds are quite strong. These bonds do not rearrange readily. As a result, the crystallization rate of silica is so slow, see Figure 9.1, that even relatively slow cooling of the liquid will result in a glass. The glassy phase of silica has dangling bonds which promote diffusion.

Data for diffusion coefficients of various elements in silica are presented on an Arrhenius plot in Figure 3.1. Atoms which occupy and move through the open spaces between the silicon and oxygen atoms of the matrix are called interstitial atoms. They diffuse rapidly, and are at the top of the chart. Substitutional atoms replace the silicon or oxygen atoms in the network matrix of the silica. These diffuse much more slowly, and are near the bottom of the chart. The interstitial diffusers are mostly small atoms or ions, but many metals also diffuse interstitially in silica, which has a relatively open structure. Others elements disrupt the network structure of silica when they are introduced. They are known as network modifiers, and include OH or water, and alkali ions such as sodium. Their diffusion coefficients increase with concentration, because higher concentrations disrupt the lattice more. They also change the diffusion rates of other elements. For example, during the growth of a field oxide on silicon, which forms by the diffusion of oxygen through the oxide layer to combine with silicon, introducing water into the ambient significantly increases the diffusion rate of oxygen through the silica layer.

Notice that silicon and oxygen diffuse at quite different rates in silica.

3.2.2
Silicon and Germanium

Both silicon and germanium can exist in three states: liquid, crystal, and amorphous. The liquid is metallic, and each atom has about nine nearest neighbors. The crystal has the diamond cubic structure, a rather open structure in which each atom has four nearest neighbors. The liquid is denser than the crystal. The amorphous phase is a randomized version of the crystal. It forms readily by vapor deposition onto any substrate at a temperature below about 400 °C. On heating, the amorphous phase will crystallize.

In the amorphous network structure, a single dangling bond, which is illustrated in Figure 3.2, can exist as a defect. These cannot exist in crystalline silicon or germanium because removing a single atom creates a vacancy and four dangling

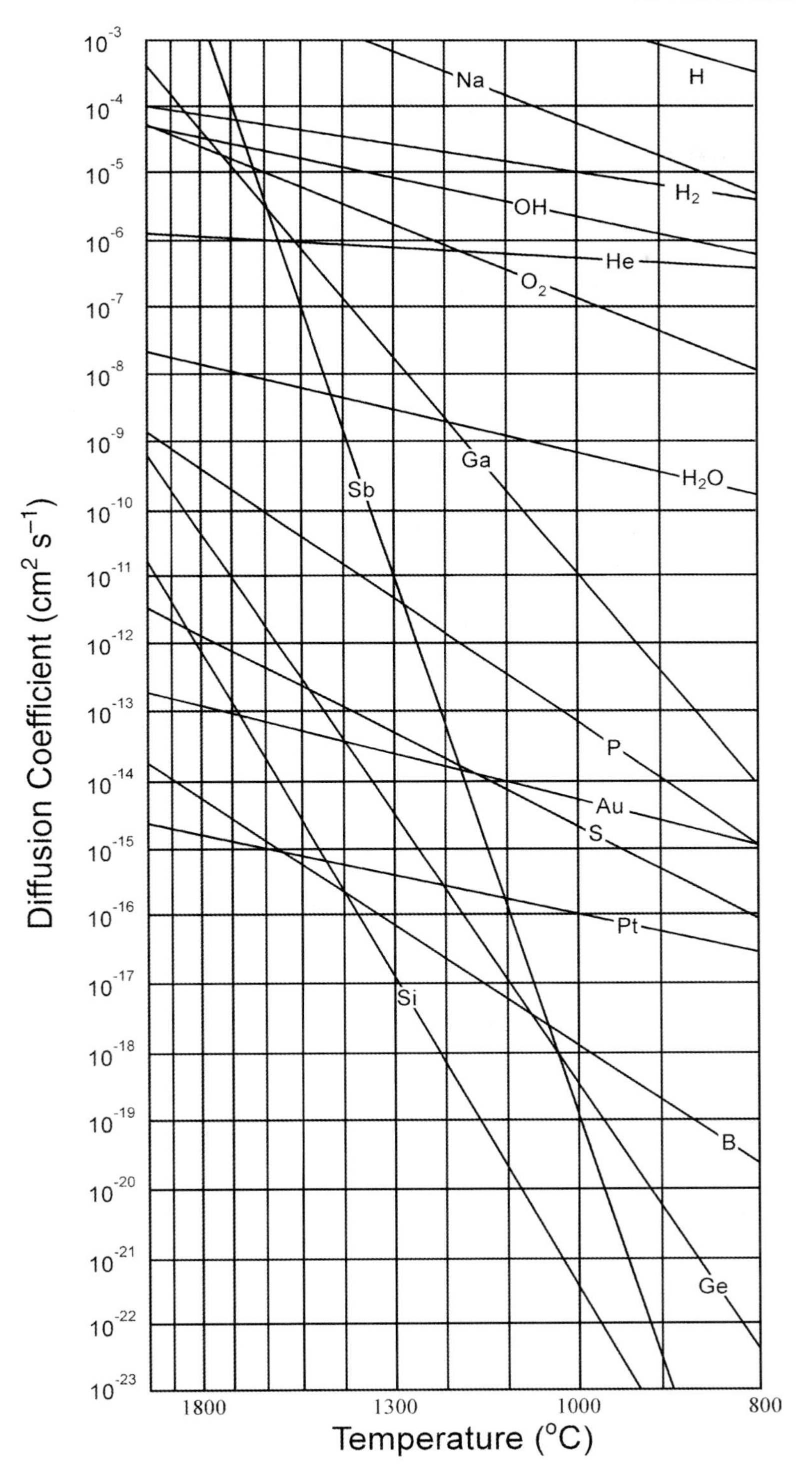

Figure 3.1 Diffusion coefficients in silica.

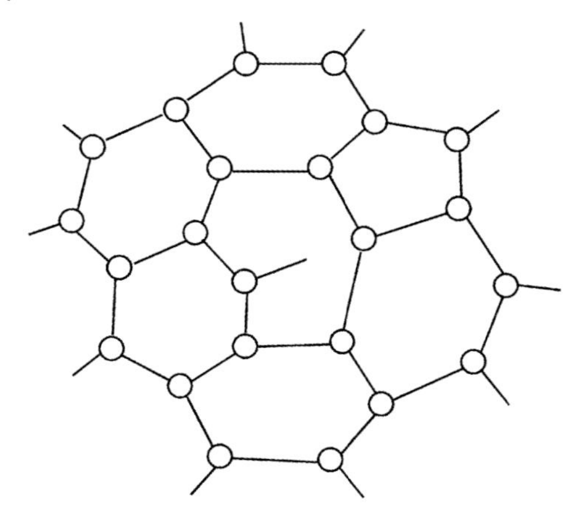

Figure 3.2 Dangling bond in a network amorphous structure.

bonds. The bonds in the vacancy pair up, leaving no dangling bonds. Dangling bonds in amorphous silicon are present in a concentration of about one in a thousand, so most of the atoms are four-fold coordinated. However, the dangling bonds are responsible for diffusion in the amorphous phase, and also for rearranging the bond structure during crystallization of the amorphous phase.

3.3
The Glass Transition

In a liquid, the atoms or molecules move around much more rapidly than in a crystal. They are constantly in motion, jiggling around relative to each other, unlike in a crystal where the atoms are bound to specific lattice sites around which each vibrates.

As a liquid is cooled, the space for the atoms to move around in decreases. A measure of this is the excess volume, which can be determined from the difference between the density of the crystal and that of the liquid. As a glass-forming material cools, this excess volume decreases, and, finally, the density of the glass approaches that of the crystal, as illustrated in Figure 3.3.

On further cooling below the glass transition temperature, the atoms can no longer jiggle around with respect to each other, and so the material becomes a solid. The thermal expansion coefficient of the glass is similar to that of the crystalline phase of the same material.

The standard definition of the glass transition temperature is where the viscosity reaches a value of about 10^{13} poise. At this viscosity, the material is essentially solid. In a glass, the diffusion process no longer depends on a lot of little jiggling motions, as in a liquid. Since the atoms are more or less locked in place, an atom must make a jump which is comparable in length to an atomic diameter. Diffusion rates in a glass are more like those in a crystal than in a liquid.

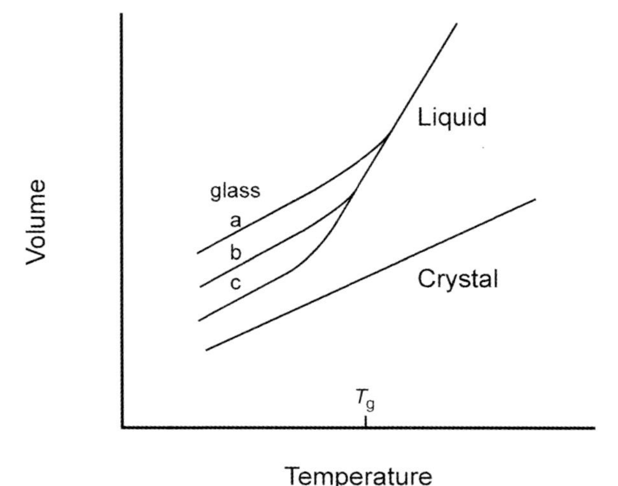

Figure 3.3 The specific volume in a liquid decreases more rapidly with temperature than that of the crystal. The thermal expansion coefficient of glass is similar to that of the crystal. The final specific volume of the glass depends on the cooling rate: (a) fast cooling, (b) normal cooling, (c) slow cooling.

3.4 The Free Volume Model

The free volume model for diffusion in amorphous materials was devised by Cohen and Turnbull [1] for cases where the diffusion process takes place by an atom jumping into an adjacent atom-size hole in the structure. The model assumes that the size distribution of the spaces in an amorphous material is a Boltzmann distribution. The probability of finding an atom-size open space is quite small, and so the diffusion rate in amorphous materials is very slow.

3.5 Fictive Temperature

The density, viscosity, diffusion coefficient and other properties of a glass depend on the thermal history of the sample [2]. On cooling and on annealing, the structure of a glass relaxes. It relaxes more quickly at a higher temperature, and the final structure, and even the specific volume, depend on the temperature at which the relaxation takes place. There is a hypothetical structure of a glass which has been quenched very rapidly to some temperature and then held there for a long time to equilibrate. This is called the fictive temperature.

In practice, the structure of a glass depends on how rapidly it was cooled through the glass transition temperature, as well as the final temperature to which it was cooled. Its specific volume depends on its thermal history, as illustrated in Figure 3.3.

The faster the glass is cooled, the higher will be its fictive temperature, the lower its density, the lower the viscosity, the larger its diffusion coefficient. Glass samples of the same composition do not all have the same properties.

3.6 Diffusion in Polymers

Polymers are long chain molecules that can be crystalline or amorphous. In many cases, they contain both crystalline and amorphous regions. The crystals in a crystalline polymer are usually small, microns or so, separated by amorphous regions. Diffusion can occur through either the crystal structure or the interface between crystals or the continuous amorphous region. The diffusion rate will generally be dominated by the properties of the amorphous region because this lower density material will generally allow faster diffusion.

The glass transition temperature and the crystalline melting temperature of most polymers is in the region between -100 and $+500\,^\circ C$, which is the region of interest for most material processing. The diffusion characteristics change significantly above and below these transformation temperatures. In an amorphous polymer the volume of the polymer increases with temperature relatively linearly until the glass transition temperature is reached. Above the glass transition temperature, the volume increases at a higher rate with increase in temperature. The diffusing molecule can occupy the regions of excess free volume, so the diffusion rate increases with the increase in volume.

In addition to being either amorphous or crystalline, polymers can be either crosslinked (thermoset) or non-crosslinked (thermoplastic), or can act somewhat in between. In a semi-crystalline thermoplastic, the crystalline domains can act like crosslinks. Above the crystalline melting temperature, the material acts more like a thermoplastic. These characteristics of a polymer affect the viscoelastic properties of the polymer. For example, a highly crosslinked polymer will be able to sustain elastic deformation, but will not be able to sustain viscous flow. A thermoplastic, on the other hand can exhibit viscous flow.

Diffusion through polymers depends on the similarity of the diffusing material and the material that it is diffusing through. This similarity is quantified as the solubility parameter, which is used to predict the solubility of a polymer in a solvent. For low molecular weight materials, the solubility parameter is approximately the square root of the molar enthalpy of vaporization. If the solubility parameters of the two materials are similar, then the bulk polymer can absorb a large volume of the permeant. If the solubility parameters of the two materials are very similar, the permeant can swell the polymer network which increases the free volume and increases the ease of diffusion through the polymer. For similar solubility parameters, the concentration of the permeant can be increased until the polymer dissolves.

The ease of diffusion of solvents or small molecules through a polymer depends on the size of the molecule as well as on the similarity of the solubility parameter.

Polymers are long chain molecules, and during the diffusion process the chains cannot move through each other. The polymer molecules can rearrange only by sliding along their lengths. This process is called "reptation" because it is like the motion of a reptile. This process also depends on the free volume of the polymer. For low free volume, only the polymer side chains can move. When the free volume is large enough to allow crank shaft motion of adjacent units in the polymer backbone, the polymer chains can rearrange much more rapidly. Reptation is slow compared to the diffusion of permeants through the polymer.

3.7 The Stokes–Einstein Relationship

Stokes studied the motion of a sphere through a viscous medium, and determined how the force required to move a sphere through a fluid at a velocity v depended on its diameter and on the viscosity of the medium. Einstein [3] postulated that this relationship should hold for an atom or molecule (a very small sphere) moving through a viscous medium, in which case the diffusion coefficient is related to the viscosity:

$$D = \frac{kT}{3\pi\eta a} \tag{3.1}$$

where η is the viscosity and a is the diameter of the atom or molecule. This is known as the Stokes–Einstein relationship [4]. It suggests that the temperature dependence of the viscosity should be related to the temperature dependence of the diffusion coefficient:

$$\begin{aligned} D &= D_0 \exp(-Q/kT) \\ \eta &= \eta_0 \exp(+Q/kT) \end{aligned} \tag{3.2}$$

with the same value of Q. The temperature dependences of the diffusion coefficient and the viscosity are usually found to be inversely related, in liquids as well as in glasses, as suggested by Equation 3.2, but the Stokes–Einstein relationship does not always correctly predict the magnitude of the pre-factor. However, it is often easier to measure the viscosity of a material than the diffusion coefficient, and so the Stokes–Einstein relationship is often used to estimate diffusion coefficients from viscosity measurements.

References

1 Cohen, M.H.and Turnbull, D. (1959) *J Chem. Phys.*, **31**, 1164.

2 Jones, G.O. (1971) *Glass*, Chapman and Hall, London, UK.

3 Einstein, A. (1956) *Investigations on the Theory of Brownian Movement* (ed. R. Furth), Dover.

4 Kittel, C.and Kroemer, H. (1980) *Thermal Physics*, 2nd edn, Freeman, p. 404.

Problems

3.1. The diffusion coefficient can be fitted to an Arrhenius expression of the form:

$$D = D_0 \exp(-Q/kT)$$

From Figure 3.1, calculate D_0 and Q for O_2 and silicon in SiO_2
Evaluate the Qs in two different units: $eV\,atom^{-1}$ and $kcal\,mol^{-1}$.
Note to the instructor: Comparing the results of this analysis from several people indicates the sensitivity of the value of D_0 to small differences in Q.

3.2. What is the viscosity as given by the Stokes–Einstein relationship of a liquid with a diffusion coefficient of $10^{-9}\,m^2\,s^{-1}$?

3.3. According to the Stokes–Einstein relationship, how will the diffusivity of helium and argon differ in the same liquid?

4
Diffusion in Crystals

4.1
Diffusion in a Crystal

The diffusion in a solid is also described by Fick's Law, Equation 2.5, but the concepts of atom velocity and mean free path, which we used to derive a form of Fick's first law for a gas or a liquid do not apply in a crystal. Instead we can consider the flux of atoms between two adjacent planes in the crystal (see Shewmon [1], separated by a distance a, such as are labeled 1 and 2 in Figure 4.1.

The flux of atoms from plane 1 to plane 2 is given by:

$$J_{1\rightarrow 2} = N_1\Gamma/2 \tag{4.1}$$

Where N_1 is the number of atoms per unit area in the plane on the left, and Γ is the jump rate of an atom. There is an equal probability that the atom will jump left or right, so the jump rate to the right is given by $\Gamma/2$.

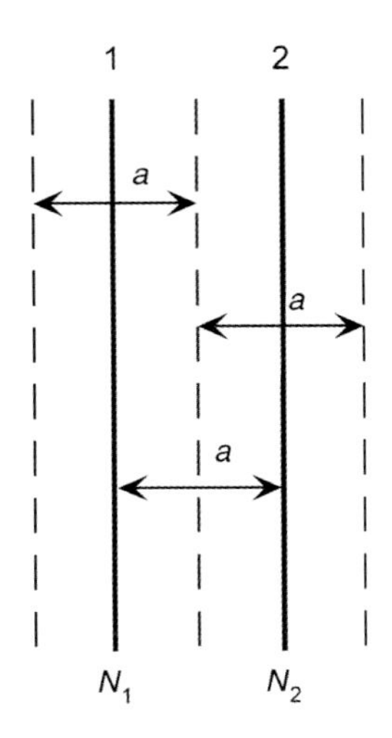

Figure 4.1 Flux of atoms between two adjacent planes in a solid.

Kinetic Processes: Crystal Growth, Diffusion, and Phase Transitions in Materials. Kenneth A. Jackson

ISBN: 978-3-527-32736-2

Similarly, the flux of atoms from plane 2 to plane 1 is given by:

$$J_{2\to 1} = N_2\Gamma/2 \tag{4.2}$$

The net flux is given by:

$$J = J_{1\to 2} - J_{2\to 1} = (N_1 - N_2)\Gamma/2 \tag{4.3}$$

N_1 is number of atoms per unit area in plane 1, and denoting C_1 as the atoms per unit volume, the concentration of atoms in plane 1 is $C_1 = N_1/a$.

The concentration gradient is given by:

$$\frac{dC}{dx} = \frac{C_2 - C_1}{a} = \frac{N_2 - N_1}{a^2} \tag{4.4}$$

So that the net flux is given by:

$$J = -a^2 \frac{\Gamma}{2} \frac{dC}{dx} \tag{4.5}$$

This is in the form of Fick's first law, $J = -D\,(dC/dx)$, so the diffusion coefficient is:

$D = a^2\Gamma/2$ in one dimension.

Γ is the probability that the atom will jump. In one dimension it can jump in either of two directions; in three dimensions it can jump in any one of six directions.

$D = a^2\Gamma/6$ in three dimensions.

This formalism assumes that the jump rate is independent of concentration. We will discuss later the case where this is not valid.

4.2 Diffusion Mechanisms in Crystals

Diffusion in simple materials such as metals or inert gas crystals is fairly simple. The atoms act more or less like spheres, and usually diffuse by the motion of vacant lattice sites. A few species can move around in the spaces between the lattice sites in some crystals, a process known as interstitial diffusion.

4.2.1 Vacancy Diffusion

In vacancy diffusion, the vacancies can move around relatively rapidly, but the motion of an atom depends on having a vacancy next to it, with which it can exchange places, as illustrated in Figure 4.2. The motion of the atoms is much slower than the motion of the vacancies.

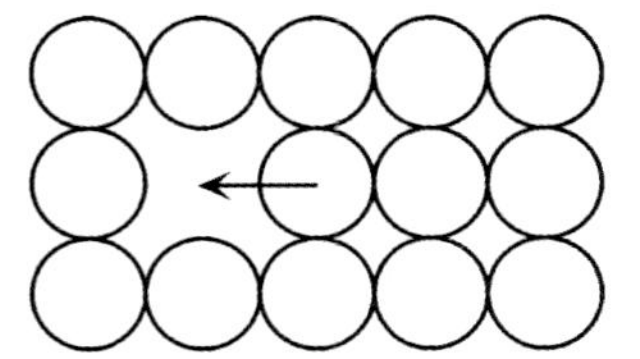

Figure 4.2 Vacant lattice site in a crystalline solid.

The equilibrium concentration of vacancies has an Arrhenius temperature dependence:

$$N_V = N \exp(-E_F/kT) \tag{4.6}$$

Where E_F is the formation energy of a vacancy. We will derive this expression later.

The motion of an atom depends on the probability of a vacancy being next to it, times the rate at which the atom can exchange places with the vacancy. This latter rate is given by a Boltzmann factor containing the energy barrier, Q_M which must be overcome in order for the atom to jump into the vacant site:

$$\exp(-Q_M/kT) \tag{4.7}$$

Q_M is called the motion energy.

So the diffusion coefficient for vacancy diffusion contains a Boltzmann factor with the sum of the formation energy of the vacancy and the motion energy:

$$D = D_C \exp[-(E_F + Q_M)/kT] \tag{4.8}$$

Since the vacancies diffuse much more rapidly than the atoms, the vacancy concentration in a sample is usually much more uniform than the chemical composition. It is often assumed that the vacancy concentration is uniform throughout the sample, even though this is not necessarily true.

The equilibrium vacancy concentration in most metals at their melting point is about 10^{-3} to 10^{-4}, and it decreases at lower temperatures. The rate at which atoms jump into vacant lattice sites is not very fast. So self-diffusion coefficients in crystals in the range of $10^{-10}\,\text{cm}^2\,\text{s}^{-1}$ or less are not uncommon at the melting point of metals. This is several orders of magnitude smaller than the diffusion coefficient in a typical liquid.

Table 4.1 contains data for the self-diffusion coefficient in the solid for several elements. All except sodium are face centered cubic crystals. Sodium is body centered cubic. The self-diffusion process occurs in these crystals by vacancy motion.

In the last column is the ratio of the melting point (MP) to the activation energy, Q. The ratio is approximately the same for the five metals which have the face centered cubic structure, and not too different for sodium, which is body centered cubic. The melting point scales with the latent heat of fusion for all these elements, and so it is not too surprising that the activation energy for self-diffusion also does.

The prefactor D_0 does not. It varies by more than a factor of 10 for the various fcc metals, with no apparent order to the values. However, it should be noted that a small

Table 4.1 Self-diffusion coefficients in the crystalline phase for some metals. The data have been fitted to $D = D_0 \exp(-Q/RT)$.

	Melting point (K)	Q (kcal mol^{-1})	D_0 (cm^2 s^{-1})	MP/Q
Cu	1358	47	0.2	29
Ag	1235	44	0.4	28
Ni	1728	67	1.3	26
Au	1337	42	0.09	32
Pb	601	24	0.28	25
Na	371	10	0.24	37

change in Q on fitting the experimental data can make a large change in the value of D_0.

4.2.2 Interstitial Diffusion

Some elements fit into the interstitial spaces in the lattice of other elements, and they move in the spaces between the atoms, as illustrated in Figure 4.3. They can move through the lattice much more rapidly than substitutional atoms since they do not require the presence of a vacancy. Examples are carbon in iron, hydrogen in platinum, and copper in silicon. The silicon lattice is much more open than the close packed structure of metals, so there are more interstitial diffusers in silicon than in metals.

Typically, interstitial diffusion rates are more like liquid diffusion rates than like vacancy-mediated diffusion rates. For example, copper in silicon has a diffusion coefficient of about 10^{-4} cm^2 s^{-1}. For that reason, it is important to keep copper from getting into the silicon during IC processing.

The rapid diffusion of carbon in iron is responsible for the ability to modify the properties of steels using heat treatment. The rapid diffusion of hydrogen through platinum is used to purify hydrogen.

Figure 4.4 shows the diffusion rates for several elements in silicon on an Arrhenius plot. The rapidly diffusing species are interstitial. Copper is at the top of the chart. The elements at the bottom are substitutional, and move by vacancy diffusion.

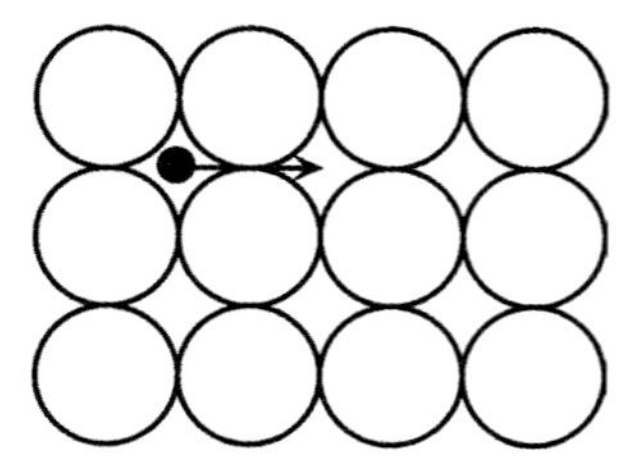

Figure 4.3 Interstitial atom in a crystal.

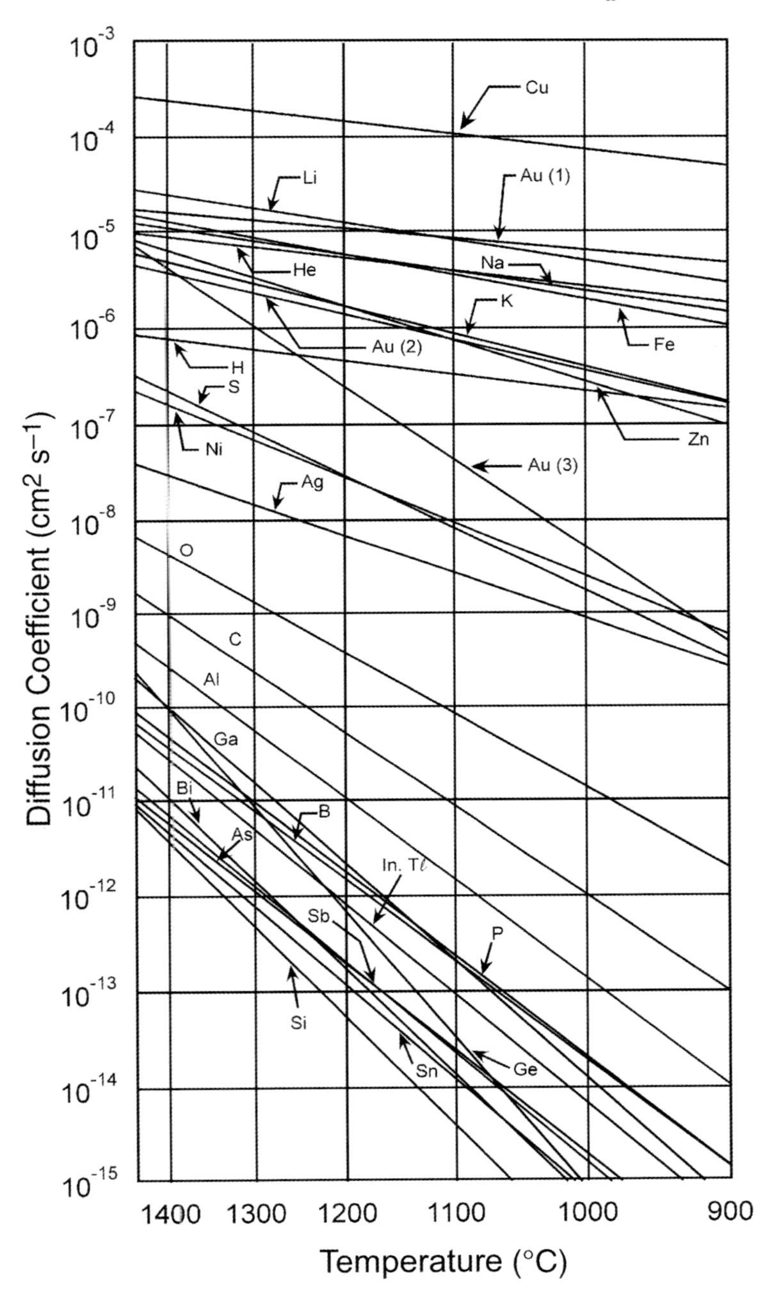

Figure 4.4 Diffusion coefficients in silicon.

The substitutional elements, such as B, Sb, As, P, are the conventional dopants in silicon, and must be on lattice sites to be electrically active. Their diffusion coefficients at 1000 °C are about ten orders of magnitude slower than copper. So during a diffusion anneal which is designed to move one of these elements 10 nm, copper will move 1 mm.

4.3
Equilibrium Concentration of Vacancies

4.3.1
Thermodynamic Analysis

If vacancies are added to a crystal, then the energy of the crystal will be increased by the energy to create a vacancy times the number of vacancies per unit volume that are formed. But the vacancies increase the entropy of the crystal (its structure is less perfect). The change in free energy of the crystal caused by adding N_V vacancies can be written:

$$F = N_V E_F - kT \ln W \tag{4.9}$$

The free energy of the crystal increases by E_F for each vacancy added, and the entropy, S, is increased by $S = k \ln W$, where W is the number of ways in which N atoms and N_V vacancies can be arranged on $N + N_V$ sites:

$$W = \frac{(N + N_V)!}{N! N_V!} \tag{4.10}$$

Using the Stirling approximation for the logarithm of a factorial:

$$\ln(N!) = N \ln(N) - N \tag{4.11}$$

we can write:

$$\ln W = (N + N_V) \ln(N + N_V) - N \ln N - N_V \ln N_V \tag{4.12}$$

This can also be written as:

$$k \ln W = -k \left(N_V \ln \frac{N_V}{N + N_V} + N \ln \frac{N}{N + N_V} \right) \tag{4.13}$$

which is a familiar form for the entropy of mixing, which we will see again when we discuss the thermodynamics of alloys. Since the logarithmic terms are negative, the entropy of mixing is positive, which corresponds to an increase in the disorder of the crystal.

The equilibrium state of the system is given by the minimum in the free energy with respect to the vacancy concentration:

$$\frac{dF}{dN_V} = 0 = E_F + kT[\ln N_V - \ln(N + N_V)] \tag{4.14}$$

or:

$$\frac{N_V}{N + N_V} = \exp(-E_F/kT) \tag{4.15}$$

The equilibrium vacancy concentration is given by a Boltzmann factor containing the formation energy of a vacancy.

4.3.2
Kinetic Analysis

The same result can be obtained in a simpler way using a kinetic model which we will discuss next.

Consider a source/sink of vacancies on or in a crystal, as illustrated in Figure 4.5. The volume of the crystal into which or from which an atom can jump into or out of the source is Aa, an effective area of the source, A, times the atomic diameter, a.

This source could be a free surface, a void in the crystal, a kink on a dislocation line, or any other defect which does not change its energy when a vacancy is removed from or added to it. The rate at which vacancies leave the source to go into the crystal is given by the number of atomic sites to which the vacancy can jump, times the rate at which it will jump. The number of atomic sites to which the vacancy can jump is given by the number of atomic sites per unit volume ($N + N_V$), times Aa, the volume of crystal accessible to the source/sink. In order to create a vacancy in the lattice, the formation energy of the vacancy, E_F, is required, plus enough energy to overcome an energy barrier, of height Q, for the vacancy to leave the source.

$$J^+ = (N + N_V)\, a\, A \exp[-(E_F + Q)/kT] \tag{4.16}$$

The rate at which vacancies will leave the crystal and enter the sink is given by the number of vacancies next to the sink, times the rate at which they jump into the sink. The number of vacancies next to the sink can be written as the number of vacancies

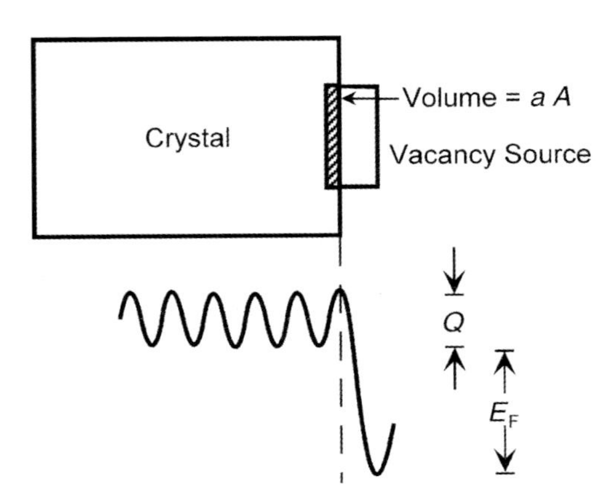

Figure 4.5 Schematic illustration of a vacancy source on or in a crystal.

per unit volume, N_V, times the same volume of crystal accessible to the source/sink, aA. The vacancy must surmount the same energy barrier of height Q to jump into the source/sink.

$$J^- = N_V\, a\, A\exp[-Q/kT] \tag{4.17}$$

There is equilibrium when the two fluxes are equal:

$$\frac{N_V}{N + N_V} = \exp(-E_F/kT) \tag{4.18}$$

This is the same result that was derived above from a statistical thermodynamics analysis.

The thermodynamic analysis implies that the vacancies appear by magic in the crystal. This is not the case. The rate analysis emphasizes that the equilibrium concentration of vacancies is achieved by vacancies joining and leaving sources and sinks.

When the temperature of the crystal changes, vacancies must leave or enter the source/sinks, and then diffuse through the crystal in order to establish the new equilibrium concentration. How rapidly this happens will depend on the density of source/sinks, on how rapidly the vacancies join and leave them, and how rapidly they diffuse through the crystal. In a typical metal, there are many dislocations which can act as sources and sinks for vacancies, and so the vacancy concentration in a metal is usually close to the equilibrium value. In silicon crystals there are few if any dislocations, and the vacancy equilibrium is established at the surfaces. The vacancy concentration can be quite far from equilibrium in a silicon crystal.

The diffusion rate for an atom due to vacancy motion depends on the probability that the atom will have a vacancy next to it, times the rate at which the atom jumps into the vacant site, as in Equation 4.8:

$$D = \frac{N_V}{N + N_V} D_V = D_V^0 \exp[-(Q_M + E_F)/kT] \tag{4.19}$$

This equation implies not only that the concentration of vacancies is at equilibrium, but also that it is uniform throughout the sample. Neither of these is necessarily true.

A small void can be a source and sink for vacancies, but the energy of the void changes when a vacancy is added or removed. We will discuss this later when we talk about nucleation, and the size dependence of the energy of small clusters of atoms or voids.

The motion energy is likely to be different for the different species in an alloy to jump into a vacant lattice site.

4.4 Simmons and Balluffi Experiment

Simmons and Balluffi [2] measured both the length and the lattice parameter of a metal sample as a function of temperature. They found that the length of the sample

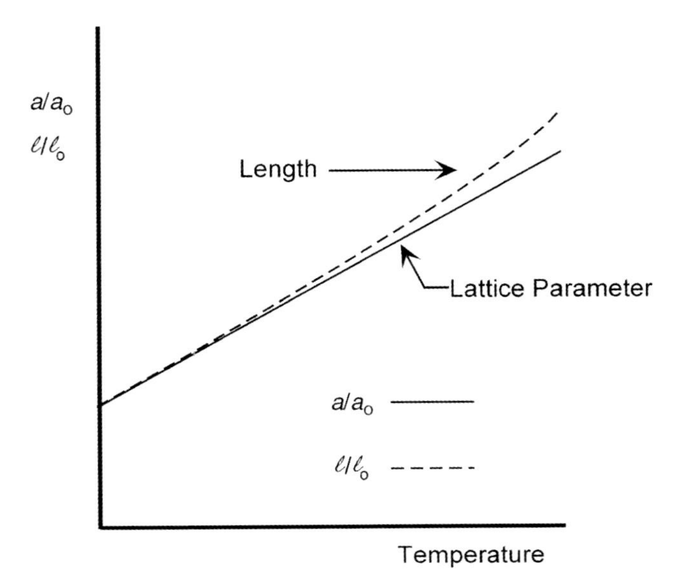

Figure 4.6 The length of a sample and the lattice parameter of the same sample at various temperatures.

increased faster with increasing temperature than did the lattice parameter, as shown in Figure 4.6.

The length of the crystal depends on both the lattice expansion and the change in volume of the crystal due to the vacancies. The vacancy concentration of their samples was given by the difference between the actual length of the sample and the lattice expansion. This measurement can be made for metals, where the vacancy concentration is 10^{-3} to 10^{-4} at the melting point, but, for silicon, the vacancy formation energy is larger, and so the vacancy concentration at the melting point is only about 10^{-5}, which is too small to be measured by this method.

4.5 Ionic and Covalent Crystals

The various elements in an ionic or covalent crystal will, in general, diffuse at different rates [3, 4]. For example, in lithium niobate ($LiNbO_3$), the lithium moves around rapidly at 125 °C, whereas the niobium does not move around much below about 1000 °C. Each element stays primarily on its own sub-lattice. The diffusion process is usually by motion of defects, such as vacancies, on the sub-lattices, or by the motion of interstitials.

The role of defects in diffusion in ionic crystals is illustrated by diffusion in lithium niobate. Lithium niobate is a non-linear optical material which is used to make SAW devices and optical modulators. Surface optical waveguides for the optical modulators are made by diffusing titanium into the crystal from the surface. The titanium substitutes for niobium in the lattice. Increasing the lithium concentration by a small amount changes the number of vacancies on the niobium sub-lattice. Changing the lithium to niobium ratio by one part in a thousand changes the diffusion coefficient of titanium by about 10%. So the stoichiometry of the crystal must be very carefully controlled in order to control the diffusion coefficient in the crystal, which determines the width of the surface waveguides.

In an ionic crystal, charge neutrality prevails, since a very small departure from charge neutrality creates a very large electrical field. If there are two differently charged point defects in a crystal, each will have its own diffusion rate, but the *effective* diffusivities of the two species will be the same. The fluxes of the two charged species will be coupled so that there is no net flux of charge. The fluxes are coupled through the electric field which is generated if the fluxes are not coupled. The electric field adds a drift component to the flux of each species.

$$J_A = -D_A \frac{dC_A}{dx} + Z_A C_A \frac{D_A}{RT} FE \tag{4.20}$$

$$J_B = -D_B \frac{dC_B}{dx} + Z_B C_B \frac{D_B}{RT} FE \tag{4.21}$$

Here the subscripts A and B refer to the two mobile species, D is the diffusion coefficient, Z is the charge state of the species, C is the concentration of the species, F is the Faraday constant, and E is the electric field. The Einstein relation has been invoked to replace the defect mobility with DF/RT in these equations. The second term in these equations is the drift term, which speeds up the slower defect and slows the faster defect, depending on the charge on the defect relative to the direction of the electric field, so that the fluxes of the two defects are the same. An initial imbalance in the net flux creates the electric field. If only one component is mobile, then the field will build up to effectively stop diffusion. On the other hand, an externally applied field will promote an ionic flux.

4.6 Stoichiometry

As mentioned before, the vacancy concentration in metals at their melting point is typically 10^{-3} to 10^{-4}. Since a deviation from stoichiometry of one part in one thousand can produce a concentration of 10^{-3} vacant sites on one of the sub-lattices, it is not surprising that variations in the stoichiometric ratio of one part per thousand can have a significant effect on diffusion rates.

This effect is used to advantage in yttrium- and calcium-stabilized zirconia. The addition of yttrium or calcium to the zirconia permits the creation of an oxygen deficient crystalline material, with many vacant sites on the oxygen sub-lattice. Oxygen ions can readily diffuse through these materials, and so they are used as oxygen sensors. The sensors are made by placing porous electrodes on the two faces of a thin sample. At an appropriate elevated temperature, oxygen ions will diffuse through the crystal, creating a voltage across the sensor. The voltage depends on the difference in oxygen concentration at the two electrodes.

Alternatively, a voltage can be applied to the two electrodes, and oxygen ions will drift through the zirconia, and come out the other side. This is used to obtain pure oxygen from air, because nothing else can move through the crystal.

4.7
Measurement of Diffusion Coefficients

A usual starting configuration is to bond together two blocks of material of differing compositions. This configuration is known as a diffusion couple. Alternatively, one material can be deposited on the surface of another to provide the starting configuration. Ion implantation can be used to provide an initial configuration.

The composition profile can be determined by slicing the final sample and measuring the composition in each slice by a wet chemical or other chemical analysis method. X-ray fluorescence, EDAX or microprobe analysis are also commonly used. Radioactive tracers are also used by introducing an appropriate radioactive isotope, and then counting the radioactivity of each slice. This is a preferred method for obtaining self-diffusion coefficients. Rutherford backscattering can be used to determine the composition distribution after ion implantation and a diffusion anneal. Secondary ion mass spectroscopy (SIMS) is also used. Here the surface is sputtered away by incident ions, and the ions which come from the surface are analyzed using a mass spectrometer.

A procedure called delta doping is used to study interactions between species in semiconductors. Silicon can be grown epitaxially on silicon. So a thin doped layer can be deposited, followed by an undoped layer, and this can be repeated for several layers. A different species can then be deposited on the surface and diffused in. The effect of this component on how the thin doped layers spread out in time gives information about the interaction between the two species. The composition of the in-diffusing species decreases with distance from the surface, so the composition dependence of the interaction can be assessed. The effects of ion implantation damage can also be explored with this method.

4.8
Surface Diffusion

Most surfaces are not clean. They have oxide layers, finger prints, films of one sort or another, foreign particles, and so on. For example, window glass is hydroscopic, and typically there is an adsorbed layer of water about five molecules thick on the surface. A thick layer of oxide spontaneously grows on aluminum when it is exposed to air. Two monolayers of oxide form almost instantly on silicon when it is exposed to air. The people doing semiconductor processing go to great lengths to clean wafers and to keep them clean during processing.

Studies of surface diffusion are usually carried out under ultra-high vacuum (UHV) conditions in order to avoid contamination [5], and these studies provide insight into surface diffusion processes under carefully controlled conditions, rather than under real world conditions. The latter are much more complex.

Studies of self-diffusion on a clean surface in a UHV system indicate that the surface diffusion process involves the motion of adatoms along the surface. Adatoms are atoms which sit on top of an otherwise flat surface. The surface diffusion rate is the product of the adatom density and the rate at which the adatoms move. The surface adatoms are less constrained than bulk atoms, and they move much more rapidly than bulk atoms. The adatoms can even make long jumps over several surface sites. At very low temperatures, the surface is relatively smooth on the atomic scale, with few adatoms, and so the surface diffusion coefficient is small. At higher temperatures, the adatom density increases, and the jump rate also increases. Near the melting point, the number of adatoms and their mobility has increased so that the surface diffusion is approximately like the diffusion that would occur if there were a one atom thick layer of liquid on the surface. This behavior is found on highly cleaned surfaces of metals.

The study of surface migration and adatom motion has been greatly facilitated by the use of scanning tunneling microscopy (STM) and atomic force microscopy (AFM). However, there is some concern that the presence of the probe tip may be altering the motion of the adatoms. One interesting phenomenon which has been observed is that an adatom can become trapped on a plateau which is bounded by a step. The atom will be less tightly bound to the surface as it goes over the step, and that corresponds to a higher energy state, which represents an energy barrier. This is known as the Schwoebel effect [6].

4.9 Diffusion in Grain Boundaries

Small angle tilt boundaries consist of an array of edge dislocations. It has been found that diffusion along the array of dislocations in such a boundary is faster than diffusion across the dislocation lines. There is more space for an atom to move along the core of a dislocation than there is for it to move through the bulk, so this is not surprising. Atoms also diffuse faster along a single dislocation. This is important in semiconductor device processing, where the diffusion of a dopant along a dislocation which penetrates a p/n junction can short out the junction. Most silicon devices are made on dislocation-free silicon for this reason. Bipolar devices are more susceptible to defects than CMOS devices.

In high angle boundaries, there are many atoms which are not tightly bound to either of the lattices on the two sides of the boundary. Atoms diffuse much more rapidly along grain boundaries than through the bulk. The diffusion along a high angle boundary in a metal is about the same as would occur if there were a one or two atom layer thickness of liquid at the grain boundary. Atoms that have diffused rapidly along a grain boundary can then diffuse out laterally from the grain boundary into the adjacent bulk. Additions are made to some alloys to reduce grain boundary diffusion by adding elements which segregate to the grain boundaries, and, once there, diffuse more slowly than the matrix atoms.

4.10
Kirkendall Effect

This effect depends on the fact that the rate at which a vacancy will change places with an atom in an alloy is different for the different components of the alloy. For example, a copper atom will jump into a vacant lattice site more rapidly than a silver atom in a copper–silver alloy. There is usually a difference in jump rates for various atom species in the same lattice. One result of this was demonstrated by Kirkendall [7] using a copper–silver diffusion couple. Copper and silver are both face centered cubic metals, and they have continuous mutual solid solubilities. Kirkendall joined a piece of copper to a piece of silver with a fine mesh of molybdenum wire between them, as illustrated in Figure 4.7.

After a diffusion anneal, the metallurgical junction, that is, the plane where the composition is 50 : 50 copper to silver, was no longer at the molybdenum wires. The metallurgical junction was displaced into the silver region. The copper exchanged places with the vacancies more rapidly than the silver, and so there was net flux of copper across the plane of the molybdenum wires into the silver side of the diffusion couple. There was a net flux of vacancies in the opposite direction. The concentration of these vacancies increased sufficiently in the region on the copper side of the

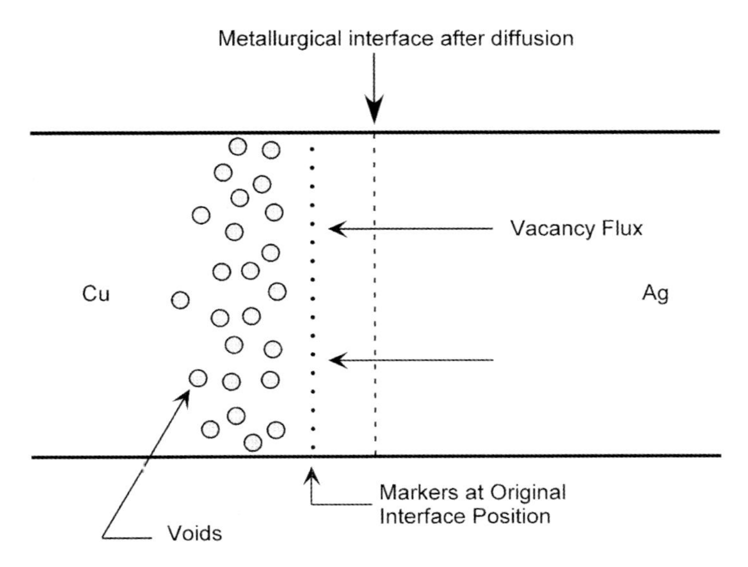

Figure 4.7 The markers indicate the initial position of the interface. The dashed line indicates the final position of the metallurgical interface after diffusion.

junction so that the vacancies precipitated out as voids [8, 9]. These are known as Kirkendall voids.

In semiconductor devices there are many places where dissimilar metals are joined. Each of these is a potential site for the creation of Kirkendall voids. These voids have often been observed to form where a lead of one metal is connected to a bond pad of a different metal. The voids weaken the strength of the connections, so that the leads tend to fall off.

4.11
Whisker Growth

It is well known that metals can deform at high temperature by grain boundary creep. The atoms can diffuse rapidly along grain boundaries, and can change the shape of a sample by moving from a grain boundary which is perpendicular to a compressive stress to a grain boundary which is parallel to a compressive stress. In doing so the sample gets shorter and wider in response to the applied stress. This phenomenon can give rise to whisker growth under appropriate conditions [10].

For deformation, a high temperature usually means a temperature above about one half of the melting point in degrees K. Tin melts at 232 °C, and so room temperature qualifies as a high temperature for deformation processes in tin.

Whisker growth can be readily observed by stacking several steel sheets which have been coated with a fairly thick layer of tin and then compressing the stack in a vice. (In the good old days, pieces cut from Pet milk can were used.) This assures a small grain size for the tin, and some suitable sites for whisker growth. The whiskers grow from their base by a process which is akin to grain boundary creep, except that the atoms deposit at a suitable grain boundary junction near the surface, and material is effectively extruded from the sample in the form of a whisker, as illustrated in Figure 4.8.

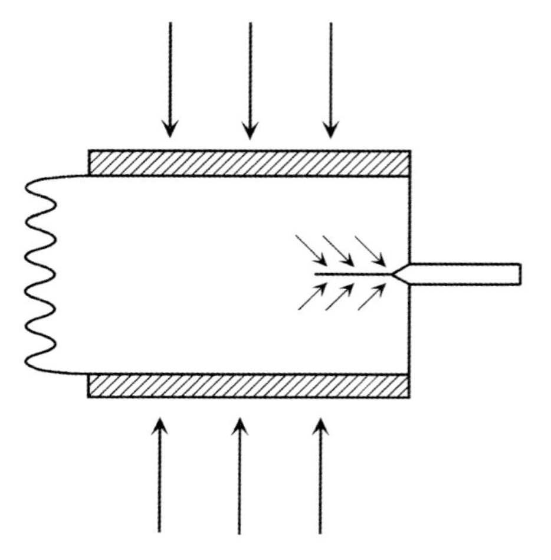

Figure 4.8 Whisker growth.

This happens at solder joints in electronic components where there is stress on the joints because of the way the devices are mounted. The whiskers tend to make short circuits.

4.12 Electromigration

This is the effect which limits the size of the conductor stripes on semiconductor devices. The minimum feature size for the devices is approaching 0.1 μm, but the width of the conductor stripes is greater than 1 μm. About half of the area on a semiconductor chip is taken up by the wiring. On advanced devices, there are several layers of metallization.

Electromigration becomes important when the current density gets above about $10^5\,\mathrm{A\,cm^{-2}}$ [11]. It is sure to be a problem above $10^6\,\mathrm{A\,cm^{-2}}$. To put this into perspective, it takes about 3 to 5 mA to switch a bipolar transistor, and about 1 mA to switch a CMOS gate. 1 mA in a 1 μm by 1 μm conductor gives a current density of $10^5\,\mathrm{A\,cm^{-2}}$, and that is for one device. The power distribution lines carry current to many devices.

Electromigration is caused by momentum transfer between the electrons and the atoms in the conductor. The electron mass is about one-thousandth the mass of an atom, and so the electron momentum cannot displace an atom which is on a lattice site. But it can bias the hopping motion of atoms in a grain boundary which are not tightly bound to any one lattice site, as illustrated in Figure 4.9.

Lower melting point metals are more susceptible to this than higher melting point metals, and so copper is more resistant to electromigration than aluminum. This is one of the reasons why copper is replacing aluminum in ICs. The other reason is that copper has lower resistivity than aluminum.

The loosely held atoms in grain boundaries can move around at very low temperatures. The motion of the atoms is effectively a displacement of charge in

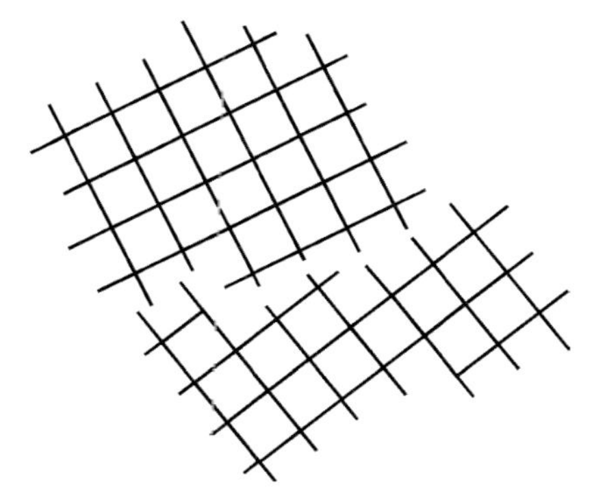

Figure 4.9 Atoms in a grain boundary are not tied strongly to sites in either lattice.

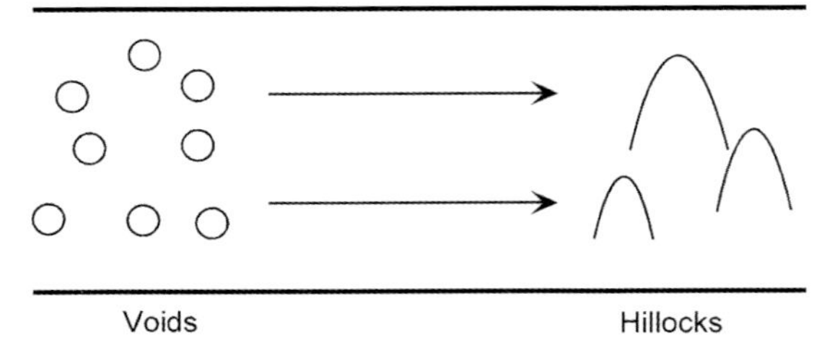

Figure 4.10 The motion of atoms caused by electromigration creates voids where the atoms leave and hillocks where they accumulate.

the metal. This motion of charge in the wire connected to the input of very high gain and well-shielded amplifiers creates noise in the amplifier. The noise has a frequency spectrum, and the amplitude of the noise is higher at lower frequencies. The amplitude of the noise is inversely proportional to the frequency, and so this is known as $1/f$ noise. It seems strange that atoms moving around in grain boundaries in a metal can make electronic noise in an amplifier. Electromigration is due to the net displacement by electron momentum of these same loosely bound atoms in the grain boundaries.

Electromigration displaces the atoms in the grain boundaries. DC currents produce a net flux of atoms, and this creates voids where the atoms leave, and makes hillocks where the atoms end up, as illustrated in Figure 4.10.

The voids decrease the cross-section of the conductor, which increases the current density, which increases the electromigration. Ultimately, an open circuit is created. Electromigration also occurs with an AC current, but the lifetime is typically about ten times longer than with DC.

The motion of material during electromigration interacts with the layers above and below the conductor, generating a stress. Putting a capping layer on top of the conductor, as illustrated in Figure 4.11, increases the stress due to migration in the conductor. This stress reduces the amount of migration, which increases the lifetime.

Electromigration occurs in grain boundaries, and the migration rate depends on the grain size. Where the grain size is small there are more atoms which can be displaced by electromigration than where the grain size is large. Hence, if the grain

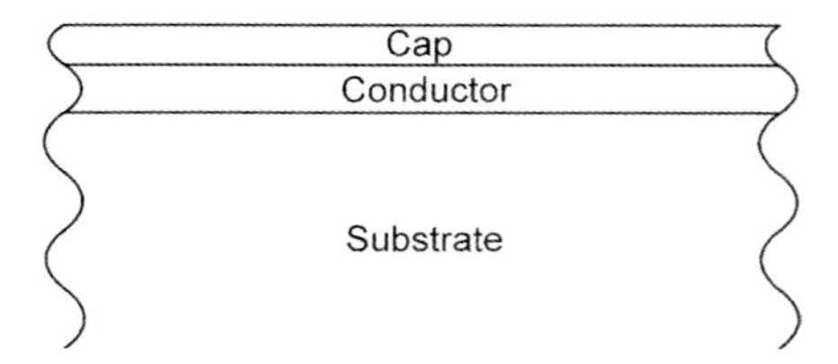

Figure 4.11 A capping layer on top of the conductor creates a back stress when the atoms migrate, and so tends to decrease the atom motion.

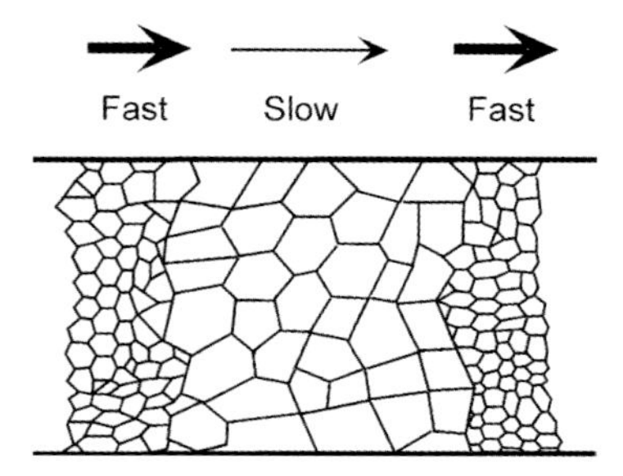

Figure 4.12 A variation in grain size produces a variation in atom flux.

size changes along a wire, the flux of atoms along the wire changes, as illustrated in Figure 4.12.

Where the atom flux decreases due to an increase in grain size, there will be a pile-up of atoms, because more atoms are arriving there than are leaving, and conversely, where the flux of atoms increases due to a decrease in grain size, there will be a net out-flux of material, creating voids. So variations in the grain size in the conductor stripes are bad for electromigration. The conductor stripes on ICs are deposited metal, and the grain size usually changes where the conductor stripe goes over irregularities in the surface, and there are often variations in grain size at a via or where the conductor stripe meets a contact pad.

Electromigration occurs in grain boundaries which lie along the axis of the conductor, but not in grain boundaries which are perpendicular to the axis. When the grain size in the conductor is about the same as the diameter of the conductor, the grain boundaries tend to be perpendicular to the axis of the conductor. This is known as a bamboo structure, illustrated in Figure 4.13.

This structure reduces electromigration significantly. As shown in Figure 4.14, the mean time to failure decreases with the width of the conductor stripe down to about 2 μm, and then increases below 2 μm because bamboo grain structures form there.

The data in the graph were obtained for the same current density in the conductor stripes of various widths, so the total current in the conductor stripes is lower for the narrower stripes in the graph.

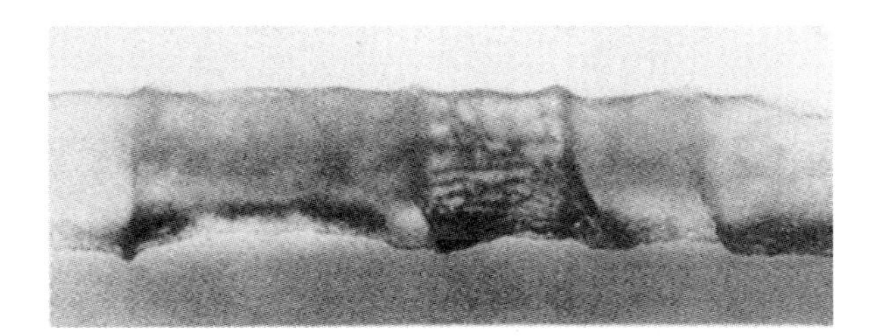

Figure 4.13 Bamboo structure in a one micron wide conductor stripe (From Vaidya *et al.* [12].)

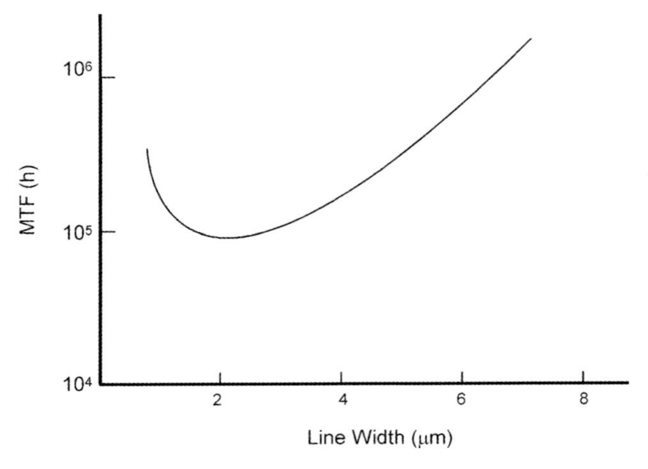

Figure 4.14 Mean time to failure for the same current density, 10^5 A cm^{-2}, for various conductor stripe widths.

References

1 Shewmon, P.G. (1989) *Diffusion in Solids*, 2nd edn, AIME, Warrendale, PA.
2 Simmons, R.O. and Balluffi, R.W. (1964) *J. Phys. Chem. Solids*, **25**, 1139.
3 Matzke, H. (1991) *Phil. Mag. A*, **64**, 1181.
4 Compaan, K. and Haven, Y. (1956) *Trans. Faraday Soc.*, **52**, 786.
5 Venables, J.A. (2000) *Surface and Thin Film Processes*, Cambridge University Press, Cambridge, UK.
6 Schwoebel, R.L. and Shipsey, E.J. (1966) *J. Appl. Phys.*, **37**, 3682.
7 Kirkendall, E.O. (1947) *Trans. AIME*, **171**, 130.
8 Huntington, H.B.and Seitz, F. (1942) *Phys. Rev.*, **61**, 325.
9 Kirkaldy, J.S. and Savva, G. (1997) *Acta Mater.*, **45**, 3115.
10 Ellis, W.C. (1958) *Growth and Perfection of Crystals* (eds R.H. Doremus, B.W. Roberts, and D. Turnbull), John Wiley & Sons, New York, NY, p. 102.
11 Murarka, S.P.and Peckerar, M.C. (1989) *Electronic Materials*, Academic Press, San Diego, CA, p. 324.
12 Vaidya, S., Sheng, T.T., and Sinha, A.K. (1980) *Appl. Phys. Lett.*, **36**, 464.

Problems

4.1. Silver and gold form a continuous series of solid solutions, and interdiffusion occurs by a vacancy mechanism. A diffusion couple was made by depositing a thick layer of silver onto a gold substrate. At 1000 K, the diffusion coefficient of silver into gold is $10^{-10}\,\text{cm}^2\,\text{s}^{-1}$, and the diffusion coefficient of gold into silver is $2 \times 10^{-11}\,\text{cm}^2\,\text{s}^{-1}$.

a Why can the two diffusion coefficients be different?
b Will the plane where the concentration is 50–50 move during diffusion? If so, which way?

c Will there be a net flux of vacancies through the interface during diffusion? If so, where will the vacancies come from and where would they go? On which side of the interface might you expect to find voids?

d Sketch the concentration profile through the junction after annealing for 3 days at 1000 K.

e Suggest an experimental method which could be used to measure the concentration profile in the annealed sample in order to determine the diffusion coefficients.

4.2. The diffusion coefficient can be fitted to an Arrhenius expression of the form:

$$D = D_0 \exp(-Q/kT)$$

From Figure 4.4, calculate D_0 and Q for copper and phosphorus in silicon. Evaluate the Qs in two different units: eV atom^{-1} and kcal mol^{-1}.
Note to the instructor: Comparing the results of this analysis from several people indicates the sensitivity of the value of D_0 to small differences in Q.

5
Diffusion in Semiconductors

5.1
Introduction

This chapter will start with a brief review of the role of donor and acceptor dopant atoms in semiconductors.

When atoms come together to form a solid, the energy levels of the electrons on the atoms spread out into bands. The total number of allowed valence states in silicon is the same as the number of valence electrons: four per atom. However, these states are spread out in energy, and this energy distribution is known as the density of states. In a semiconductor, there is an energy gap between the valence states where the electrons on neighboring atoms interact to hold the solid together, and the conduction states where the electrons can move through the crystal. At low temperatures, all the electrons are in the valence states, and so silicon does not conduct electricity. At higher temperatures, some electrons have enough energy to cross the band gap to the conduction state. These electrons can move through the crystal, and conduct electricity. When an electron leaves the valence band for the conduction band, it leaves behind a hole. A hole is effectively a positive charge, and it can also move through the crystal and conduct electricity. Dopant atoms in the crystal can also help to create conduction electrons or holes in the crystal.

We can define an effective density of states N_C for the conduction band, and N_V for the valence band. The occupancy of these states depends on temperature. However, unlike a collection of atoms where many atoms can have the same energy, no two electrons in a crystal can occupy the same state. The occupancy of the electronic states in a crystal is described by a Fermi function, rather than the Boltzmann function which describes the energy distribution of atoms.

The Fermi function has the form:

$$f(E) = \frac{1}{1+\exp[(E-E_F)/kT]} \tag{5.1}$$

where E_F is the Fermi energy from 1 to 0 depending on the sign of $E - E_F$. $f(E)$ is 1 for energies lower that E_F, so that all the states up to E_F in energy are filled, and $f(E)$ is 0 for energies greater than E_F, so all the states above E_F in energy are empty. At higher

Kinetic Processes: Crystal Growth, Diffusion, and Phase Transitions in Materials. Kenneth A. Jackson

ISBN: 978-3-527-32736-2

temperatures, there is not a sharp transition between the filled and empty states. The transition spreads out by about kT in energy, so that some of the states below E_F are empty, and some of the states above E_F are filled.

The concentration of electrons in the conduction band, n, can be written as:

$$n = N_C \exp[(E_F - E_C)/kT] \tag{5.2}$$

and the concentration of holes in the valence band, p, can be written as:

$$p = N_V \exp[(E_V - E_F)/kT] \tag{5.3}$$

For an intrinsic semiconductor, that is one containing no dopants, the number of holes and electrons must be the same, that is, $n_i = p_i$. The Fermi level in an intrinsic semiconductor is E_F^i, so we can turn Equations 5.2 and 5.3 around in order to express the effective densities of states in terms of the intrinsic carrier concentration and the Fermi level of the intrinsic material:

$$\begin{aligned} N_C &= n_i \exp[(E_C - E_F^i)/kT] \\ N_V &= n_i \exp[(E_F^i - E_V)/kT] \end{aligned} \tag{5.4}$$

The carrier concentrations in non-intrinsic material can then be related to its Fermi level and to the Fermi level and carrier concentration of intrinsic material, rather than to the density of states and the positions of the band edges:

$$\begin{aligned} n &= n_i \exp[(E_F - E_F^i)/kT] \\ p &= n_i \exp[(E_F^i - E_F)/kT] \end{aligned} \tag{5.5}$$

From Equation 5.5 a very important relationship follows directly:

$$np = n_i^2 \tag{5.6}$$

Equations 5.5 and 5.6 provide simple expressions for the carrier concentrations in doped semiconductors in terms of the carrier concentrations and the Fermi level of the intrinsic material, rather than in terms of the density of states and the positions of the band edges.

Dopant atoms, which are intentionally introduced into silicon, are usually shallow donors (a Group V element with an energy level close to the conduction band) or shallow acceptors (a Group III element with an energy level close to the valence band). These donor and acceptor sites are usually charged at room temperature, that is, the extra electron on the donor atoms has left the donor atom and is in the conduction band, so that the donor atom is positively charged. Similarly, the acceptor atoms have picked up an electron from the valence band, creating a hole, and the acceptor atom is negatively charged. At room temperature, the electrons and holes can migrate through the lattice, but the charged dopants cannot move, so they create a charge which has a fixed position. The charged dopant atoms create fixed internal electrical fields in the semiconductor.

Overall, the crystal must maintain a neutral charge, so that the excess charge created by the dopant atoms is compensated by a change in the number of electrons and holes.

$$p - n = N_A - N_D \tag{5.7}$$

N_A and N_D are the number of acceptor and donor atoms per unit volume, respectively.

Combining Equations 5.7 and 5.6 results in a quadratic equation for n:

$$n^2 - n(N_D - N_A) - n_i^2 = 0 \tag{5.8}$$

If all of the dopants are ionized, which is likely to occur at high doping levels, the concentration of ionized dopants should be used in Equation 5.8.

With no dopants, $N_A = N_D = 0$, so $n = n_i$, as it should be for intrinsic material.

For a large excess of donor atoms,

$$N_D - N_A \approx N_D \gg n_i; \quad n \approx N_D, \quad p \approx n_i^2/N_D \tag{5.9}$$

For a large excess of acceptor atoms,

$$N_A - N_D \approx N_A \gg n_i; \quad p \approx N_A, \quad n \approx n_i^2/N_A \tag{5.10}$$

The carrier concentrations, n and p, as determined from Equation 5.8 can be inserted into Equation 5.5 to obtain the position of the Fermi level in the doped material. The Fermi level rises above the intrinsic level for net donor concentrations, and falls below the intrinsic level for net acceptor concentrations.

5.2 Diffusion in Silicon

5.2.1 Vacancy Diffusion in Silicon

Vacancy diffusion in silicon is much more complex than in metals, because the motion of the vacancies depends on the position of the Fermi level [1]. It is even more complex in compound semiconductors, because vacancies can exist on the different sub-lattices in the crystal. In the metals, the motion of the vacancy does not involve charge effects, and the barrier to motion is basically the elastic interaction of an atom squeezing past its neighbors in order to move into the vacant site. However, in silicon the vacancy can exist in various charge states, as illustrated in Figure 5.1, which shows the energy levels of vacancies in various charge states.

For example, the energy level of a vacancy with a single negative charge is 0.44 V below the conduction band. The number of vacancies in each charge state depends on the Fermi level, which depends on the dopant concentration.

The concentration of singly charged vacancies in doped silicon can be related to that in intrinsic silicon:

$$\frac{V^-}{V_i^-} = \exp\left(\frac{E_F - E_F^i}{kT}\right) = \frac{n}{n_i} \tag{5.11}$$

Where V^- is the concentration of singly charged vacancies in doped silicon, V_i^- is the concentration of singly charged vacancies in intrinsic silicon, E_F is the Fermi

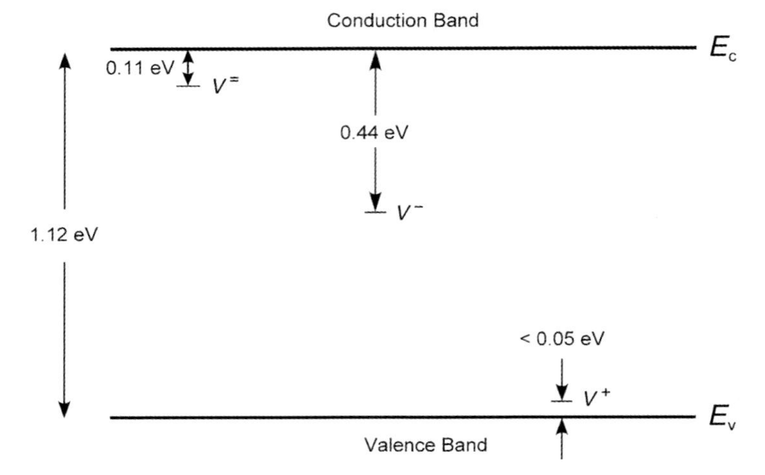

Figure 5.1 Energy levels for a vacancy in various charge states in silicon. (After Fair [2].)

level in the doped silicon, E_F^i is the Fermi level in intrinsic silicon, n is the electron concentration in the doped sample, and n_i is the electron concentration in intrinsic silicon.

For n-type silicon, $E_F > E_F^i$, so $n > n_i$, and conversely in p-type silicon. For a dopant concentration of $10^{18}/cm^3$, which is about 20 ppm, $n/n_i \cong 1.2$ in n-type, and $n/n_i \cong 0.8$ in p-type silicon.

Similar expressions give the concentrations of vacancies in other charge states.

Table 5.1 lists the activation energy, Q, as well as D_0 for self-diffusion by vacancies for various charge states of the vacancy, as well as the values for vacancy diffusion of common substitutional dopants. The diffusion coefficient is different for each charge state and, in general, has a different temperature dependence for each charge state.

Table 5.1 Activation energy, Q, and D_0 for diffusion by vacancies in silicon. (After Fair [2].).

		Q (eV)	D_0 (cm^2 s^{-1})
Si	V^o	3.9	0.015
	V^-	4.54	16
	$V^=$	5.1	10
	V^+	5.1	1180
As	V^o	3.44	0.066
	V^-	4.05	12.0
B	V^o	3.46	0.037
	V^+	3.46	0.76
P	V^o	3.66	3.85
	V^-	4.0	4.44
	$V^=$	4.37	4.42
Ge		5.28	6.25×10^5

For intrinsic silicon, the diffusion coefficient can be written as the sum of the concentration of each type of vacancy times the diffusion coefficient in each charge state:

$$D_\mathrm{i} = \sum_r [V^r]_\mathrm{i} D_\mathrm{i}^r \tag{5.12}$$

Here V^r is the concentration of vacancies in charge state r, and D_i^r is the corresponding diffusion coefficient.

For doped silicon, the self-diffusion coefficient for vacancy diffusion can be written as:

$$D_\mathrm{Si} = D_\mathrm{Si}^\mathrm{o} + D_\mathrm{Si}^- \left(\frac{n}{n_\mathrm{i}}\right) + D_\mathrm{Si}^{--} \left(\frac{n}{n_\mathrm{i}}\right)^2 + D_\mathrm{Si}^+ \left(\frac{n_\mathrm{i}}{n}\right) + D_\mathrm{Si}^{++} \left(\frac{n_\mathrm{i}}{n}\right)^2 \tag{5.13}$$

5.2.2
Diffusion of Phosphorus in Silicon

The diffusion coefficient for phosphorus in silicon can be written:

$$D_\mathrm{p} = h\, D_\mathrm{i}^\mathrm{o} + D_\mathrm{i}^- \left(\frac{n}{n_\mathrm{i}}\right) + D_\mathrm{i}^{--} \left(\frac{n}{n_\mathrm{i}}\right)^2 \tag{5.14}$$

where h is a constant. The contribution of positively charged vacancies can be ignored because phosphorus is a donor. But phosphorus combines with a vacancy to form a negatively charged phosphorus–vacancy complex by the reaction:

$$\mathrm{P}^+ + \mathrm{V}^= \rightleftharpoons [\mathrm{PV}]^- \tag{5.15}$$

These complexes can diffuse, but do so more slowly than unassociated phosphorus atoms. During the in-diffusion of phosphorus, the phosphorus combines with vacancies to form a complex at high concentrations near the surface. Farther into the wafer where the phosphorus concentration drops off to a low enough value so that the Fermi level passes through the energy level of the doubly-negatively-charged vacancy at 0.11 eV below the conduction band, then the phosphorus-vacancy complexes break up, and the phosphorus diffuses more rapidly. This creates a kink in the phosphorus concentration profile, as illustrated in Figure 5.2. The open circles represent the total phosphorus content. The black dots represent the electrically active phosphorus, which is substitutional on the lattice sites. At the high concentrations, most of the phosphorus is tied up in vacancy complexes, and so is not electrically active.

This is also responsible for an effect called *emitter push*, illustrated in Figure 5.3. When phosphorous is diffused into silicon to form an emitter after a boron diffusion, the boron under the phosphorus emitter diffuses further into the silicon because of the increased vacancy concentration resulting from the break-up of the phosphorus–vacancy complexes.

The diffusion coefficient of arsenic in silicon has the same dependence on the various charge states of the vacancy as phosphorus. At high concentrations,

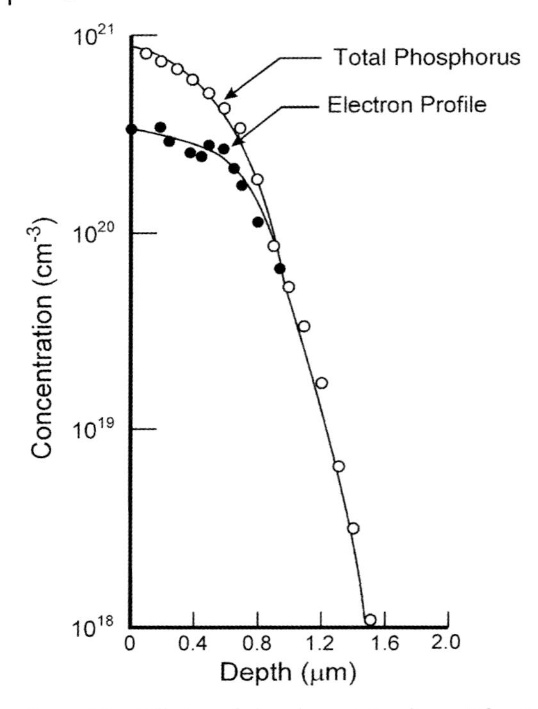

Figure 5.2 Diffusion of phosphorus into silicon. (After Fair and Tsai [3].)

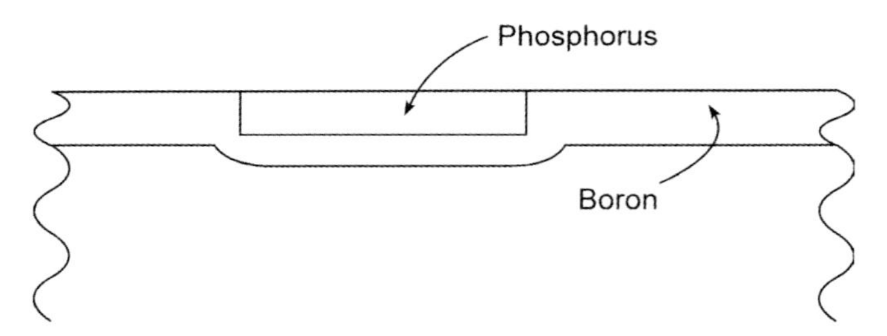

Figure 5.3 Emitter push.

two arsenic atoms can combine with a vacancy to form a complex which diffuses slowly.

5.2.3 Diffusion of Arsenic in Silicon

At lower concentrations, one of the arsenic atoms leaves the complex:

$$\begin{aligned} &[VAs_2]^o \rightleftharpoons [VAs]^o + As^+ + e^- \\ &V^- + As^+ \rightleftharpoons [VAs]^o \end{aligned} \tag{5.16}$$

The V–As complex is the dominant diffusing species in arsenic-doped silicon.

5.2.4 Diffusion of Boron in Silicon

Unlike phosphorus and arsenic in silicon, boron is rather well behaved. Boron is an acceptor, and so the dominant vacancy in boron-doped silicon is positively charged. The diffusion coefficient for boron is given by:

$$D_{\mathrm{B}} = D_{\mathrm{i}}^{+}\frac{n_{\mathrm{i}}}{n} = D_{\mathrm{i}}^{+}\left(\frac{p}{n_{\mathrm{i}}}\right) \tag{5.17}$$

And so the diffusion coefficient for boron in silicon increases approximately linearly with boron concentration, as shown in Figure 5.4.

The diffusion coefficients for dopants in silicon are strongly dependent on dopant concentration. The diffusion coefficient for a dopant depends not only on its concentration, but also on the concentration of other dopants.

This behavior is quite different from metals, where the diffusion coefficients in alloys are relatively independent of concentration.

Diffusion processes in compound semiconductors are even more complicated than in silicon. There has probably been more work done studying diffusion in silicon than in all of the compound semiconductors put together, and so there is a lot that is not known about diffusion in compound semiconductors.

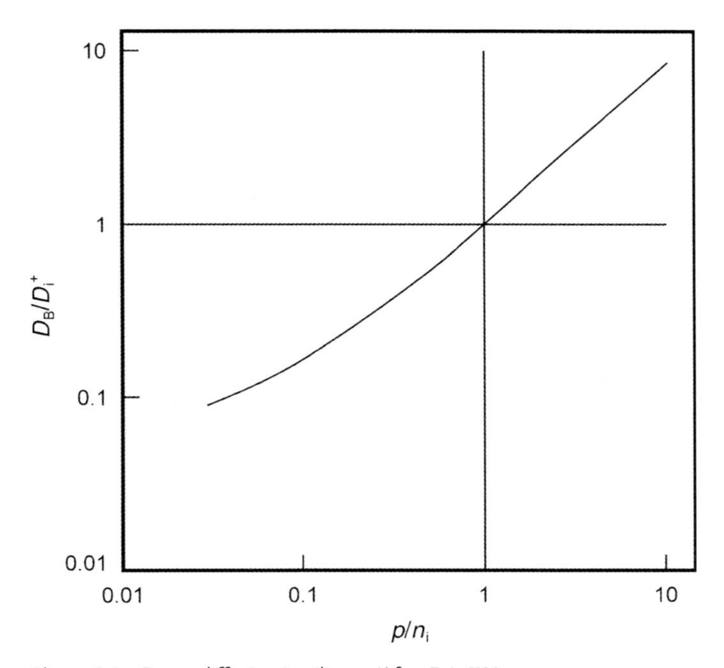

Figure 5.4 Boron diffusion in silicon. (After Fair [2].)

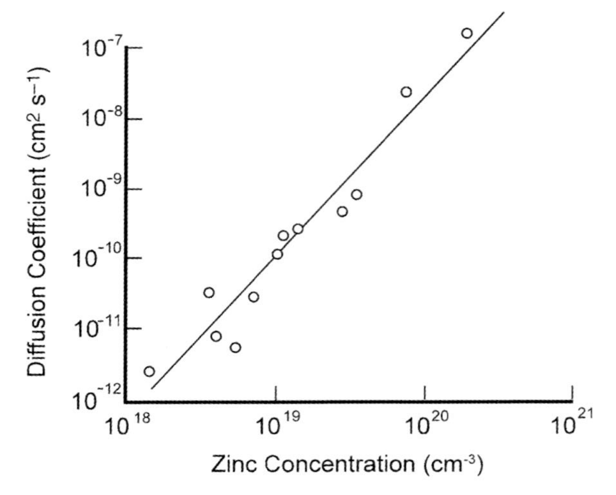

Figure 5.5 Zinc diffusion in GaAs. (After Tuck [4].)

5.3
Diffusion of Zinc in GaAs

An example of complex behavior is zinc diffusion in GaAs. This has been studied extensively. When GaAs is doped with zinc, most of the zinc atoms substitute for gallium on gallium sites, where the zinc acts as an acceptor. However, a small fraction of the zinc atoms are interstitial, and these diffuse much more rapidly than the substitutional zinc atoms, and so the diffusion process is dominated by interstitial zinc. The interstitial zinc atoms are donors, and so are positively charged. The creation of a zinc interstitial leaves a vacancy on a gallium site which becomes positively charged. The result of all this is that the diffusion coefficient for zinc in GaAs increases with the square of the zinc concentration, as illustrated in Figure 5.5.

5.4
Recombination Enhanced Diffusion

The diffusion process can also be influenced by recombination events [5, 6]. When an electron in the conduction band recombines with a charged dopant, the energy given off by the recombination can generate a photon in a direct gap material, or it can dissipate as heat, that is by generating phonons. In an indirect gap material such as silicon, it will always dissipate as heat. The heat increases the kinetic energy of the dopant atom as well as the kinetic energy of other atoms around where the recombination occurs. Some very small fraction of the time, this extra energy results in the dopant atom making a diffusion jump. The activation energy for the diffusion jump is decreased by the recombination energy, E_R, which is illustrated in Figure 5.6.

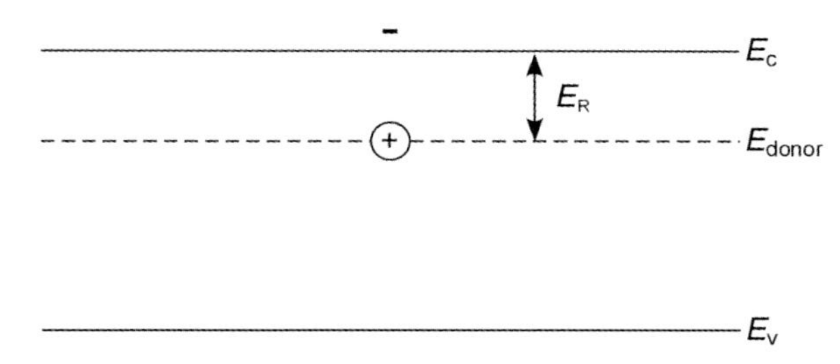

Figure 5.6 Energy diagram for recombination at a donor.

This process is known as recombination enhanced diffusion (RED). Even though the dopant atom does not jump every time there is a recombination event, the jump rate for the dopant atom can be much faster than the thermal diffusion jump rate.

Recombination events occur when the device is turned on, so that current is flowing through it. So RED can significantly increase the diffusion rate of dopants in operating devices over the diffusion rate in devices which are not in operation. RED was responsible for the failure of early semiconductor lasers. The devices could sit around for a long time at room temperature, but failed fairly quickly, by defect diffusion in the active region of the laser, when they were turned on.

5.5 Doping of Semiconductors

Traditionally, the doping of semiconductors was done by gas-phase diffusion. This was a two-step process. In the first step, called the *pre-dep*, the wafers were exposed to appropriate gas-phase dopants until a sufficient amount of dopant was deposited on or into the surface of the wafers. The second step, known as the *drive-in*, was a diffusion anneal designed to diffuse the dopant into the wafer so that a p/n junction was created at the desired depth below the surface of the wafer.

In order to dope wafers as n-type, the wafers were loaded into a rack alternating with wafers of p-glass, a phospho-silicate glass. The rack was then inserted into a tube furnace, and phosphorus, which has a high vapor pressure, diffused out of the p-glass, into the gas, and deposited on the wafer. For p-type doping, boron trichloride gas, BCl_3, which decomposes and deposits boron on the wafer, was used.

These vapor deposition processes are difficult to control precisely. They are sensitive to surface contamination on the wafer, and they are subject to non-uniformity due to variations in gas flow, non-uniformities in temperature, and so on.

The concentration profile in the wafer during pre-dep is an error function, (erf), because the surface concentration is constant, and it becomes Gaussian during drive-in, because there is no longer a constant supply of dopant at the surface. The total amount of dopant in the wafer does not change during drive-in. We will discuss the mathematics of these diffusion profiles in more detail later.

5.6
Point Defect Generation in Silicon during Crystal Growth

The formation of defects in silicon crystals during crystal growth illustrates a case where the point defects (vacancies and interstitials) are not in thermal equilibrium, and so they can precipitate to form larger defects, which affect device properties [7, 8].

In silicon, the self-interstitials move rapidly, and have a fairly large formation energy. Vacancies, on the other hand, move much more slowly, but have a smaller formation energy. The interface between solid and liquid silicon is expected to be an ideal source and sink for both self-interstitials and vacancies, so the concentration of each type of defect can be assumed to be at its equilibrium value at the interface. Since the vacancies have a lower formation energy than interstitials, there are many more vacancies than interstitials at the interface.

During rapid growth, when the interface is moving rapidly compared to how fast the point defects can move by diffusion, the defect concentrations are fixed at the interface, but, in the crystal, the interstitials and vacancies annihilate each other so that at some distance from the interface there are only vacancies left, as illustrated in Figure 5.7.

The crystal temperature drops in the crystal away from the interface, and so the equilibrium vacancy concentration also decreases. At some distance behind the interface the actual vacancy concentration becomes sufficiently greater than the equilibrium concentration, and the vacancies precipitate out as tiny voids.

For slow growth, on the other hand, the interstitials can diffuse faster than the interface is moving. The initial concentrations at the interface are the same, but as the interstitials are eliminated by combining with vacancies, they are replaced by more

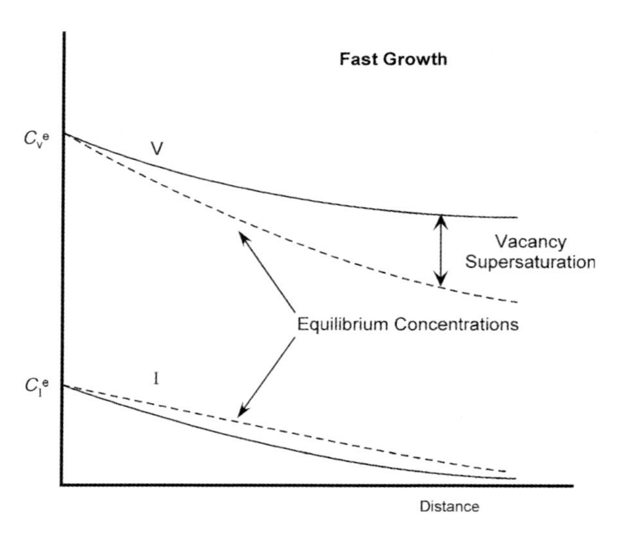

Figure 5.7 Defect concentrations in a silicon crystal behind the interface during rapid growth. The dashed lines indicate the equilibrium concentrations.

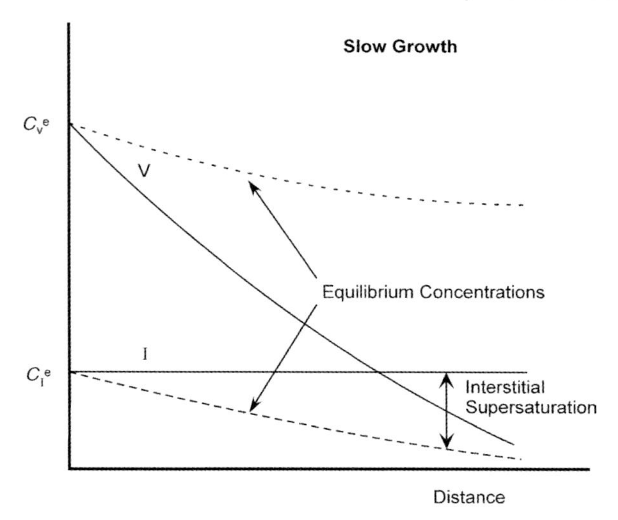

Figure 5.8 Defect concentrations in a silicon crystal behind the interface during slow growth. The dashed lines indicate the equilibrium concentrations.

interstitials diffusing from the interface. So, at some distance ahead of the interface, all of the vacancies have disappeared, and there are only interstitials left, as illustrated in Figure 5.8.

As the crystal cools locally, the interstitial concentration exceeds the equilibrium value, and the interstitials precipitate out as planar stacking faults on (1 1 1) planes.

As illustrated schematically in Figure 5.9, the concentration of defects in the crystal is dominated by interstitial precipitates at slow growth rates, and by vacancy precipitates at rapid growth rates.

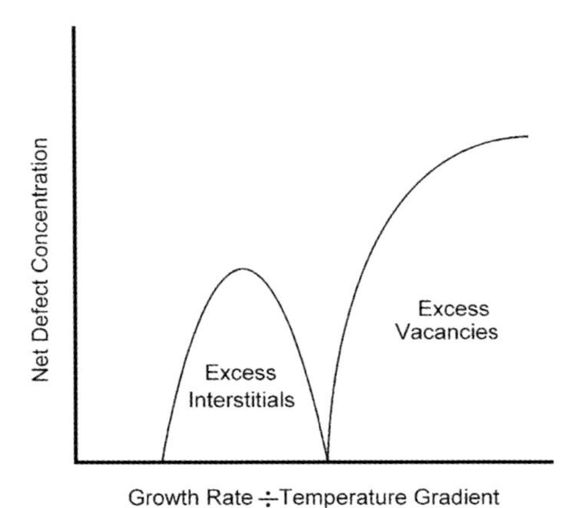

Figure 5.9 Net defect concentration in silicon plotted against the growth rate divided by the temperature gradient.

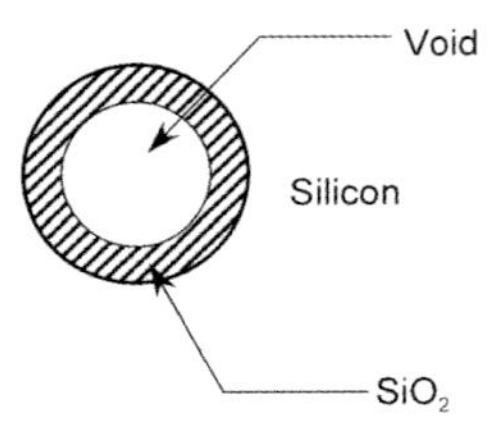

Figure 5.10 Microvoid in silicon formed by the precipitation of vacancies. These microvoids are typically only several tens of nanometers in diameter. Oxygen precipitation forms a layer of SiO_2 inside the microvoid.

The precipitation process depends on the temperature gradient, and the precipitate concentration is found experimentally to scale with the growth rate divided by the temperature gradient, as in Figure 5.9. The preferred growth condition is in the gap where the density of both types of defect is low [9, 10].

The stacking faults formed by interstitial precipitation tend to be relatively large because of the mobility of the interstitials, so their presence and the damage they do to device yields have been apparent for a long time. The vacancies precipitate out as microvoids, and the oxygen which is present in Czochralski silicon migrates to these small voids, and forms a stabilizing layer of silica inside the void, as illustrated in Figure 5.10.

These voids went undetected until the devices got so small that these defects began to reduce the breakdown voltage of CMOS gates if they were in the silicon under the gate. Researchers then went looking for the source of the problem and found them.

5.7
Migration of Interstitials (and Liquid Droplets) in a Temperature Gradient

One would expect that point defects would migrate from the hot end of a sample towards the cold end, that is, they would diffuse towards lower temperatures in a temperature gradient, because the defects move more rapidly at higher temperatures, and the temperature gradient would bias their motion. However, interstitials in silicon go the other way; they migrate towards higher temperatures.

Liquid droplets in a solid alloy also do this. The liquid droplets which form are rich in a second component which lowers the melting point of the alloy. The composition of the liquid droplet will be fairly uniform throughout a small droplet because of rapid liquid diffusion. The liquid will be below the melting point of the alloy at the lower temperature in the temperature gradient, and above the melting point at the higher temperature. So the liquid droplet will melt into the solid at the higher temperature, and the liquid will freeze at the lower temperature. The droplet will move towards higher temperatures.

5.8 Oxygen in Silicon

Czochralski silicon, which is commonly used for device fabrication, contains about 10^{18} oxygen atoms per cubic centimeter. There are 5×10^{22} silicon atoms per cubic centimeter in silicon. During Czochralski growth of silicon, the silicon is melted and held in a crucible lined with high purity silica. Some of the silica dissolves into the liquid, which introduces oxygen into the liquid, and some of the oxygen then ends up in the crystal. The concentration of oxygen which gets into the crystal is in equilibrium at about 1300 °C. Below that temperature, it will tend to precipitate. Microvoids can form due to an excess of vacancies during crystal growth, and these are stabilized by the formation of a layer of oxide inside the microvoid [11]. However, the oxygen can precipitate without the aid of vacancies, in the form of small crystoballite crystals in the silicon. The oxygen diffuses interstitially to join these precipitates. An SiO_2 molecule in the precipitate occupies the same volume as two silicon atoms in the matrix, and so when precipitation occurs the excess silicon atoms are displaced to interstitial sites. The interstitials precipitate out as extrinsic stacking faults, which are bounded by partial dislocations. These defects are one atom layer thick sheets of extra atoms, lying on a (1 1 1) plane, which can grow to be hundreds of microns in diameter. The SiO_2 precipitate can also produce enough stress to punch out dislocation rings. These dislocations and the partial dislocations bounding the stacking faults provide nucleation sites for the precipitation of unwanted impurity elements, such as Fe, Ni, Cu, and so on, which may be present.

5.9 Gettering

Getter is a strange word. It originated with vacuum tube technology, where a metal, such as titanium, was deposited on the interior surface of the glass envelope. This metal would absorb or “get” unwanted ions and atoms inside the vacuum tube, extending the life of the tube. The deposited metal came to be known as a getter. This terminology has carried over into semiconductor technology, where it refers to a material or process intended to “get” unwanted impurities in the semiconductor.

The precipitation of oxygen is used to “getter” unwanted impurities during silicon processing. This is known as intrinsic gettering [12]. Oxygen is intentionally caused to precipitate deep in the wafer so that unwanted impurities will precipitate on the defects created by the oxygen precipitation. The key to doing this is that growing an oxide layer on silicon at a temperature below 1300 °C reduces the oxygen concentration in the Czochralski silicon. Oxide layers are typically grown in air or in an oxygen atmosphere at temperatures of 1050 to 1100 °C or so. The layers grow by the diffusion of oxygen through the oxide layer to the silicon interface, where new oxide forms. Some of the oxygen already in the silicon also goes into the oxide layer, reducing the oxygen concentration in the silicon next to the oxide to the equilibrium value at the annealing temperature.

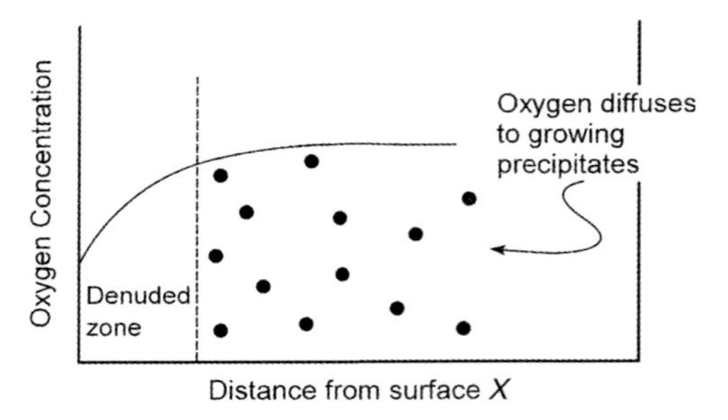

Figure 5.11 Denuded zone for intrinsic gettering.

So a layer which has a lower oxygen concentration, called a denuded zone, can be created at the surface of the wafer, as illustrated in Figure 5.11. The wafer is then heat treated so that the oxide in the bulk of the wafer precipitates, creating defects, which getter impurities, as outlined above.

There are also processes which are referred to as extrinsic gettering, which involve doing something to the back side of the wafer to induce precipitation there. These include scratching or abrading the back side of a wafer to create damage. The backside of a wafer can also be damaged with a high velocity jet of water. Polysilicon, which contains many grain boundaries, is also deposited on the back side of wafers to provide a sink for impurities. Intrinsic gettering usually has a cost advantage because it can be done in a batch mode in an annealing furnace.

5.10
Solid-State Doping

As devices have decreased in size, the diffusion anneal to remove ion implantation damage and to make the dopants electrically active can result in excessive motion of the dopant. The ion implantation process can create too much damage for critical features.

Solid-state doping is used to control the small scale features at the base region in bipolar transistors, and in the gate region in CMOS transistors, as illustrated in Figure 5.12 for a bipolar transistor. First, a patterned layer of field oxide is grown into the silicon to provide isolation between adjacent devices. p+-doped polysilicon is deposited on the silicon, and then oxide is either deposited or grown on the poly. The poly and the oxide are patterned to define the active region, and annealed so that the p+-dopant diffuses into silicon. Next, a thin oxide layer is either deposited or grown on the surface. Boron is implanted through the thin oxide to make the lightly doped p base region under the emitter. The implanted boron does not penetrate through the thick oxide. The thin layer of oxide is then removed to expose the silicon in the emitter region, and n+-doped poly is deposited in the opening. The n-type dopant is diffused into the silicon from the n+-poly to make the emitter. The diffusion distances are

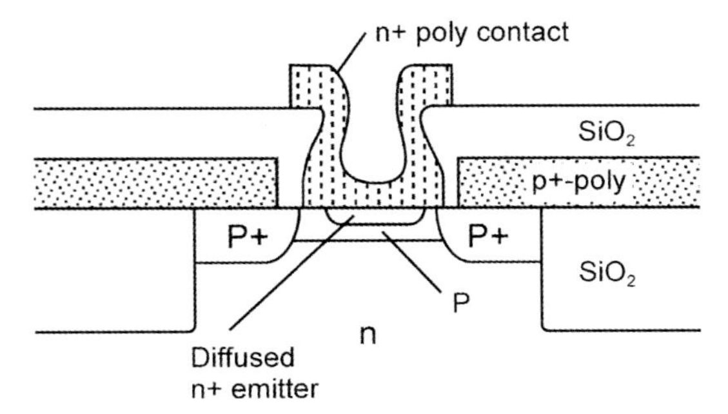

Figure 5.12 Solid state doping from polysilicon in bipolar device fabrication. Both the extrinsic base and the emitter are diffused from doped polysilicon.

extremely small, and the process can be accurately controlled with solid-state doping. The process described above is called "self-aligned", because the location of all the features is controlled by the initial opening which was made in the silica and p + -poly layers. The doped poly layers provide contacts to the silicon.

References

1 Murarka, S.P. and Peckerar, M.C. (1989) *Electronic Materials*, Academic Press, San Diego, CA, p. 182.
2 Fair, R.B. (1981) *Silicon Integrated Circuits, Part B* (ed. D. Kahng), Academic Press, New York, NY, p. 1.
3 Fair, R.B. and Tsai, J.C.C. (1975) *J. Electrochem. Soc.*, **122**, 1689.
4 Tuck, B. (1974) *Introduction to Diffusion in Semiconductors*, Peter Peregrinus, Herts, UK.
5 Kimerling, L.C. (1978) *Solid State Electron.*, **21**, 1391.
6 Kimerling, L.C. (1983) *Physica B + C*, **116**, 1.
7 Voronkov, V.V. and Falster, R. (1999) *J. Appl. Phys.*, **86**, 5975.
8 Voronkov, V.V.and Falster, R. (2001) *J. Cryst. Growth*, **226**, 192.
9 Brown, R.A., Wang, Z.H.,and Mori, T. (2001) *J. Cryst. Growth*, **225**, 97.
10 Wang, Z.H.and Brown, R.A. (2001) *J. Cryst. Growth*, **231**, 442.
11 Nakamura, K., Saishoji, T.,and Tomioka, J. (2002) *Solid State Phenom.*, **82–84**, 25.
12 Wilkes, J.G., Benson, K.E.,and Lin, W. (2000) *Handbook of Semiconductor Technology*, Vol. 2, *Processing of Semiconductors* (ed. K.A. Jackson), Wiley-VCH, Weinheim, p. 1.

Further Reading

Murarka, S.P.and Peckerar, M.C. (1989) *Electronic Materials*, Academic Press, San Diego, CA.
Plummer, J.D., Deal, M.D.,and Griffin, P.B. (2000) *Silicon VLSI Technology*, Prentice-Hall, Upper Saddle River, NJ.
Campbell, S.A. (1996) *The Science and Engineering of Microelectronic Fabrication*, Oxford, New York, NY.

Problems

5.1. Single crystals of silicon grown by the Czochralski method contain oxygen in excess of their room temperature solubility limit. The oxygen solubility in silicon at 1100 °C is $5 \times 10^{16}/\text{cm}^3$, and the diffusion coefficient for oxygen in silicon is given by:

$$D_{\text{Oxygen}} = 0.19\exp(-2.54\ \text{eV}/kT)\ \text{cm}^2\ \text{s}^{-1}$$

Plot the oxygen concentration profile in a wafer with an initial oxygen concentration of 20 ppma after a 48 h anneal at 1100 °C in an inert ambient.

6
Ion Implantation

6.1
Introduction

Ions of widely different energy ranges are used in various applications in semiconductor technology, as illustrated in Figure 6.1.

In ion beam deposition, low energy ions are used to minimize damage to the substrate. For comparison, an atom with thermal energy from an evaporative source has an energy which is a fraction of an electron volt. An ion incident at a surface with about 30 eV energy has enough energy to displace an atom in most substrates. For sputtering and cleaning of surfaces, argon ions with an energy of the order of 1 keV are frequently used. This is enough energy to remove (sputter) atoms from the

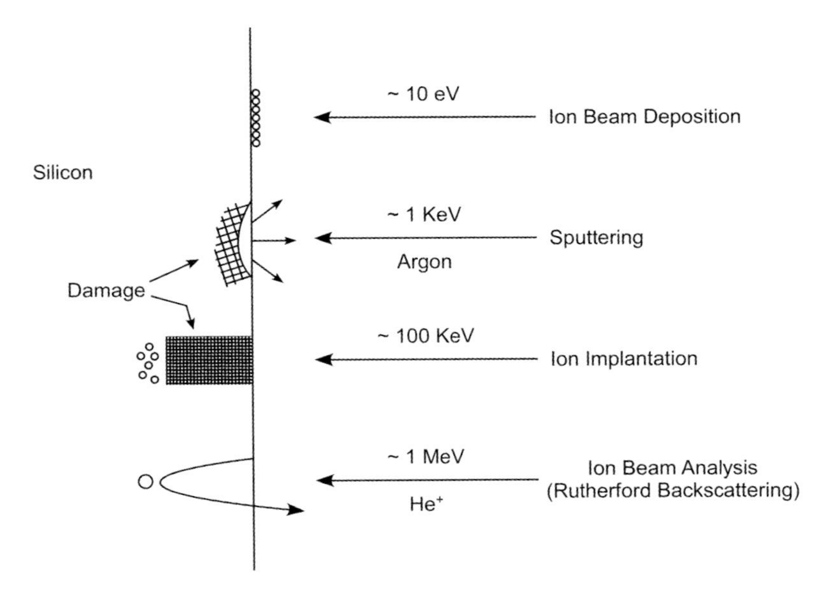

Figure 6.1 Ions of various energies incident on silicon.

Kinetic Processes: Crystal Growth, Diffusion, and Phase Transitions in Materials. Kenneth A. Jackson

ISBN: 978-3-527-32736-2

surface, without doing too much sub-surface damage. For ion implantation, ions with energies of about 100 keV are used. They create significant damage in the substrate. High energy helium ions (alpha particles) are used for ion beam analysis (Rutherford backscattering) which will be discussed below.

Today, most of the dopants are introduced into semiconductor wafers by ion implantation. This method is much more costly than vapor deposition, but the dopant concentration can be carefully controlled since it depends on the ion current. The dopant distribution can be made very uniform by the use of planetaries, which rotate the substrate in a controlled pattern through the ion beam. The ions are implanted with enough energy so that they go right through minor amounts of surface contamination. The ions are typically implanted with about 70 keV energy, and they end up about 100 nm into the sample.

6.2 Ion Interactions

The ions slow down gradually along their path by interaction with electrons in the substrate, and they can also be scattered by making direct collisions with the nuclei of the atoms in the substrate. The strength of these two types of interaction depends on the type of incident ion, on its energy, and on the substrate material. The energy loss due to these two processes is shown in Figure 6.2 for boron, phosphorus and arsenic implanted into silicon.

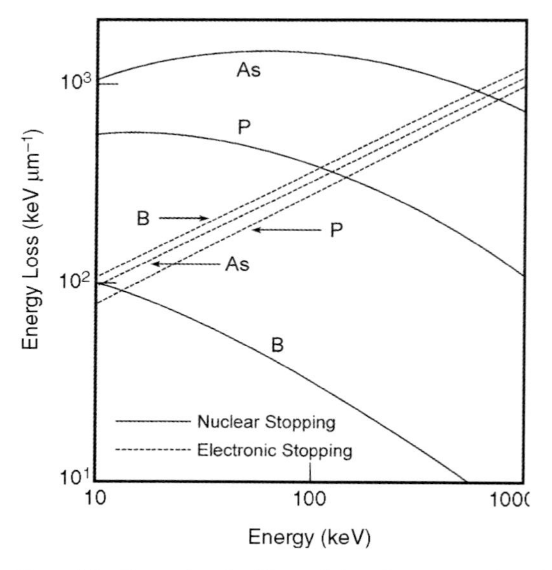

Figure 6.2 Energy loss in silicon of phosphorus, arsenic and boron ions due to nuclear collisions and to electronic interactions. (After Smith [1].)

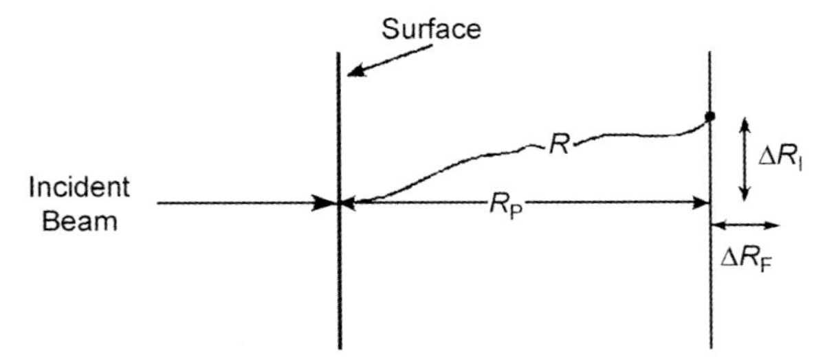

Figure 6.3 Incident ions interact with the electrons in the substrate, and are scattered by collisions with nuclei, and come to rest at the end of range, R_P.

The probability of an incident ion colliding with the nucleus of an atom is about what is to be expected for a sphere the size of the ion nucleus hitting a sphere the size of an atomic nucleus. The effective area within which the ion will make a collision with the nucleus is known as the nuclear collision cross section.

The interaction with the substrate increases as the ions slow down, and so the deceleration rate of the ions increases as they penetrate into the substrate. During implantation, most of the ions end up more or less at the same distance into the substrate, which is known as the end of range, as illustrated in Figure 6.3.

Values for the end of range boron, phosphorus, and arsenic ions implanted into silicon are shown in Figure 6.4.

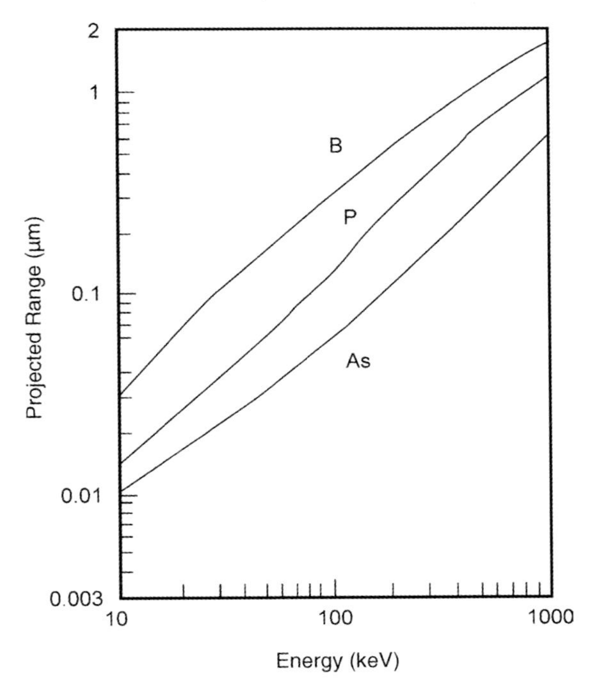

Figure 6.4 End of range for boron, phosphorus, and arsenic ions implanted into silicon. (After Smith [1].)

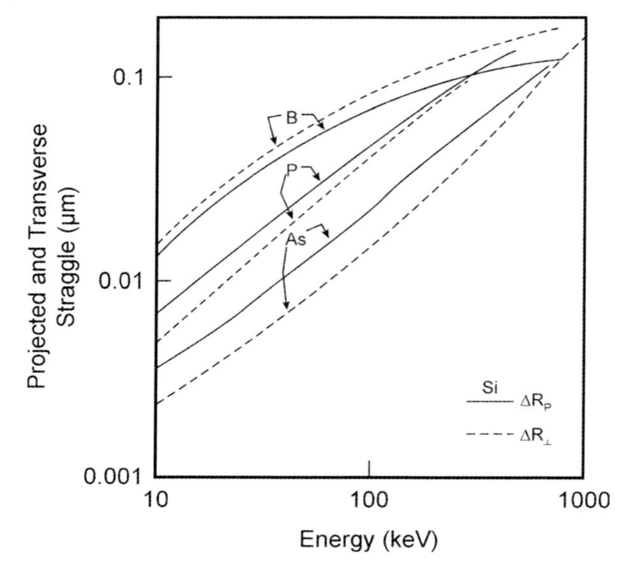

Figure 6.5 Straggle for ions implanted into silicon. (After Smith [1].)

There is some scatter in where the ions are located at the end of range. This is known as straggle. There is straggle both parallel and perpendicular to the incident path of the ions, as illustrated in Figure 6.5.

The straggle both parallel and perpendicular to the incident path of the ions is important in determining where the ions will finally locate. The lateral straggle is especially important for patterned implants, as illustrated in Figure 6.6.

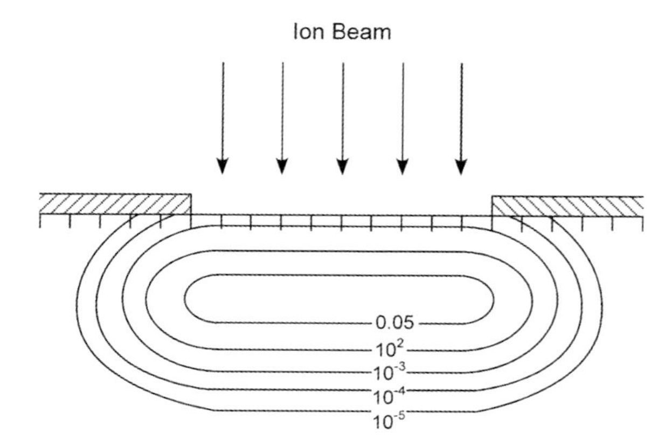

Figure 6.6 Position of ions resulting from parallel and perpendicular straggle.

6.3 Implantation Damage

The collisions displace atoms in the substrate, and create defects. The amount and type of damage which the ion creates in the crystal depends on both the energy of the incident ion and on its mass, as illustrated in Figure 6.7.

The incident ions can create interstitials and vacancies, some of which recombine. There is a standard Monte Carlo program called TRIM which is used to evaluate how many point defects are generated by nuclear collisions. However, the residual damage is usually much less than predicted by the TRIM program. A lot of the initial damage caused by the ions anneals out because of the high local temperature. Recent molecular dynamics studies suggest that most of the damage is concentrated at the end of range. Unfortunately, the MD simulations have only been done for relatively low energy incident ions, because the disturbed volume for higher energy ions is so large that it can only be modeled with large multiprocessor machines.

During ion implantation, the collision of the ion with the nucleus of a substrate atom can result in a focusing collision. This happens when the incident ion hits the end of a row of atoms, where the atoms are closely spaced in the row. The whole row of atoms is displaced, so that a vacancy is created at one end of the row, and an interstitial is created at the other end of the row. This creates a vacancy and an interstitial which are well separated in space, and so are much less likely to recombine than if they were created next to each other.

At the end of range, the residual energy of the ion can make a local hot spot, called a thermal spike, which can be sufficient to melt a small volume of the substrate. On rapid cooling the material in the hot spot can become amorphous. The damage to the substrate depends on the total dose, as illustrated in Figure 6.8.

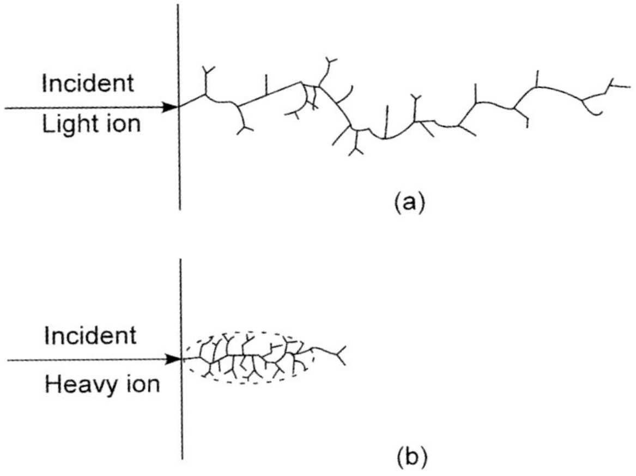

Figure 6.7 Light ions (a) penetrate farther into the substrate than heavy ions (b) with the same energy. Heavy ions create more local damage.

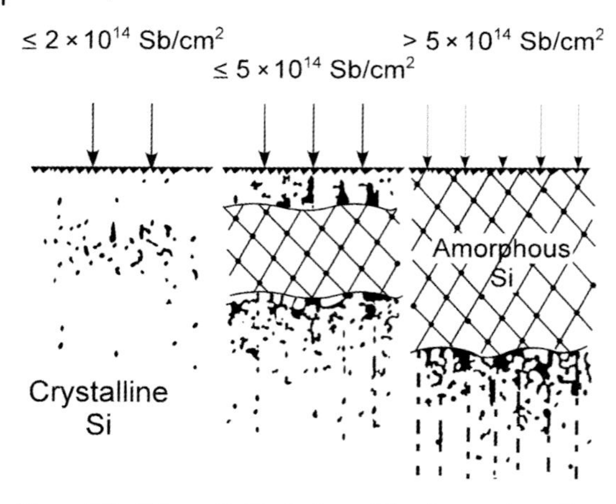

Figure 6.8 Schematic illustration of the damage in a silicon substrate after various doses of antimony ions.

A heavy dose can create an amorphous layer. Figure 6.9 is a TEM picture of a buried amorphous layer which is about 100 nm thick in silicon after implantation with argon. The damage above the amorphous layer is largely microtwins. Below the amorphous layer the damage in the substrate consists of small dislocation loops and point defect clusters.

The critical dose which is sufficient to create a buried amorphous layer depends on the type of ion and the implant temperature. Heavier ions create an amorphous layer at a lower dose, and there is a temperature for each type of ion above which an amorphous layer will not form.

One interesting recent result from modeling of ion implantation into a metal is that the molten zone caused by the thermal spike at the end of range expanded the lattice so much that the surface of the sample was lifted one or two monolayers above the rest of the surface. When the hot spot cooled down, the surface did not relax back to its original position, but stayed in the elevated position. This created a lot of vacancies where the thermal spike had been. This will not happen in silicon, which contracts on melting.

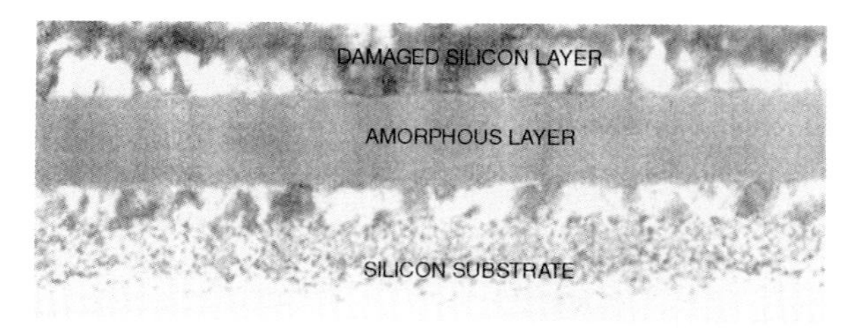

Figure 6.9 Buried amorphous layer in silicon created by 200 keV argon ions. (Courtesy Marcus [2].)

In silicon processing, after implantation, the substrate is annealed to remove the implantation damage, and to get the dopants onto substitutional sites in the lattice where they will be electrically active. Often a low temperature anneal is used to remove the defects, and a shorter, higher temperature anneal is used to make the dopants electrically active. It is more difficult to anneal out heavy damage than lighter damage. Solid phase epitaxial regrowth of an amorphous layer requires a longer anneal than is required to remove lighter damage.

The defects created by ion implantation can play a major role in the diffusion process and defect interactions make it much more complex. When the annealing time is long, these defects anneal out relatively quickly, so that they are not present during most of the anneal, and so they do not significantly affect diffusion. However, as device feature sizes have decreased, the distance the dopants have to move decreases, and annealing times decrease. The defects are present for more of the total annealing time, so their effect on dopant diffusion must be taken into account. The total annealing schedule must be very carefully worked out, because the final positions of the dopants will depend on the total time at temperature which they experience.

The dopant diffusion process depends on the Fermi level, defect interaction, and complex formation, and can only be described by a messy set of coupled equations. The details of many of these interactions are unknown, and are the subject of current research.

There is a project at Stanford to develop the mathematics of these interactions, and to develop computer programs to predict the resulting dopant distributions. The computer program is Stanford University Process Engineering Module (SUPREM), and the current version is SUPREM V. The program is continuously updated to improve the accuracy, and to take into account more sophisticated understanding of defect interactions on the diffusion process. The computer program can also be used to calculate oxide layer growth. Information about commercial versions of SUPREM can be found on the internet.

6.4 Rutherford Backscattering

Rutherford backscattering (RBS) provides a convenient tool for determining the location of implanted dopants. RBS can also be used to estimate the degree of damage to a crystal and to locate buried amorphous layers created by the implantation process. The method consists of impinging a beam of helium ions onto the substrate, and then measuring the energy distribution of the ions which are backscattered from the substrate. The energy which is lost by a helium ion when it is scattered by the nucleus of an atom in the substrate depends on the mass of the nucleus, and on the scattering angle.

Both the total energy and the total momentum are conserved in the collision, as illustrated in Figure 6.10. For example, if the incident ion and the nucleus it strikes have exactly the same mass, then in a direct collision the incident ion will stop and the

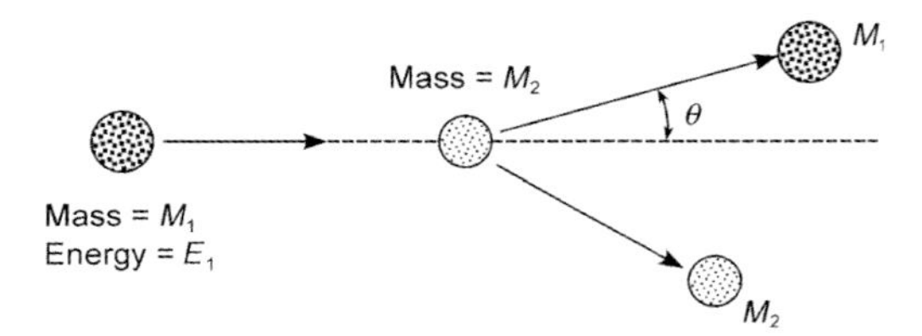

Figure 6.10 Scattering of an ion of mass M_1 by a nucleus of mass M_2.

nucleus will depart with the velocity of the incident ion. This preserves both energy and momentum. If the incident ion has a larger mass than the target nucleus, both will continue in the direction of the incident ion. If the incident ion has a smaller mass than the target nucleus, then the incident ion bounces backward, that is, it will be backscattered. Helium ions are backscattered from everything except hydrogen. Of course, there is an angular variation in the scattering, depending on how the incident ion strikes the target nucleus but, in RBS, the detector for the backscattered helium ions is placed so that only the ions which are backscattered more or less straight back are counted, and the backscattered angle is fixed.

The more massive the target nucleus, the greater will be the speed of the backscattered helium ion. The difference between the energy of the incident ion and its energy when it is backscattered is an energy loss. This energy loss is greater for scattering from light atoms than from heavy atoms.

Silicon is a relatively light atom, and so an incident helium ion loses a fairly large amount of energy on backscattering from a silicon nucleus.

So there is a specific amount of energy which a helium ion can lose on being backscattered from silicon at a given angle. A helium ion can also lose energy, for example by electron interactions, if it penetrates into the silicon, scatters from a silicon nucleus, and then comes back out. So an ion which is backscattered by a silicon atom which is below the surface will have lost more energy than an ion backscattered from a silicon atom at the surface of the sample. This means that there is a silicon edge in the energy loss spectrum, and similarly for other elements.

The only way for a helium ion to have an energy loss which is less than the silicon edge is to be backscattered from a heavier atom, and thus the presence of heavier atoms can be detected in the energy loss region before the silicon edge.

The energy loss due to scattering from any particular type of atom at the surface is known, but the helium ion will also lose extra energy due to electronic interactions if the heavier atom is not at the surface. This loss is known as a function of depth. The depth distribution of a particular species of dopant atom can thus be determined.

The energy loss spectrum from a silica sample which was implanted with germanium is shown in Figure 6.11. The energy of the incident helium ions was 2 MeV. The peak at about 1.6 MeV is scattering from the implanted germanium, the silicon edge is at about 1.2 MeV, and the oxygen edge is at about 750 keV.

In order to make a measurement like this, the sample is usually oriented in a non-channeling direction.

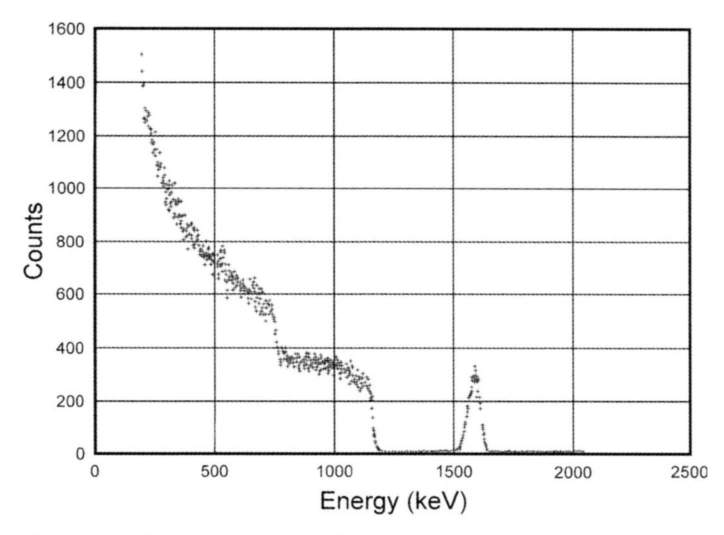

Figure 6.11 Energy loss spectrum for germanium implanted into silica. (Courtesy Minke [3].)

6.5 Channeling

Ions can "channel" down the open spaces in a lattice, such as those illustrated in Figure 6.12, if they are incident in the right direction. There is a steering effect due to the atoms along the sides of the channel which tends to keep the ion in the channel.

When a silicon crystal is irradiated in a channeling direction, there will be scattering from the first atom in each column of atoms, but the incident ions which do not strike a surface atom will tend to proceed down the open channels, although they will still lose energy due to electronic interactions. Typical back-scattering channeling yield from a perfect crystal is shown in Figure 6.13. The data labeled Virgin Crystal Channeling Orientation has a peak due to scattering from silicon atoms at the surface.

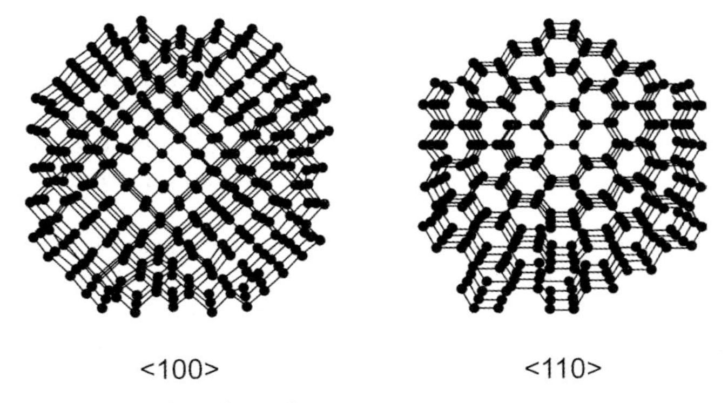

Figure 6.12 Channels in silicon.

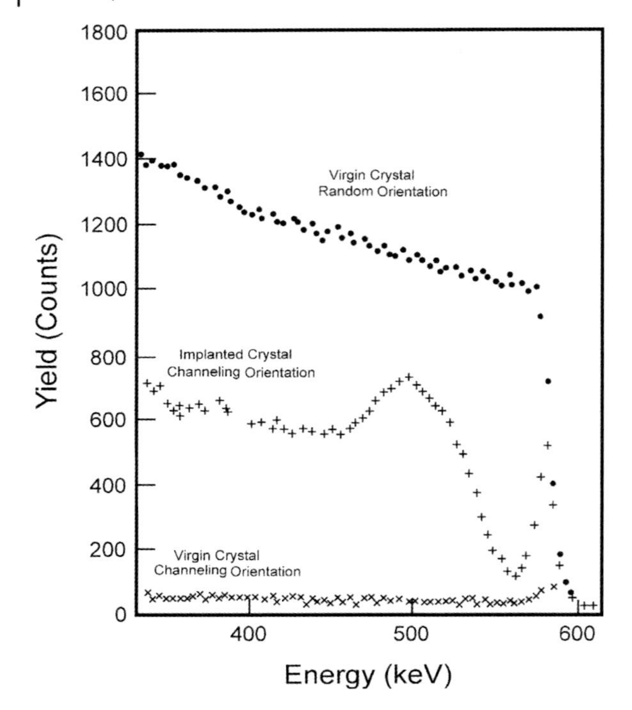

Figure 6.13 Energy loss spectrum from a perfect crystal in a channeling and in a random direction, and from a crystal which was damaged by boron ion implantation. (After Morehead and Chowder [4].)

The rest of the ions disappeared down the channels and were not scattered back to the detector. So it appears that there were silicon atoms only at the surface.

By contrast, the backscattering data from the Virgin Crystal randomly oriented sample in Figure 6.13 indicates that ions were scattered from atoms throughout the crystal. This illustrates the difference in yield between a channeling and a random incident direction.

Channeling can be used to evaluate the perfection of a crystal, since the distance a channeling ion can penetrate into a crystal depends on its perfection. The center curve in Figure 6.13 was taken in the channeling direction, after implantation with boron. The backscattered yield has increased significantly over that from the virgin crystal, due to the damage to the crystal structure created by the implant, but it is still significantly less than scattering from the randomly oriented sample. The difference provides information about the degree of damage.

An ion in a channel is unlikely to have a nuclear collision, and it samples a lower electron density, so it penetrates much more deeply into the sample than an ion which is incident in a random direction. Even with a random incident direction, some of the ions will be scattered into a channel, and so there are some ions which penetrate much more deeply into the sample than the average end of range. This creates a spread in the end of range where the implanted ions end up.

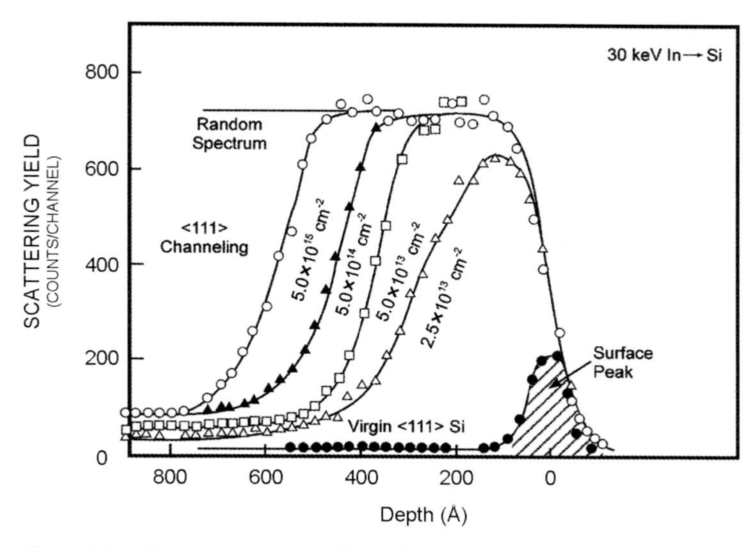

Figure 6.14 Energy loss spectrum from a buried amorphous layer. (After Thompson *et al.* [5]). Some of the atoms appear to be outside of the surface due to instrument broadening.

If there is a buried amorphous layer, the backscattering from the amorphous region will be similar to scattering from a randomly oriented crystal, as illustrated in Figure 6.14. The location of a buried amorphous layer can be detected using channeling ions backscattered from the amorphous layer.

The energy loss spectrum in Figure 6.14 is from a sample implanted with various doses of indium which have created amorphous layers of various thickness in the silicon. Only the surface peak is present in the virgin sample, since the helium ions were incident in the ⟨1 1 1⟩ direction and most of them channeled into the silicon. The backscattering in the channeling direction increases dramatically with the implantation dose as the amorphous layer thickens. The backscattering from the amorphous layer in the indium-implanted sample in Figure 6.14 is similar to the yield from a random sample. This is unlike the backscattered yield from the boron-implanted sample in Figure 6.13, where the yield from the implanted sample is significantly below the yield in a random orientation. This indicates that the boron implantation resulted in a damaged crystal, rather creating an amorphous layer.

6.6 Silicon-on-Insulator

A thin layer of silicon, about 0.1 to 1 μm thick, on top of an insulating layer provides a substrate with superior properties for the fabrication of many semiconductor devices. This configuration reduces leakage currents, provides improved electrical characteristics, and removes the necessity for a deep trench isolation, which is used to eliminate latch-up.

There are three methods currently used to make SOI, and two of them involve ion implantation. silicon implanted with oxygen (SIMOX) involves the implantation of a wafer with a heavy dose of oxygen. The wafer is then annealed, producing a buried layer of SiO_2. The silicon above the buried oxide layer, which is where the devices will be made, retains significant damage from the implantation.

Figure 6.15 shows a buried silica layer produced by SIMOX. A layer of epi silicon has been deposited on the top silicon layer. There are heavily damaged regions both above and below the oxide.

The method called Smart Cut uses hydrogen ion implantation. After implantation, a layer of oxide is grown, then a thick layer of polysilicon is deposited to provide mechanical support. The wafer is then annealed, so that the hydrogen precipitates as gas bubbles. This splits off a thin layer of silicon from the original wafer. Devices are then fabricated in this layer. The thin layer of silicon, which split off from the wafer, is now supported by the polysilicon, but separated from it by an oxide layer. This method also produces a damaged silicon layer on an insulator.

The third method is called wafer bonding. A p-type layer is created at the surface of an n-type wafer by in-diffusion of an appropriate dopant. The wafer is oxidized, and then bonded to another oxidized wafer. The n-type silicon is then removed from the first wafer, leaving the p-type layer on top of the oxide, supported by the second wafer. This is done with mechanical polishing, followed by etching with an etch that removes n-type silicon faster than p-type silicon. For a wafer which is initially 600 μm thick, stopping within 1 μm of the other side is tricky. Although the selective etch helps, there is still some variability in the final thickness of the layer, as well as in the uniformity of its thickness.

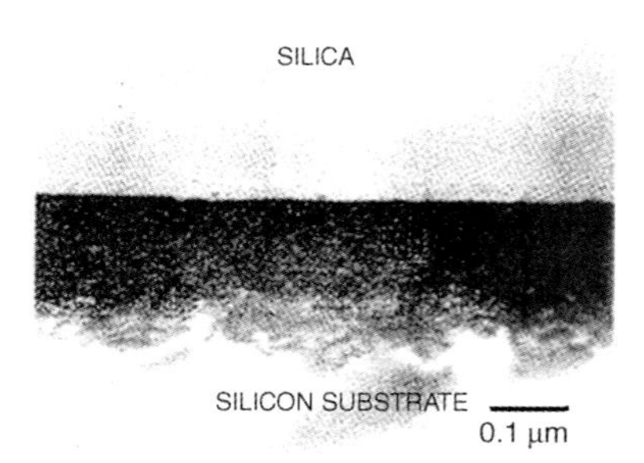

Figure 6.15 SIMOX silicon-on-insulator. (After Pinizzotto [6].)

None of these are ideal methods. Two produce defective layers having a well-controlled thickness, and the third produces a good quality wafer, but has poor thickness control. All three of these methods are used in production by different manufacturers of ICs.

References

1 Smith, B. (1977) *Ion Implantation Range Data for Silicon and Germanium Device Technologies*, Oxford, New York, NY.
2 Marcus, R.B. *Personal communication.*
3 Minke, M. (2002) PhD Thesis, Univ. of Arizona.
4 Morehead, F.F. and Chowder, B.L. (1971) *Proceedings of the First International Conference on Ion Implantation* (eds F. Eisenand L. Chadderton), Gordon and Breach, New York, NY.
5 Thompson, D.A., Golanski, A., Haugen, H.K., Stevanovic, D.V. Cater, G., and Cristodoulides, C.E. (1980) *Radiat. Effects*, **52**, 69.
6 Pinizzotto, R.F. (1984) *Mater. Res. Soc. Symp. Proc.*, **27**, 265.

Further Reading

Murarka, S.P. and Peckerar, M.C. (1989) *Electronic Materials*, Academic Press, San Diego, CA.
Poate, J.M., Foti, G., and Jacobson, D.C. (eds) (1983) *Surface Modification and Alloying*, Plenum, New York, NY.
Mayer, J.W., Erickson, L., and Davies, J.A. (1970) *Ion Implantation in Semiconductors*, Academic Press, New York, NY.

Problems

6.1. Discuss the energy losses experienced by an ion incident on a crystal.
6.2. Discuss the damage generated by ions incident on a crystal.
6.3. Write a report of six pages on the preparation of silicon-on-insulator wafers.

7 Mathematics of Diffusion

7.1 Random Walk

The diffusion of atoms on a lattice occurs by the random motion of atoms from lattice site to lattice site. Insight into this process can be gained by examining the motion of a random walker in one dimension. In one dimension, a random walker makes steps of equal length to the left or to the right in a random sequence. The probability distribution for where a walker ends up after a given time will be derived in this section. In the following section, the diffusion equation will be derived and then a solution to the diffusion equation will be shown to provide this same distribution.

Consider a random walker making steps randomly to the left or to the right. Each step is of length a. After N steps, the final displacement of the walker from its initial position is:

$$R_N = a_1 + a_2 + a_3 + a_4 + \cdots + a_N \tag{7.1}$$

where each step, a_i, is of length a, and either to the right, $+a$, or to the left, $-a$. Plus or minus steps are equally probable, so that on average, the walker moves neither to the left nor to the right. The average value of the final displacement is zero:

$$\overline{R_N} = 0 \tag{7.2}$$

Although the average final position is zero, not all of the walkers will end up precisely at zero. There will be a spread in their final positions. A measure of this spread can be obtained from the square of R_N, which is not zero:

$$R_N^2 = Na^2 + \sum_{i \neq j} a_i a_j \tag{7.3}$$

The first term comes from multiplying each term in Equation 7.1 by itself, which gives a positive term in each case. The second term is the sum of the products of each term with all the other terms. These terms are randomly positive and negative, since there are only four possible combinations of plusses and minuses: $+\,+$, $-\,-$, $+\,-$, and $-\,+$. The first two result in a positive term, and the second two in a negative term.

Kinetic Processes: Crystal Growth, Diffusion, and Phase Transitions in Materials. Kenneth A. Jackson

ISBN: 978-3-527-32736-2

For random positive and negative steps, each of these is equally probable, and so the summation term is zero on average. So the average value of R_N^2 is:

$$\overline{R_N^2} = Na^2 \tag{7.4}$$

and the root mean square displacement is:

$$\sqrt{\overline{R_N^2}} = \sqrt{N}a \tag{7.5}$$

The r.m.s. displacement is proportional to the square root of the number of steps.

Writing the number of steps as the jump rate, Γ, times the time, t: $N = \Gamma t$, the final displacement can be related to the diffusion coefficient. For diffusive motion in one dimension (see Equation 4.5),

$$D = \frac{1}{2} a^2 \Gamma = \frac{a^2 N}{2t} = \frac{\overline{R_N^2}}{2t} \tag{7.6}$$

In three dimensions, $N = 6Dt/a^2$, so that

$$D = \frac{1}{6} a^2 \Gamma = \frac{a^2 N}{6t} = \frac{\overline{R_N^2}}{6t} \tag{7.7}$$

The final position of the walker can be analyzed in more detail, as illustrated in Table 7.1.

The walker moves horizontally, starting at 0, and the probability of its being at any position, R, after each time step is shown in each successive row. The location of each walker is Ra, where R is the number in the first row of the table, and a is the length of each step. After one step, the walker moved either left or right one step, so the probability that the walker is one step to the right is $\frac{1}{2}$, and the probability that it is one step to the left is also $\frac{1}{2}$. After two steps, there is $\frac{1}{2}$ probability that the walker which moved one step to the right at the first step will move a further step to the right, and $\frac{1}{2}$ probability that it will step to the left, returning to the starting position. Similarly, there is $\frac{1}{2}$ probability that the walker which moved one step to the left at the first step will move a further step to the left, and $\frac{1}{2}$ probability that it will step to the right, returning to the starting position. So there is $\frac{1}{4}$ probability that R is equal to $+2$ or -2, and $\frac{1}{2}$ probability that $R = 0$. The probabilities after each step are given by the

Table 7.1 The probability that a walker which starts at position 0 will be in various positions to the left or right of its starting position after making steps randomly to the left or right.

R	**−4**	**−3**	**−2**	**−1**	**0**	**+1**	**+2**	**+3**	**+4**
Starting position					1				
After 1 step				1/2		1/2			
After 2 steps			1/4		2/4		1/4		
After 3 steps		1/8		3/8		3/8		1/8	
After 4 steps	1/16		4/16		6/16		4/16		1/16

binomial expansion, divided by 2^N, where N is the number of steps. The normalizing factor, $(\frac{1}{2})^N$, assures that the total probability that the walker is somewhere is equal to 1, which is given by the sum of the numbers in each row.

The probability distribution given by the table after N steps is given by the normalizing factor, $(\frac{1}{2})^N$, times the standard binomial expansion:

$$P = \frac{1}{2^N} \frac{N!}{\left(\frac{N+R}{2}\right)!\left(\frac{N-R}{2}\right)!} \tag{7.8}$$

Using $\ln N! \approx (N+1)\ln N - N + (\ln 2\pi)/2$, which is more accurate than the usual Stirling approximation, and using $\ln\,(1 \pm R/N) \approx \pm R/N - (R/N)^2/2$, after some algebra, Equation 7.8 becomes:

$$P = \sqrt{\frac{1}{2\pi N}}\exp\left(-\frac{R^2}{2N}\right) \tag{7.9}$$

Below, this result will be compared with the corresponding solution to the diffusion equation.

7.2 The Diffusion Equation

The diffusion equation is derived from Fick's first law by adding conservation of matter [1].

The flux across a plane is given by Fick's first law, Equation 2.5, 4.5, as:

$$J = -D\frac{dC}{dx} \tag{7.10}$$

The number of atoms crossing an area A of a plane in time Δt is:

$$A\Delta t J = -A\Delta t\, D\frac{dC}{dx} \tag{7.11}$$

The difference between the number of atoms crossing plane 1 and plane 2 in Figure 7.1 in time Δt is:

$$A\Delta t\, D\left[\left(\frac{dC}{dx}\right)_2 - \left(\frac{dC}{dx}\right)_1\right] \tag{7.12}$$

This difference must equal the net change in the number of atoms in the volume between the two planes, which is $\Delta CA\Delta x$, where ΔC is the change in the number of atoms per unit volume:

$$A\Delta t\, D\left[\left(\frac{dC}{dx}\right)_2 - \left(\frac{dC}{dx}\right)_1\right] = \Delta CA\Delta x \tag{7.13}$$

which can be rewritten as:

$$\frac{D\left[\left(\frac{dC}{dx}\right)_2 - \left(\frac{dC}{dx}\right)_1\right]}{\Delta x} = \frac{\Delta C}{\Delta t} \tag{7.14}$$

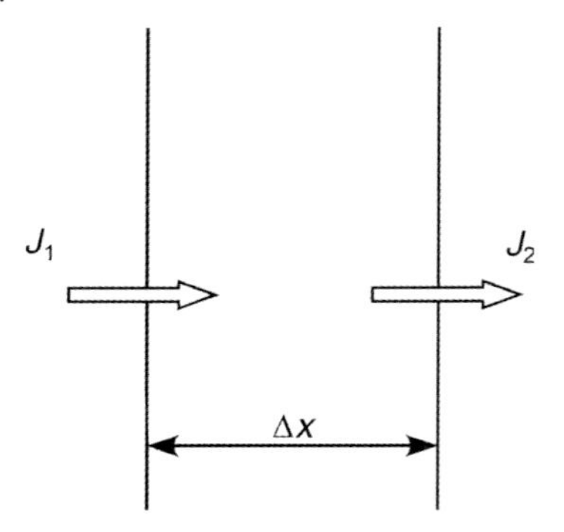

Figure 7.1 Flux of atoms across two planes of area A separated by Δx.

Going to the limit of small Δx and small Δt, this becomes:

$$D\frac{\partial^2 C}{\partial x^2} = \frac{\partial c}{dt} \tag{7.15}$$

This is the time-dependent diffusion equation for a one-dimensional flux. In three dimensions, this becomes:

$$D\nabla^2 C = \frac{\partial c}{dt} \tag{7.16}$$

where ∇^2 is the Laplacian operator, defined as:

$$\nabla^2 = \frac{\partial^2}{\partial x^2} + \frac{\partial^2}{\partial y^2} + \frac{\partial^2}{\partial z^2} \tag{7.17}$$

Heat flow is also described by the diffusion equation. The heat flux in a temperature gradient is given by:

$$J = -k\frac{\mathrm{d}T}{\mathrm{d}x} \tag{7.18}$$

Here k is the thermal conductivity in units such as cal $\mathrm{cm^{-1}\,s^{-1}\,K^{-1}}$. This differs from Fick's first law in that the temperature gradient is given in $\mathrm{K\,cm^{-1}}$, say, whereas the heat flux is in cal $\mathrm{cm^{-2}\,s^{-1}}$. When the diffusion equation, Equation 7.16, is used for thermal diffusion, the thermal diffusivity $\varkappa$ replaces D. Both $\varkappa$ and D have the same dimensions, such as $\mathrm{cm^2\,s^{-1}}$. Instead of the conservation of mass as in the analysis above for chemical diffusion, for heat flow, energy is conserved. This requires that the difference between the heat flux into and out of a volume element raises the temperature by an amount that depends on the specific heat per unit volume, and so the relationship between the thermal conductivity and the thermal diffusivity is $\varkappa = k/C\varrho$, where C is the specific heat and ϱ is the density.

7.3
Solutions to the Diffusion Equation

7.3.1
Gaussian Concentration Distribution

For simple geometries, there are analytical solutions to the diffusion equation. For example, there is a book by Carlslaw and Jaeger [2] entitled "The Conduction of Heat in Solids", which contains analytical solutions to the diffusion equation in a variety of geometries. The problems are posed as heat flow problems, and so the solutions are stated in terms of thermal diffusion, but the solutions can be applied to mass diffusion problems, simply by replacing the thermal diffusivity with the chemical diffusivity.

7.3.1.1 Gaussian Distribution in One Dimension

The partial differential equation for time-dependent diffusion in one dimension, Equation 7.15, can be solved for the case where the starting condition is a layer of atoms of some species in a plane at $x = 0$, and zero concentration everywhere else, out to infinity in plus and minus x. The formal solution has the form:

$$C(x,t) = \frac{A}{2\sqrt{\pi Dt}} \exp\left(-\frac{x^2}{4Dt}\right) \tag{7.19}$$

where A is the total number of diffusing atoms per unit area in the initial layer. This distribution is known as a Gaussian. There is a finite concentration at $x = 0$ which changes with time, and the concentration is zero at $x = \pm\infty$. This solution is relatively difficult to derive, but it can be verified readily that this is a valid solution. Since

$$\int_{-\infty}^{\infty} \exp\left(-\frac{x^2}{4Dt}\right) \mathrm{d}x = 2\sqrt{\pi Dt} \tag{7.20}$$

the total amount of the diffusing component is:

$$\int_{-\infty}^{\infty} C(x,t)\,\mathrm{d}x = A \tag{7.21}$$

The total number of diffusing atoms can also be written as $C_0 l$, where C_0 is the concentration in atoms cm^{-3} in a layer which has a thickness l. The Gaussian solution is valid in this case after the diffusion process has proceeded long enough so that $\sqrt{(2Dt)} \gg l$.

Figure 7.2 illustrates the initial condition and the Gaussian profile at two subsequent times. The concentration at $x = 0$ decreases with time, and is proportional to $1/\sqrt{(Dt)}$. The width of the concentration distribution increases with time, and is proportional to $\sqrt{(Dt)}$. The area under the curve is the total number of atoms in the sample, and it stays constant in time.

When a thin layer of composition C_0 atoms cm^{-3} and thickness l is deposited on a surface, as illustrated in Figure 7.3, and all the deposited material diffuses into the substrate, and the total amount of material in the substrate stays constant.

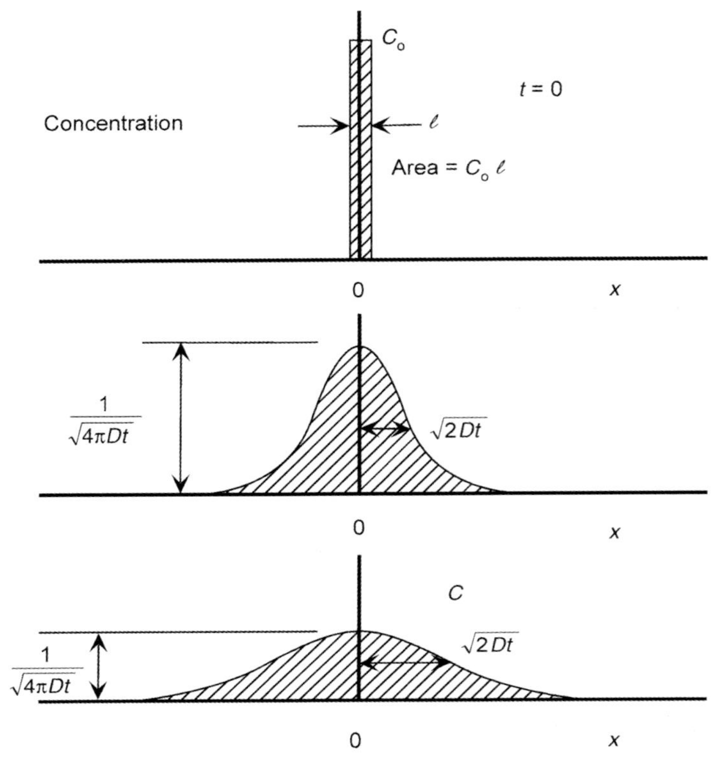

Figure 7.2 Gaussian diffusion profile. The hatched area indicates the total amount of dopant in the sample, which remains constant.

The concentration profile in this case will also be Gaussian after enough time has elapsed. The mathematical solution applies only for $x > 0$. The Gaussian solution has zero slope at $x = 0$, implying that there is no diffusion flux across the plane at $x = 0$.

In this case, since all of the deposited material diffuses into the substrate, the concentration profile in the substrate is:

$$C(x,t) = \frac{C_0 l}{\sqrt{\pi Dt}} \exp\left(-\frac{x^2}{4Dt}\right) \tag{7.22}$$

Since

$$\int_0^{\infty} \exp\left(-\frac{x^2}{4Dt}\right) \mathrm{d}x = \sqrt{\pi Dt} \tag{7.23}$$

The total amount of material deposited on the substrate, $C_0 l$ atoms cm^{-2} is the total amount of material in the diffusion profile.

For analysis of the dopant distribution following ion implantation, $C_0 l$ is replaced with Q, the total number of atoms cm^{-2} implanted into the sample. The Gaussian

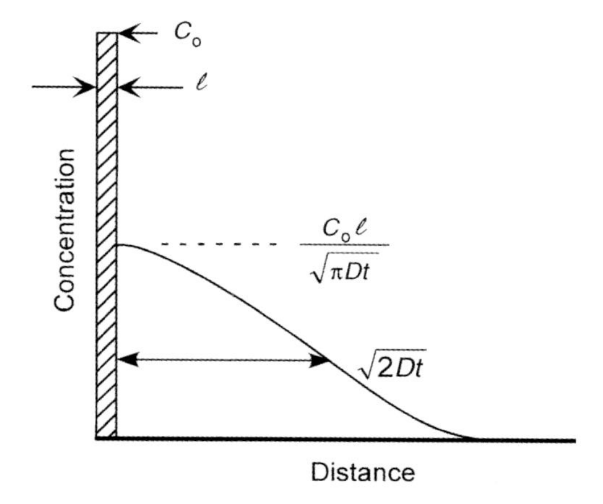

Figure 7.3 Diffusion in from a surface.

solution is valid after a time which is long enough so that $\sqrt{(2Dt)}$ is large compared to the implantation depth.

Equation 7.22 is the formal solution for the concentration profile during a diffusion *drive-in* anneal.

It is valid when the annealing time is long compared to the time it takes for all the original deposited or implanted material to diffuse into the sample. This formal solution assumes that the diffusion coefficient is constant, that there are no effects of clustering, and no variations in defect density. If any of these conditions are not met, a more complex analysis must be done, for example, using the SUPREM program.

7.3.1.2 Cylindrical Coordinates

The concentration profile in a cylinder or sphere which results from an initial concentration located at $x = 0$ also has a Gaussian shape. For a circle or a cylinder, the solution is two-dimensional, and is independent of position along the z-axis, as illustrated in Figure 7.4.

In cylindrical coordinates, $r^2 = x^2 + y^2$, and the diffusion equation becomes:

$$\frac{\partial^2 C}{\partial r^2} + \frac{1}{r}\frac{\partial C}{\partial r} = \frac{\partial C}{\partial t} \tag{7.24}$$

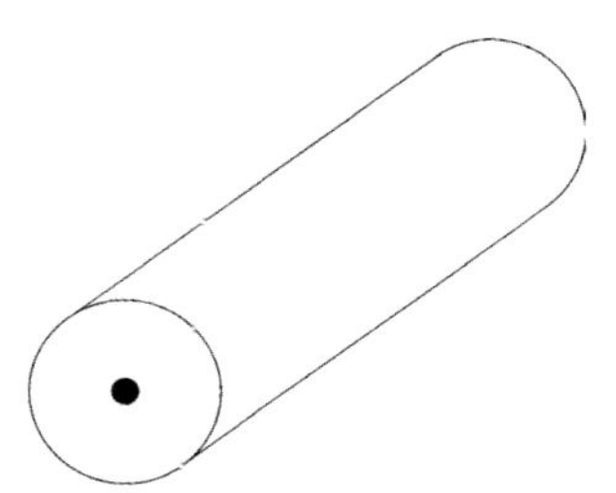

Figure 7.4 Diffusion in cylindrical coordinates, from a line of atoms initially at the center of the cylinder.

For a total amount of a second component, Q atoms cm^{-1}, initially along the axis of the cylinder, the solution takes the form:

$$C = \frac{Q}{4\pi Dt} \exp\left(-\frac{r^2}{4Dt}\right) \tag{7.25}$$

7.3.1.3 Spherical Coordinates

In spherical coordinates, Figure 7.5, $r^2 = x^2 + y^2 + z^2$, the diffusion equation becomes:

$$\frac{\partial^2 C}{\partial r^2} + \frac{2}{r}\frac{\partial C}{\partial r} = \frac{\partial C}{\partial t} \tag{7.26}$$

For a total number, Q, of atoms of the second component initially at the center of the sphere, the solution takes the form:

$$C = \frac{Q}{(4\pi Dt)^{3/2}} \exp\left(-\frac{r^2}{4Dt}\right) \tag{7.27}$$

Both of these profiles preserve the total number of atoms, and have a Gaussian shape, similar to the diffusion of a fixed total number of atoms diffusing into a sample from its surface.

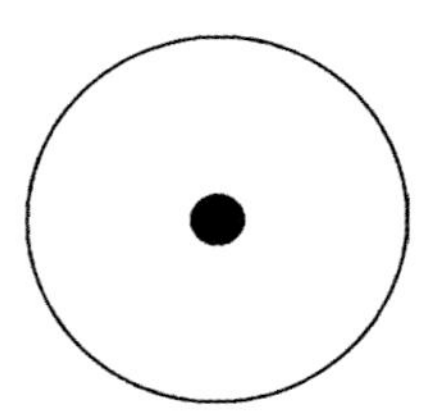

Figure 7.5 Diffusion in spherical coordinates, of atoms initially at the center of the sphere.

7.3.2 Error Function Concentration Distribution

The other important solution to the one-dimensional diffusion equation is used when the surface concentration stays constant, as during deposition onto the surface, that is during *pre-dep*. The solution in this case is given by the complementary error function:

$$C(x,t) = C_S \operatorname{erfc}\left(\frac{x}{2\sqrt{Dt}}\right) \tag{7.28}$$

Where the complementary error function is defined in terms of the error function as:

$$\operatorname{erfc}(z) = 1 - \operatorname{erf}(z) \tag{7.29}$$

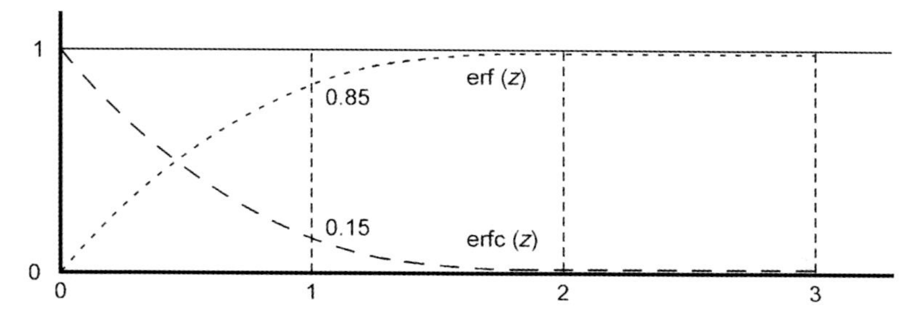

Figure 7.6 The error function and the complementary error function.

and the error function is an integral of the Gaussian probability distribution function:

$$\mathrm{erf}(z) = \frac{2}{\sqrt{\pi}} \int_0^z \exp(-\eta^2)\, d\eta \tag{7.30}$$

These functions are important in statistical analysis, which is where the name comes from.

Figure 7.6 illustrates the error function and the complementary error function. For our case, the argument z in Equation 7.29 is $x/2\sqrt{(Dt)}$.

The flux into the surface depends on the concentration gradient at the surface, and this decreases with time:

$$J_{x=o} = -D\left(\frac{dC}{dx}\right)_{x=o} = \sqrt{\frac{D}{\pi t}}\, C_s \tag{7.31}$$

The same formal solution applies for a diffusion couple, where the two dissimilar materials are joined at $x = 0$, as illustrated in Figure 7.7.

The concentration at the interface is initially half way between the concentrations on either side, and stays there (if we ignore the Kirkendall effect). So the concen-

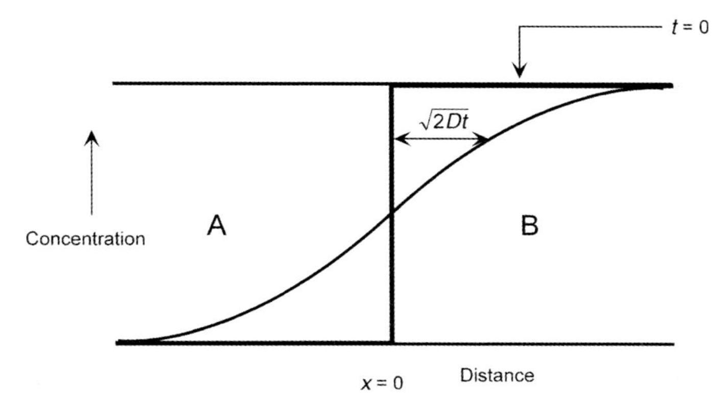

Figure 7.7 Diffusion profile for a diffusion couple.

tration profile in each half of the couple is given by an error function. The concentration to the right in the figure is a complementary error function, starting at $C = 0.5$ at $x = 0$, and increasing to $C = 1$ at $x = \infty$. On the left-hand side the concentration profile is a complementary error function with a negative argument, starting at $C = 0.5$ at $x = 0$, and decreasing to $C = 0$ at $x = -\infty$.

Both the Gaussian and the error function solutions are valid only if the concentrations at infinity do not change with time.

The arguments for both the Gaussian and the error function solutions involve x^2/Dt, or equivalently, $x/\sqrt{(Dt)}$. This group of variables is dimensionless. For any diffusion problem, always look at its value *before* starting the analysis.

$\sqrt{(Dt)}$ is a length, known as the diffusion distance. If $\sqrt{(Dt)}$ is large compared to the distances involved, such as the sample thickness, then the concentration will be uniform throughout the thickness after time t. If, and only if, $\sqrt{(Dt)}$ is small compared to the sample thickness, so that the other side of the sample is effectively at infinity, then a Gaussian or error function solution, if appropriate, may be used.

Alternatively, if the distance in the problem is known, then the time for diffusion over that distance, x^2/D, should be considered. x^2/D is a time, and if the available time is long compared to this time, then the concentration will be uniform over the distance x. On the other hand, if the time for diffusion is short compared to the time available, then the concentration distribution will not be uniform throughout the distance x, and so a Gaussian or error function solution, if appropriate, may be used.

7.3.3
p/n Junction Depth

When a dopant is diffused into a substrate which is uniformly doped at a concentration C_b with dopant of the opposite carrier type, a p/n junction will be formed where the concentration of the in-diffusing dopant equals the concentration of the substrate dopant. Traditionally, a p/n junction is formed during the drive-in anneal of the semiconductor. In this case, a Gaussian concentration profile, as given by Equation 7.22 is developed. A p/n junction is formed where the concentration of the in-diffusing dopant is C_b. The junction depth x_j is given by:

$$C_b = C(x,t) = \frac{Q}{\sqrt{\pi Dt}} \exp\left(-\frac{x_j^2}{4Dt}\right) \tag{7.32}$$

So that:

$$x_j = 2\sqrt{Dt \ln\left(\frac{Q}{C_b\sqrt{\pi Dt}}\right)} \tag{7.33}$$

Where Q is the total amount of dopant per square centimeter diffusing into the sample.

7.3.4 Separation of Variables

7.3.4.1 Concentration at the Surface is Specified

For diffusion of a dopant into a wafer from the surface with a constant composition at the surface, the concentration profile in from the surface is given by the complementary error function, provided that the diffusion time is short enough so that $\sqrt{(2Dt)} \ll l$, where l is the thickness of the wafer. For long times, $\sqrt{(2Dt)} \gg l$, the concentration in the wafer will be uniform. In between these two limits, where $\sqrt{(2Dt)} \sim l$, we need a different solution.

The solution for this case is formally the same as the solution for outgassing of a slab, where there is initially a uniform concentration of a second component in the slab, but the concentration at the surface of the slab is zero, as illustrated in Figure 7.8.

The rate at which the second component leaves the slab is controlled by diffusion of the second component to the surface of the slab. This is a one-dimensional problem, and the diffusion process is described by Equation 7.15:

$$D\frac{\partial^2 C}{\partial x^2} = \frac{\partial C}{\partial t} \tag{7.34}$$

The method of separation of variables consists of looking for a solution of the form:

$$C(x,t) = X(x) \cdot T(t) \tag{7.35}$$

where $X(x)$ is a function only of x, and $T(t)$ is a function only of t.

So that:

$$\frac{\partial C}{\partial t} = X(x) \cdot \frac{\mathrm{d}}{\mathrm{d}t} T(t) \tag{7.36}$$

and

$$\frac{\partial^2 C}{\partial x^2} = T(t) \cdot \frac{\mathrm{d}^2}{\mathrm{d}x^2} X(x) \tag{7.37}$$

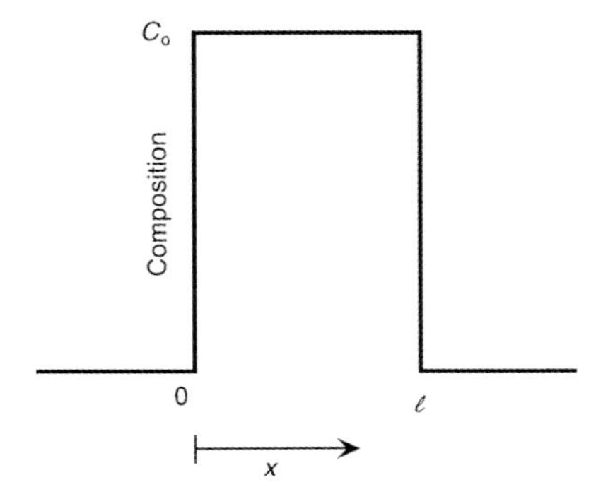

Figure 7.8 Initial composition profile in slab.

The Gaussian and error functions which were discussed above are *not* solutions of this type.

The differential equation becomes:

$$D\frac{d^2X(x)}{dx^2}\cdot T(t) = X(x)\,\frac{dT(t)}{dt} \tag{7.38}$$

which can be rewritten as:

$$\frac{1}{X(x)}\frac{d^2X(x)}{dx^2} = \frac{1}{D}\left(\frac{1}{T(t)}\frac{dT(t)}{dt}\right) = -k^2 \tag{7.39}$$

Since the $X(x)$ is independent of time, and the $T(t)$ is independent of x, the two must be equal to a constant. The constant we will take to be $-k^2$, and so the partial differential equation splits into two ordinary differential equations:

$$\frac{d^2X(x)}{dx^2} + k^2X(x) = 0 \tag{7.40}$$

and

$$\frac{dT(t)}{dt} = -k^2DT(t) \tag{7.41}$$

which have formal solutions:

$$X(x) = A\cos(k\,x) + B\sin\,(k\,x) \tag{7.42}$$

$$T(t) = T_0\exp\,(-k^2Dt) \tag{7.43}$$

The constants A, B, T_0, and k can be chosen to fit the boundary conditions.

Any k provides a valid solution so we write a formal solution as a sum of terms with all possible values of k:

$$C(x,t) = \sum_k\{[A_k\cos(kx) + B_k\sin\,(kx)]\exp\,(-k^2Dt)\} \tag{7.44}$$

In this form, the constant T_0 is redundant. For our outgassing problem, the concentration at the two surfaces of the slab is zero at all times, so that $C(0,t)=0$, which means that there can be no cosine terms in our solution, so $A=0$. We can make $C(l,t)=0$ by setting $k=n\pi/l$, where n is an integer, which will make all the sine terms equal to zero at $x=l$.

$$C(x,t) = \sum_n B_n\sin\left(\frac{n\pi x}{l}\right)\exp\left[-\left(\frac{n\pi}{l}\right)^2Dt\right] \tag{7.45}$$

The essence of the method of separation of variables is that the initial concentration profile at $t=0$ can be fitted as a boundary condition, and then the $X(x)$ part of the formal solution does not change in time. All of the time dependence is in $T(t)$. We can represent any form of initial concentration in the range 0 to l as a Fourier series, that is, as a summation of sine and/or cosine terms, as illustrated in Figure 7.9.

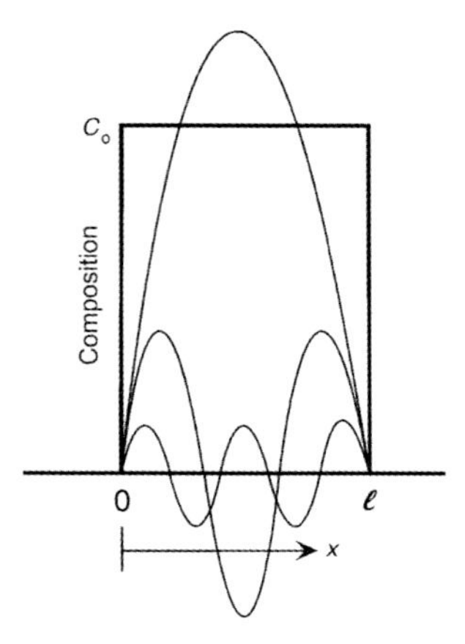

Figure 7.9 Representation of the original concentration profile as a sum of sine terms.

The trick to doing this is outlined below. For any arbitrary function $\Phi(x)$ defined between 0 and l, we can write formally:

$$\Phi(x) = \sum_{i=1}^{\infty} G_i \sin\left(\frac{i\pi x}{l}\right) \tag{7.46}$$

In order to determine the G_is, note that:

$$\int_0^l \sin^2\left(\frac{n\pi x}{l}\right) \mathrm{d}x = \frac{l}{2} \tag{7.47}$$

and

$$\int_0^l \sin\left(\frac{n\pi x}{l}\right) \sin\left[\frac{m\pi x}{l}\right] \mathrm{d}x = 0 \quad \text{for } m \neq n \tag{7.48}$$

so if we multiply both sides of Equation 7.46 by $(2/l)\sin(m\pi x/l)$, and integrate between 0 and l, then all of the terms in the series are zero except for the one containing m, and that term is equal to G_m. So that:

$$\frac{2}{l}\int_0^1 F(\mathrm{x}) \sin\left(\frac{m\pi x}{l}\right) \mathrm{d}x = G_m \tag{7.49}$$

If the integral can be evaluated, then we obtain the value of G_m.

For our outgassing problem, $\Phi(x) = C_0$, is a constant, so the integral can be readily evaluated.

$$\begin{aligned} B_n &= \frac{2C_0}{1}\int_0^1 \sin\int\left(\frac{n\pi x}{1}\right) \mathrm{d}x \\ &= -\frac{2C_0}{n\pi}\left[\cos\left(\frac{n\pi x}{1}\right)\right]_o^1 \end{aligned} \tag{7.50}$$

For n even, $\cos(n\pi x/l) = 1$ for both $x = 0$ and for $x = l$, so $B_n = 0$ for even values of n.

For n odd, $\cos(n\pi x/l) = 1$ for $x = 0$ and $\cos(n\pi x/l) = -1$ for $x = l$, so the square bracket is equal to 2, and $B_n = 4C_0/n\pi$ for odd values of n. We can replace n with $2i + 1$, which makes $i = 0, 1, 2, 3, \ldots$ correspond to $n = 1, 3, 5, 7, \ldots$, so that a summation over i will contain only the terms where n is odd.

$$B_n = \frac{4\,C_0}{n\pi} = \frac{4\,C_0}{\pi\,(2\,i+1)} \tag{7.51}$$

And our final solution becomes:

$$C(x,t) = \frac{4\,C_0}{\pi}\sum_{i=0}^{\infty}\frac{1}{2\,i+1}\,\sin\left[\frac{(2i+1)\pi x}{l}\right]\exp\left[-\left(\frac{(2i+1)\pi}{l}\right)^2 Dt\right] \tag{7.52}$$

The shorter wavelength terms in the series decay faster. The term for $i = 0$ decays as $\exp[-(\pi/l)^2 Dt]$, and the second term, in which $i = 1$, decays as $\exp[-9(\pi/l)^2 Dt]$, which is nine times faster. After some time, only the $i = 0$ term is left:

$$C(x,t) = \frac{4C_0}{\pi}\,\sin\left(\frac{\pi x}{l}\right)\exp\left[-\frac{\pi^2 Dt}{l}\right] \tag{7.53}$$

So the concentration profile in the slab starts off with a steep concentration rise at the surface. As diffusion proceeds, the corners of the profile become rounded, as described by an error function. Then, after some time, the profile becomes a simple sine function, as shown in Figure 7.10, which decays in time until the profile is flat.

For the diffusion into a wafer which has an initial concentration of zero in the wafer but a constant surface concentration, C_0, of the diffusing species, the concentration profile has the same mathematical form, and the solution will be $C_0 - C(x, t)$, where $C(x, t)$ is given by Equation 7.52.

7.3.4.2 Flux Specified at Surface

A diffusion problem is well defined if the concentration or flux is known at all the surfaces. The separation of variables method can also be used if the initial boundary condition describes the flux into the surface of the sample. The diffusion flux into the sample can be written as: $J = -D(\mathrm{d}C/\mathrm{d}x)$, so that the derivative of the general solution for the concentration, Equation 7.44, can be taken,

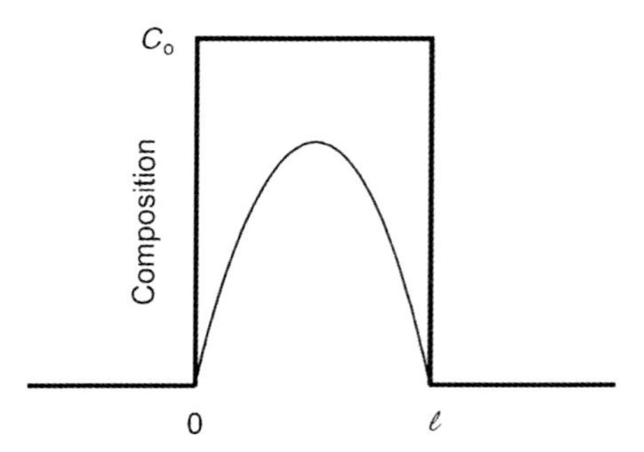

Figure 7.10 Composition profile after some time, when only one sine term is left.

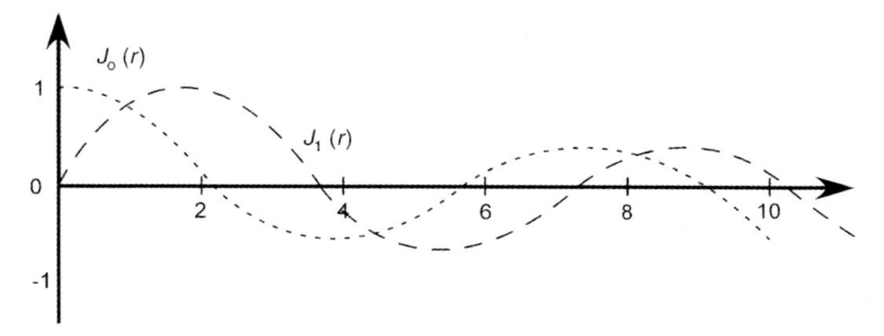

Figure 7.11 Bessel functions.

$$\frac{dC(x)}{dx} = k \sum_{k=o}^{\infty}\{[-A_k \sin(k\,x) + B_k \cos\,(k\,x)]\exp[-k^2\,Dt]\} \tag{7.54}$$

and then fitted with a Fourier series.

7.3.4.3 **Cylinder**

Formal solutions can be obtained for diffusion into or out of a cylinder using the separation of variables method. The radial part of the solution is expanded in terms of Bessel functions, rather than sines and cosines. Figure 7.11 contains a sketch of two Bessel functions.

They are rather similar to sines and cosines, except that the amplitude decreases with distance from the origin. They have zeros between the zeros of sines and cosines.

7.3.4.4 **Sphere**

Formal solutions can also be obtained for diffusion into or out of a sphere using the separation of variables method. The radial part of the solution is expanded in terms of Legendre polynomials. The first few Legendre polynomials are:

$$P_0(x) = 1$$
$$P_1(x) = x$$
$$P_2(x) = (3x^2 - 1)/2$$
$$P_3(x) = (5x^3 - 3x)/2$$
$$P_4(x) = (35x^4 - 30x^2 + 3)/8$$
$$P_5(x) = (63x^5 - 70x^3 + 15x)/8$$
$$\vdots$$

In general, formal solutions can be found only for simple geometries, such as a plane, a cylinder or a sphere. The time-dependent solutions for even these cases can be quite complex. Mel Lax, a formidable mathematician, observed that after he had found a formal solution to a problem in terms of an infinite series of Bessel functions, that as much computer time was required to evaluate the series as to evaluate the solution numerically directly from the differential equations.

Numerical solutions will be discussed in the next section.

Unless there are sources or sinks inside a sample, there can be no extrema in the concentration field inside the sample. This can be proven mathematically, but physically it is just a consequence of the fact that diffusive motion is due to atoms jumping around randomly, which is an averaging process, which precludes the possibility of getting a bump or a depression inside the sample.

7.4 Numerical Methods

Numerical methods can be used for solving differential equations. This is done by using finite increments for time and distance, rather than the infinitesimal increments used in the differential equation.

There are two principal schemes for numerical analysis, finite difference and finite element. In finite difference analyses, the grid size (increment size) is made small enough so that a valid solution can be obtained. Usually the mesh size is kept the same everywhere in the solution. The solution is either propagated in time from initial conditions, or interpolated between boundary conditions. We are going to go through a finite difference calculation below.

Finite element analysis is a more complex scheme which is usually used only for large problems, because it is more difficult to set up. It makes use of the fact that solutions to problems often have some regions where the variables are changing rapidly, and other regions where they are changing slowly. And so a different size of grid is used in different regions. But this scheme requires defining the grid carefully, and the calculations are more complex where the grid size changes. Often, a great deal of effort is expended on designing the grid. There are now computer programs to help with the task of grid design. Finite element methods also make use of higher order approximations for calculating values at grid points, so that the mesh can be made coarser, while still retaining accuracy of the calculation. Finite elements methods are used, for example, for accurate stress analyses, and for time-dependent fluid flows.

The book *Numerical Recipes* [3] contains a large variety of numerical solutions to various equations. The book outlines the mathematics behind each equation, and presents code for solving the equations numerically. There are versions for fortran, C and basic. CDs containing the code are available.

7.4.1 Finite Difference Method for Diffusion

Using $x_j = j\Delta x$ and $t_n = n\Delta t$, the one-dimensional time-dependent diffusion equation is:

$$D\frac{\partial^2 C}{\partial x^2} = \frac{\partial C}{\partial t} \tag{7.55}$$

This can be converted into a difference equation:

$$D\left[\frac{C(x_{j+1},t_n)-2C(x_j,t_n)+C(x_{j-1},t_n)}{(\Delta x)^2}\right]=\left[\frac{C(x_j,t_{n+1})-C(x_j,t_n)}{(\Delta t)}\right] \tag{7.56}$$

This is just the formal definition of the derivatives written in terms of finite increments, as was done in the derivation of Equation 7.14, rather than in the form of the differential equation, which implies infinitesimal increments.

This equation can be rearranged:

$$C(x_j,t_{n+1})=C(x_j,t_n)+\frac{D\Delta t}{(\Delta x)^2}\left[C(x_{j+1},t_n)-2C(x_j,t_n)+C(x_{j-1},t_n)\right] \tag{7.57}$$

The term on the left-hand side is for time t_{n+1} and all the terms on the right-hand side are for time t_n, so that the configuration for the next time increment, $n+1$, can be calculated from values in the previous time increment, n.

Stable solutions for this difference equation are obtained for:

$$D\Delta t/(\Delta x)^2 \leq 1/2 \tag{7.58}$$

For the special case:

$$D\Delta t/(\Delta x)^2 = 1/2 \tag{7.59}$$

the equation above takes the simple form:

$$C(x_j,t_{n+1})=\frac{[C(x_{j+1},t_n)+C(x_{j-1},t_n)]}{2} \tag{7.60}$$

7.4.2 Initial Surface Concentration Boundary Conditions

The concentration as a function of time can be calculated using a spreadsheet by writing the concentration profile at $t=0$ into the first row, and then using Equation 7.57 or 7.60 to calculate the concentration profiles at subsequent times. The same equation can be copied into each cell of the spreadsheet except for the boundaries. For example, for the outgassing problem discussed above, the concentration is zero at $x=0$ and at $x=l$ for all times.

The concentration at $t=0$ everywhere inside the slab has the same value, so this is arbitrarily set to 1 in Table 7.2. This value is inserted in the row at $t=0$. The value of either Δx or Δt can be chosen arbitrarily to provide a fine enough mesh for the calculation, and then the other increment can be calculated so that Equation 7.58 is satisfied.

For example, for $D=10^{-5}\,\text{cm}^2\,\text{s}^{-1}$, if we are asked to find the composition in a 1 mm thick slab after 40 s, the diffusion distance after 40 s will be $\sqrt{(Dt)}=\sqrt{(10^{-5}\times 40)}=0.02\,\text{cm}=0.2\,\text{mm}$. We can expect that the composition will change well into the slab, which is 1 mm thick. We can choose $\Delta x=0.001\,\text{cm}$, so there will be 100 data points in thickness of the sample. Then we can choose Δt to satisfy Equation 7.59:

Table 7.2 Illustrating the method for solving the one-dimensional time dependent diffusion equation using a spread sheet.

x▶ t▼	0	Δx	$2x$	$3\Delta x$	$4\Delta x$	$5\Delta x$	...	...	...	$l-\Delta x$	l
0	0	1	1	1	1	1	1	1	1	1	0
Δt	0										0
$2\Delta t$	0										0
$3\Delta t$	0										0

$\Delta t = (\Delta x)^2/2D = 10^{-6}/2 \times 10^{-5} = 0.05$ s. So the data for $t = 40$ s will be in the row for which $n = 40/0.05 = 800$.

This spread sheet solution will generate a complementary error function (erfc) profile at short times when $\sqrt{(Dt)} \ll l = 1$ mm, and a sine profile when $\sqrt{(Dt)} > 1$ mm.

Notice that if there is a 1 in only at one position in the first row of Table 7.2, say at the column labeled $5\Delta x$, and if the algorithm of Equation 7.59 is used to generate the subsequent rows, then the numbers in Table 7.1 will be generated. The calculation illustrated in Table 7.2 takes the concentration in each box in the $t = 0$ row, and spreads it out as time proceeds.

The calculation proceeds similarly for any arbitrary composition profile at $t = 0$. The initial profile does not need to be analytic or integrable, as is required to carry out a Fourier analysis, for example, as in Equation 7.49. An example of an arbitrary initial profile is discussed in the next section.

7.4.3
Implanted Concentration Profile

It is usual for all the dopant ions which are ion implanted into silicon to stay in the crystal. None are lost at the surface by evaporation. Since none of the dopants pass through the surface, the concentration gradient of the dopant at the surface is zero. The concentration at the surface rises as atoms diffuse there. The total amount of dopant in the crystal is constant.

The initial as-implanted concentration profile after an ion implantation is usually represented by:

$$N(x) = \frac{Q}{\sqrt{2\pi}\,\Delta R_p} \exp\left[-\frac{1}{2}\frac{(x - R_p)^2}{(\Delta R_p)^2}\right] \tag{7.61}$$

where Q is the total number of implanted ions per unit area, R_P is the end of range, and ΔR_P is the straggle. The implanted ions penetrate into the sample a distance which is R_P on average, with a spread about the average depth given by ΔR_P.

The formal analytical solution for the evolution of the concentration in time during an anneal is complex, but it can be obtained readily by inserting this initial concentration distribution into the first row of the spread sheet and using a no-flux boundary condition in the first column which represents the surface.

7.4.4
Zero Flux Boundary Condition

A zero flux condition at $x = 0$ can be imposed by making the composition at $j = -1$ the same as that at $j = +1$, so that there is no net flux through the cells in the column $j = 0$. Equation 7.57 for column 0 then becomes:

$$C(x_0, t_{n+1}) = C(x_0, t_n) + \frac{1}{2}\frac{D\Delta t}{(\Delta x)^2}[2C(x_1, t_n) - 2C(x_0, t_n)] \tag{7.62}$$

The factor 1/2 is introduced because the phantom column at $j = -1$ does not contribute flux to the concentration in column $j = 0$. For the special case where

$$D\Delta t/(\Delta x)^2 = 1/2 \tag{7.63}$$

this equation reduces to the simple form:

$$C(x_0, t_{n+1}) = \frac{[C(x_0, t_n) + C(x_1, t_n)]}{2} \tag{7.64}$$

This equation can be copied into each cell of the first column of the spreadsheet. For example, putting a 1 in the (0,0) cell, and 0s in the rest of the first row, with the zero flux boundary condition in the rest of the first column will generate a Gaussian distribution, centered at $x = 0$, and with unit area.

This scheme can follow the evolution of an initial implant profile, such as Equation 7.61, into a Gaussian distribution at times which are diffusion times which are long compared to the time R_p^2/D, which the time it takes for the initial implant profile to smooth out.

7.5
Boltzmann–Matano Analysis

The Boltzmann–Matano analysis [4, 5] provides a method for determining the diffusion coefficient from concentration profiles when it is concentration dependent, as in Figures 5.2, 5.4 or 5.5, for example.

$$D = D(C) \tag{7.65}$$

The flux in the diffusion field becomes:

$$J = -D(C)\frac{\mathrm{d}C}{\mathrm{d}x} \tag{7.66}$$

So that the diffusion equation in one dimension takes the more complex form:

$$\frac{\partial}{\partial x}\left(D\frac{\partial C}{\partial x}\right) = \frac{\partial C}{\partial t} \tag{7.67}$$

If we define a new variable, $z = x/t^{1/2}$, then this partial differential equation can be rewritten as an ordinary diffusion equation.

The derivatives with respect to x and t can be written in terms of derivatives with respect to z.

$$\frac{\partial}{\partial x} = \frac{\partial z}{\partial x} \cdot \frac{\partial}{\partial z} = \frac{1}{t^{1/2}} \frac{\partial}{\partial z}$$
$$\frac{\partial}{\partial t} = \frac{\partial z}{\partial t} \cdot \frac{\partial}{\partial z} = -\frac{x}{2t^{3/2}} \frac{\partial}{\partial z} \tag{7.68}$$

And so Equation 7.67 can be written:

$$\frac{1}{t^{1/2}} \frac{\partial}{\partial z} \left(D(C) \frac{1}{t^{1/2}} \frac{\partial C}{\partial z} \right) = -\frac{x}{2t^{3/2}} \frac{\partial C}{\partial z} \tag{7.69}$$

which can be written as an ordinary differential equation, since it contains only the variable z.

$$\frac{\mathrm{d}}{\mathrm{d}z} \left(D(C) \frac{\mathrm{d}C}{\mathrm{d}z} \right) = -\frac{z}{2} \frac{\mathrm{d}C}{\mathrm{d}z} \tag{7.70}$$

Integrating once gives:

$$D(C) \frac{\mathrm{d}C}{\mathrm{d}z} = \frac{1}{2} \int_{C}^{C_1} z \mathrm{d}C \tag{7.71}$$

or:

$$D(C) = \frac{\int_{C}^{C_1} z \mathrm{d}C}{2 \left(\frac{\mathrm{d}C}{\mathrm{d}z} \right)} \tag{7.72}$$

For a diffusion couple, where the starting concentrations at $t = 0$ are: $C = C_1$ for $x < 0$, and $C = C_2$ for $x > 0$, these two concentrations are the values at $z = -\infty$ and $z = +\infty$, respectively. Experimental data can be plotted as shown in Figure 7.12 as a function of z.

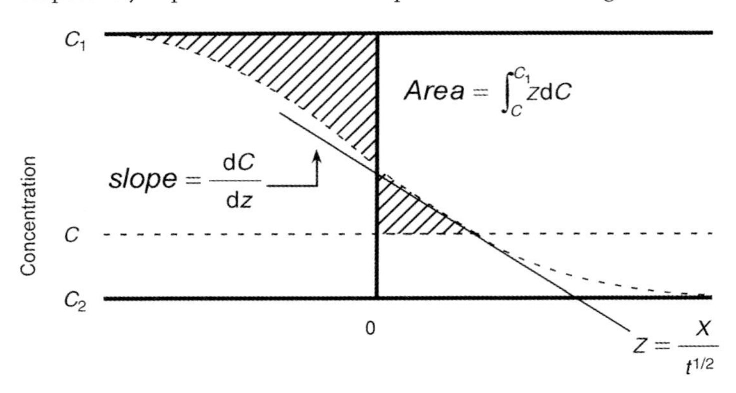

Figure 7.12 Boltzmann–Matano plot.

Concentration data should be taken as a function of distance at several different times. These data should all superimpose when plotted against z if the variations in D are due to differences in composition. The hatched area in Figure 7.12 corresponds to the integral of z as it goes from $-\infty$ at C_1 to the value of z where the composition is C. The derivative in the denominator of Equation 7.72 is obtained from the slope of the curve at C. The values of the integral and the slope are then inserted into Equation 7.72 to obtain the value of the diffusion coefficient at the composition C. The procedure can be repeated for other compositions, and measurements can be made at a various temperatures to determine the temperature dependence of D.

7.6 Diffusion During Phase Separation

The discussion of diffusion processes in this chapter is based on the standard equations for diffusion. The diffusion fluxes are assumed to depend only on the composition gradients. This is an appropriate assumption when the volume elements are large compared to atomic dimensions. For diffusion on the atomic scale, the interactions between atoms should be taken into account. The diffusion fluxes then depend on activity gradients rather than composition gradients. In Chapter 23, difference equations for diffusion are developed that take interatomic interactions into account. These equations include up-hill diffusion for phase separation or ordering in a regular solution. They reduce to the standard equations for diffusion when neighboring volume elements do not interact, when composition gradients are small on the atomic scale, or when the temperature is far above the critical temperature.

References

1 Shewmon, P. (1989) *Diffusion in Solids*, 2nd edition, Minerals, Metals & Materials Society.
2 Carlslaw, H.S. and Jaeger, J.C. (1959) *Conduction of Heat in Solids*, Clarendon Press, Oxford.
3 Press, W.H., Teukolsky, S.A., Vetterlong, W.T., and Flannery, B.P. (1992) *Numerical Methods*, Cambridge.
4 Matano, C. (1933) *Jpn. J. Phys.*, **8**, 109.
5 Murarka, S.P. and Peckerar, M.C. (1989) *Electronic Materials*, Academic Press, p. 156.

Problems

7.1. Discuss the relationship between random walk and diffusion.

7.2. (a) Write a computer program for a one-dimensional random walk, where each walker makes 100 steps which are randomly $+1$ or -1.
Run the program for 500 walkers.
Plot the final positions of the 500 walkers as a histogram (bar graph).
(b) For one-dimensional diffusion, the diffusion coefficient is given by $D = a^2\Gamma/2$, where a is the jump distance, and Γ is the jump rate. What is the

relationship between the parameters for the simulation in question 1 and x, Dt, and N_o in the equation in question 3?

(c) Compare the distribution of final positions of the walkers from part 1 with:

$$N(x) = \frac{aN_0}{\sqrt{4\pi Dt}} \exp\left(\frac{-x^2}{4Dt}\right)$$

Note that with an even number of jumps, a walker cannot end up on an odd numbered site

7.3. A wafer was annealed for 30 min at a temperature where the diffusion coefficient in silicon of a dopant deposited on the surface is $10^{-8}\,\text{cm}^2\,\text{s}^{-1}$. Assuming that the surface concentration of the dopant remains fixed at 1.0,

(a) Use a finite difference method to plot the composition profile after the anneal.
(b) Compare your result with the erfc solution for the composition profile.

7.4. 10^{15} atoms cm^{-2} of boron were implanted into a silicon wafer at 70 keV. For this implantation voltage the end of range, R_P, for boron is about 0.2 μ, and the straggle, ΔR_P, is about 0.05 μ. The as-implanted distribution is given by:

$$C = \frac{Q_0}{\sqrt{2\pi}\Delta R_P} \exp\left[-\frac{(x-R_P)^2}{2(\Delta R_P)^2}\right]$$

where Q_0 is the implanted dose in atoms cm^{-2}.

(a) Plot the as-implanted concentration profile.
(b) Using the numerical method in a spread-sheet, determine the concentration profiles after 30 s, 1 min and 5 min at 1200 °C, where the diffusion coefficient of boron in silicon is $10^{-12}\,\text{cm}^2\,\text{s}^{-1}$.

7.5. (a) A 1 μm thick layer of phosphorus was deposited onto a 0.6 mm thick silicon wafer. The wafer was then annealed for 30 min at a temperature of 1250 °C, where the diffusion coefficient of phosphorus in silicon is $6 \times 10^{-12}\,\text{cm}^2\,\text{s}^{-1}$. The solubility of phosphorus in silicon at 1250 °C is 1.2×10^{21} atoms cm^{-3}. Sketch the resulting phosphorus concentration profile. A cubic centimeter of silicon contains 5×10^{22} atoms.

(b) A monolayer of copper was accidently deposited on the surface of the same wafer before the diffusion anneal. Sketch the concentration distribution of the copper after the diffusion anneal. The diffusion coefficient of copper in silicon is $2 \times 10^{-4}\,\text{cm}^2\,\text{s}^{-1}$ at 1250 °C.

7.6. A wafer was annealed for 1 h at a temperature where the diffusion coefficient in silicon of a dopant deposited on the surface is $10^{-8}\,\text{cm}^2\,\text{s}^{-1}$. Assuming that the surface concentration of the dopant remains fixed at 1.0:

(a) Use a finite difference method to determine the composition profile after the anneal.

(b) Compare a plot of your result with the erfc solution for the composition profile.

7.7. A slab of copper 1 mm thick containing 1% of oxygen is put into a vacuum chamber at a temperature where the diffusion coefficient of oxygen in copper is $10^{-8}\,cm^2\,s^{-1}$. Assuming that the oxygen concentration at the surface is zero, determine analytically the distribution of oxygen in the sample after 24 h.

7.8. Write a paper of about five pages on finite difference methods and their application to diffusion problems.

8
Stefan Problems

During phase transformations, the compositions in the two phases are usually different, as indicated by the relevant phase diagram. These composition differences must be applied as boundary conditions at the interface, in order to describe the diffusion field. But the interface between the two phases, where the boundary condition is to be applied, moves in time. And the concentrations at the boundary influence the rate of motion of the boundary. The equations describing this state of affairs are non-linear, and the solutions can be quite complex. This class of moving boundary problems is known as Stefan problems. The mathematical solutions to these problems are of great interest to some applied mathematicians.

We will discuss some important approximate solutions for these problems. Applied mathematicians do not approve of these solutions, but they provide simple answers for cases where the approximations are valid, which can be of great practical value. One commonly used simplification scheme is to assume that the concentration profile is given by a steady state solution to the diffusion equation.

8.1
Steady State Solutions to the Diffusion Equation

The steady state solutions assume that the interface moves sufficiently slowly so the same diffusion profile exists as would be found at a stationary interface. The procedure is to derive the concentration field assuming that the interface is stationary. From the concentration field the flux of atoms to the interface can be calculated. Then this flux is used to calculate how fast the interface is moving.

As an example, we will analyze the growth of an oxide layer on silicon.

Oxygen combines readily with silicon to form SiO_2. At room temperature, a few atomic layers of oxide form very quickly on a clean silicon surface. Oxide layers are grown on silicon by heating wafers up to 1050 or 1100 °C in air or in an oxygen atmosphere, where an oxide layer which is a few microns thick will grow in half an hour or so. This process was very important in the early days of silicon manufacture because the oxide could be patterned using photolithography and etching, and then the patterned oxide was used as a mask to selectively deposit dopants. The oxide

Kinetic Processes: Crystal Growth, Diffusion, and Phase Transitions in Materials. Kenneth A. Jackson

ISBN: 978-3-527-32736-2

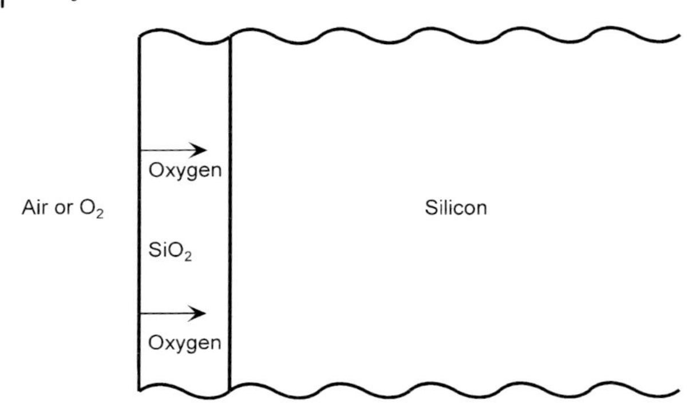

Figure 8.1 Oxide growth on silicon.

withstands the annealing temperatures for the dopant drive-in, whereas the photomask material does not. This process is used today to grow field oxides, which are used to isolate the active devices from each other on integrated circuits. Silicon rather than germanium came into widespread use because SiO_2 is stable and inert, whereas germanium oxide is volatile.

Another advantage of the growth of the oxide layer on silicon is that the growth occurs by the diffusion of oxygen through the oxide to the interface between the oxide and the silicon, and combines with silicon there to make more oxide, as illustrated in Figure 8.1. So the interface between the silica and silicon is clean, and any dirt which is on the outer surface stays there.

If the interface were stationary, the concentration profile would not change with time, and would be as illustrated in Figure 8.2.

The concentration at the surface of the oxide layer, C_S, is determined by the oxygen concentration in the atmosphere over the wafer. The concentration at the interface,

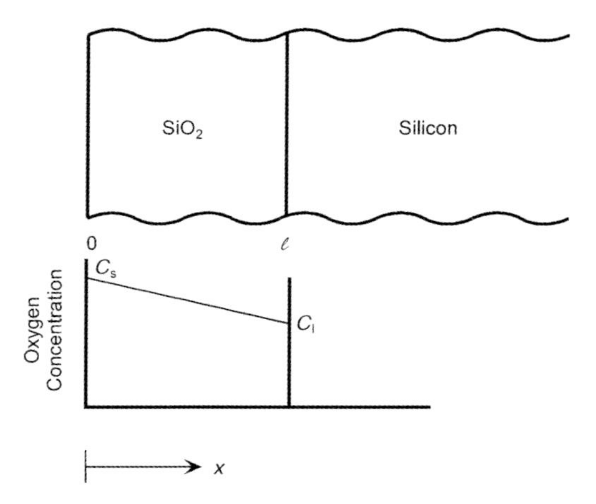

Figure 8.2 Oxygen concentration in the oxide layer.

C_I, is approximately the value corresponding to the equilibrium concentration of oxygen, C_e, in silicon which is in equilibrium with silica at the oxidizing temperature.

The steady state solution is independent of time:

$$\frac{\partial C}{\partial t} = 0 \tag{8.1}$$

The diffusion equation reduces to:

$$\frac{d^2C}{dx^2} = 0 \tag{8.2}$$

which has the simple solution:

$$C = ax + b \tag{8.3}$$

The concentration field is linear with distance through the oxide layer. Applying the boundary conditions, $C = C_S$ at $x = 0$, and $C = C_e$ at $x = l$, gives:

$$C = C_s - \frac{(C_s - C_e)}{l} x \tag{8.4}$$

where l is the thickness of the oxide layer. This concentration profile produces a flux of oxygen to the interface which is given by Fick's first law:

$$J = -D\frac{dC}{dx} = D\frac{C_s - C_I}{l} \tag{8.5}$$

C_S and C_I are small excesses in the oxygen concentration of the oxide over the stoichiometric value, which we will denote as C_{oxide}. This flux provides oxygen to make the oxide layer grow at a rate v. So that:

$$C_{oxide} v = J \tag{8.6}$$

Since $v = dl/dt$, Equations 8.5 and 8.6 can be combined to give a differential equation for the growth rate of the oxide:

$$C_{oxide} \frac{dl}{dt} = D\frac{(C_s - C_e)}{l} \tag{8.7}$$

or

$$l dl = D\frac{C_s - C_e}{C_{oxide}} dt \tag{8.8}$$

which has the solution:

$$\frac{1}{2} l^2 = D\frac{C_s - C_e}{C_{oxide}} t + B \tag{8.9}$$

For $l = 0$ at $t = 0$, then $B = 0$.

So the thickness of the oxide increases with the square root of time:

$$l = \left[2\frac{C_s - C_e}{C_{oxide}} Dt\right]^{\frac{1}{2}} \tag{8.10}$$

This describes the experimentally observed rate of growth of oxide layers on silicon after an initial transient period.

This solution assumes that C_s and C_I are constant, and that the oxidation reaction, combining oxygen and silicon to make oxide, is fast. It ignores the change in amount of excess oxygen in the oxide (the diffusing species) with time. In order to obtain a solution, the method assumes that the interface is not moving, and then uses the solution to calculate how fast the interface is moving. No wonder purists do not like this solution. But it works.

8.2
Deal–Grove Analysis

For very thin layers, the diffusion flux of atoms to the interface from the surface is very rapid, and so the rate at which oxygen combines with silicon to make SiO_2 can be the rate limiting process. A widely used analysis which includes this reaction rate is known as the Deal–Grove model [1]. In addition to using the steady state solution for diffusion, as above, they included a reaction rate term by assuming that the reaction rate is proportional to the excess concentration of oxide at the silicon/oxide interface:

$$k_S(C_I - C_e) \tag{8.11}$$

The flux of atoms to the interface is given, as in Equation 8.5, by:

$$D\frac{C_S - C_I}{l} \tag{8.12}$$

At steady state, these two rates must be equal.

$$k_S(C_I - C_e) = D\frac{C_S - C_I}{l} \tag{8.13}$$

If the rates are not equal, and the diffusion flux is greater, then C_I will increase. This increases the reaction rate, and decreases the diffusion flux. Conversely, if the reaction rate is faster, this will decrease C_I, which will reduce the reaction rate and increase the diffusion flux.

Equation 8.13 can be rearranged to give:

$$C_S - C_e = C_S - C_I + C_I - C_e = (C_S - C_I)\left(1 + \frac{D}{k_S\,l}\right) \tag{8.14}$$

As in Equation 8.7, the rate at which the oxide layer thickens is given by:

$$\frac{dl}{dt} = \frac{D}{l}\frac{(C_S - C_I)}{C_{\text{oxide}}} \tag{8.15}$$

Replacing $C_S - C_I$ with $C_S - C_e$ from Equation 8.14:

$$\frac{dl}{dt} = \frac{k_S(C_S - C_e)}{C_{\text{oxide}}\left(1 + \frac{k_S l}{D}\right)} \tag{8.16}$$

or

$$\left(1+\frac{k_S l}{D}\right)\mathrm{d}l = \frac{k_S(C_S - C_e)}{C_{\mathrm{oxide}}}\mathrm{d}t \tag{8.17}$$

which has the solution:

$$l^2 + \frac{2D}{k_S}l = \frac{2D(C_S - C_e)}{C_{\mathrm{oxide}}}t + \mathrm{constant} \tag{8.18}$$

This is usually written as:

$$l^2 + Al = B(t + \tau) \tag{8.19}$$

where:

$$B = 2D\frac{(C_S - C_e)}{C_{\mathrm{oxide}}}$$
$$\frac{B}{A} = k_S\frac{(C_S - C_e)}{C_{\mathrm{oxide}}} \tag{8.20}$$

For long times, l is large, so the term in l^2 in Equation 8.19 will be large compared to the linear term Al. The growth rate then depends on diffusion, and the thickness of the layer will increase as the square root of time, as in Equation 8.10 in the previous analysis.

For short times, the term in l^2 will be small compared to the term Al in Equation 8.19. The growth rate of l is then proportional to B/A, and the thickness of the oxide layer increases linearly with time. This is the reaction rate limited regime.

The constant τ provides for an initial layer thickness which is not zero at $t=0$.

A and B are known as the Deal–Grove coefficients, and are usually presented as values of B and B/A. B depends on the diffusion coefficient, and B/A depends on the reaction rate constant. Both are temperature dependent, with a temperature dependence given by:

$$C\exp(-E/kT) \tag{8.21}$$

Experimentally determined values of C and E for a variety of growth conditions and temperatures are presented in Table 8.1.

The Deal–Grove analysis, which is commonly used to estimate the time required to grow an oxide layer with a desired thickness, is based on the steady state solution for the diffusion field. Although it is not mathematically correct, this solution is very simple, and this analysis has proved to be very useful.

8.3 Diffusion Controlled Growth of a Spherical Precipitate

We will now use the steady state solution for the diffusion field to calculate the growth rate of a spherical precipitate, for the case where growth rate is limited by diffusion to the precipitate [3].

Table 8.1 Deal–Grove coefficients, B and B/A (From [2]).

Ambient	B	B/A
Dry O_2	$C = 7.72 \times 10^2\,\mu^2\,h^{-1}$ $E = 1.23\,eV$	$C = 6.23 \times 10^6\,\mu^2\,h^{-1}$ $E = 2.0\,eV$
Wet O_2	$C = 2.14 \times 10^2\,\mu^2\,h^{-1}$ $E = 0.71\,eV$	$C = 8.95 \times 10^7\,\mu^2\,h^{-1}$ $E = 2.05\,eV$
H_2O	$C = 3.86 \times 10^2\,\mu^2\,h^{-1}$ $E = 0.78\,eV$	$C = 1.63 \times 10^8\,\mu^2\,h^{-1}$ $E = 2.05\,eV$

In spherical coordinates, the steady state diffusion equation is:

$$D\left(\frac{\partial^2 C}{\partial r^2} + \frac{2}{r}\frac{\partial C}{\partial r}\right) = \frac{\partial C}{\partial t} \tag{8.22}$$

For steady state, the solution does not change in time, so $\partial C/\partial t = 0$. The solution to this equation takes the form:

$$C = \frac{A}{r} + B \tag{8.23}$$

This solution is unattainable exactly in nature, because the integral of the concentration field is infinite, which means that an infinite amount of material would be required to reach steady state. For example, for the sphere:

$$\int_{r_0}^{\infty} 4\pi r^2\left(\frac{A}{r} + B\right)\mathrm{d}r = \infty \tag{8.24}$$

In practice, the outer limit on the concentration field can be taken as a few times the diffusion distance, which is $\sqrt{(Dt)}$. The concentration there will actually be small enough so that it has little influence on the composition at the interface.

The initial concentration of the solution before precipitation, remains the concentration far from the precipitate, C_∞, the concentration in the precipitate particle is C_P, and the concentration in the matrix at the interface is C_I, as illustrated in Figure 8.3. We will assume that the concentrations in the precipitate particle and in the matrix at the interface are the equilibrium values, as given by the phase diagram, as illustrated in Figures 8.4 and 8.5. In other words, we are assuming that the reaction rate at the interface is rapid, so that the growth rate is limited by the diffusion process.

Usually there will be many precipitate particles, and each particle will draw material only from the region around itself. The density of precipitate particles depends on the nucleation process, and the background concentration decreases as the precipitation proceeds. In our analysis, we will ignore nucleation effects and the interactions between precipitate particles. We will assume that the background concentration remains constant, and that our particle starts out at $t = 0$ with radius $a = 0$.

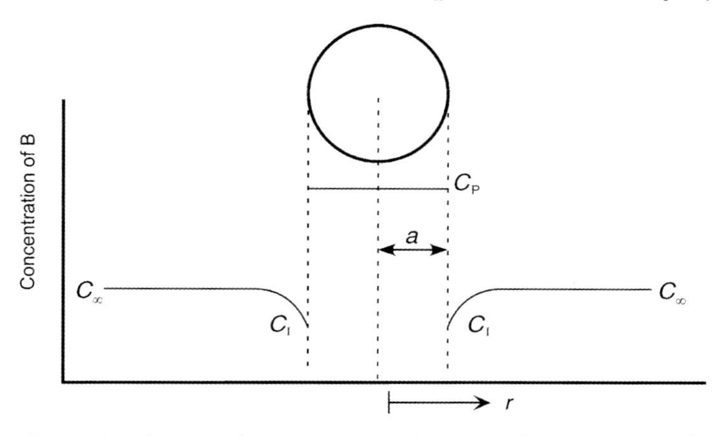

Figure 8.3 Schematic of a precipitate particle and the diffusion field around it.

At steady state, the concentration field has the form given in Equation 8.23, and after applying the boundary conditions that $C = C_\infty$ far from the particle and $C = C_I$ in the matrix at the interface, this solution becomes:

$$C = (C_I - C_\infty)\frac{a}{r} + C_\infty \tag{8.25}$$

This concentration profile is illustrated in Figure 8.3.

The flux of atoms in the diffusion field at the interface is:

$$J = -D\left(\frac{dC}{dr}\right)_{r=a} = (C_I - C_\infty)\frac{D}{a} \tag{8.26}$$

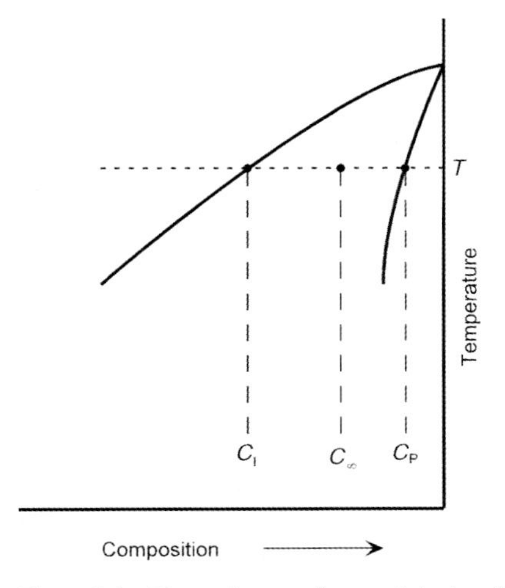

Figure 8.4 Phase diagram for precipitation from a liquid.

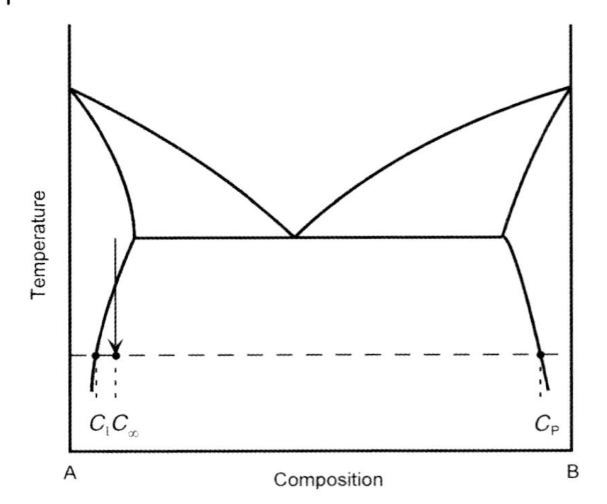

Figure 8.5 Phase diagram for precipitation from a solid solution.

The rate at which atoms arrive at the surface of the particle is the flux of particles to the surface times the surface area of the particle:

$$-4\pi a^2 J = 4\pi a D\,(C_\infty - C_I) \tag{8.27}$$

The rate of growth of the volume, V, of the particle can be written in terms of the rate at which the radius increases:

$$\frac{dV}{dt} = \frac{dV}{da} \cdot \frac{da}{dt} = \frac{d}{da}\left(\frac{4}{3}\pi a^3\right)\frac{da}{dt} = 4\pi a^2 \frac{da}{dt} \tag{8.28}$$

The amount of material in the precipitate particle is VC_P, where C_P is the number of atoms per unit volume in the precipitate, as given by the phase diagram. The rate of growth of the particle is given by the rate at which atoms arrive at it:

$$C_P \frac{dV}{dt} = -4\pi a^2 J \tag{8.29}$$

or

$$C_p 4\pi a^2 \frac{da}{dt} = 4\pi Da(C_\infty - C_I) \tag{8.30}$$

which can be rewritten as:

$$a da = D \frac{C_\infty - C_I}{C_p} dt \tag{8.31}$$

which, for $a = 0$ at $t = 0$ has the solution:

$$\frac{1}{2}a^2 = \frac{C_\infty - C_I}{C_p} Dt \tag{8.32}$$

So the radius of the precipitate particle increases with the square root of time. It is surprising that this is the same relationship we found for the diffusion limited growth of a layer in one dimension, Equation 8.9.

A limitation on the validity of this solution can be estimated by examining the size of the diffusion field around the particle. We assumed that the interface was stationary when we used the steady state equation for the diffusion field. If the radius of the particle is small compared to the radius of the diffusion field, then we are justified in this assumption that the motion of the interface is not important in determining the shape of the diffusion field.

From Equation 8.32, the time it takes a particle to grow to a radius a is:

$$t_a = \frac{C_P}{C_\infty - C_I}\frac{a^2}{2D} \tag{8.33}$$

The distance which an atom can diffuse during this same time is:

$$R^2 = Dt_a = \frac{C_P}{C_\infty - C_I}\frac{a^2}{2} \tag{8.34}$$

For $C_P \gg C_\infty - C_I$, R will be large compared to a. Our solution will be of questionable validity if this condition is not met. This suggests that our solution is more likely to be valid for solid state precipitation, as illustrated by the phase diagram of Figure 8.5, than for precipitation in a liquid, where a typical phase diagram is more likely to resemble Figure 8.4.

In our analysis, we have ignored the change in concentration which takes place in the diffusion field. That is, as the radius of the particle changes, the diffusion field around it also changes. So there is a correction to the flux which we calculated which goes into changing the diffusion field. Our solution is valid if the diffusion flux to the interface changes slowly as the particle grows. This will not be true if the radius of the particle is similar to the radius of the diffusion field. This suggests how our steady state solution should be modified for the case where the radius of the particle is not small compared to the radius of the diffusion field. An analysis of the diffusion-controlled growth of a precipitate which does not rely on the steady state diffusion filed was presented by Ham [4].

8.4 Diffusion Limited Growth in Cylindrical Coordinates

In cylindrical coordinates, the diffusion equation is:

$$D\left(\frac{\partial^2 C}{\partial r^2} + \frac{1}{r}\frac{\partial C}{\partial r}\right) = \frac{\partial C}{\partial t} \tag{8.35}$$

For steady state, the solution does not change in time, so $\partial C/\partial t = 0$. The solution to this equation takes the form:

$$C = A\ln(r) + B \tag{8.36}$$

with the boundary conditions:

$$C = C_\infty \text{ at } r = R_\infty;\ C = C_\mathrm{I} \text{ at } r = a \tag{8.37}$$

Equation 8.36 give a concentration which is infinite at $r = \infty$, so the boundary condition must be applied at some finite radius, which we will call R_∞.

The solution becomes:

$$C = \frac{C_\infty \ln\left(\frac{r}{a}\right) + C_\mathrm{I} \ln\left(\frac{R_\infty}{r}\right)}{\ln\left(\frac{R_\infty}{a}\right)} \tag{8.38}$$

which is messier than the solution in either 1D or for a sphere.

The concentration gradient at $r = a$ is:

$$\left(\frac{\mathrm{d}C}{\mathrm{d}r}\right)_{r=a} = \frac{C_\infty - C_\mathrm{I}}{\ln\left(\frac{R_\infty}{a}\right)} \tag{8.39}$$

And the rate of change of volume is:

$$\frac{\mathrm{d}V}{\mathrm{d}t} = 2\pi\, a \frac{\mathrm{d}a}{\mathrm{d}t} \tag{8.40}$$

So the differential equation for the growth rate of the cylinder is;

$$a(\ln R_\infty - \ln a)\mathrm{d}a = D\frac{C_\infty - C_\mathrm{I}}{C_\mathrm{p}}\mathrm{d}t \tag{8.41}$$

which has a solution:

$$\frac{a^2}{2}\left(\ln R_\infty - \ln a - \frac{1}{2}\right) = D\frac{C_\infty - C_\mathrm{I}}{C_\mathrm{P}}t \tag{8.42}$$

The growth rate of the radius of the cylinder is only approximately proportional to the square root of time.

The steady state solutions above are all approximate solutions to the diffusion problem. We ignored time dependent variations to the diffusion field, but we also assumed that the compositions at the interface and far from the interface did not change in time. We ignored surface tension effects which are important for small spheres and cylinders, we ignored stress effects which can be important for precipitation in a solid, and we ignored nucleation effects by assuming that the precipitate started with zero radius at $t = 0$.

But these solutions have simple forms, and they provide valuable approximations to the full solutions.

8.5 Diffusion Controlled Growth of a Precipitate

An alternate method of modeling the diffusion-controlled growth of a precipitate is to use the modified diffusion equation developed in Chapter 23. This requires a computer program to solve these equations. The thermodynamic properties of a regular solution, which are often used to fit the boundaries of the solid–solid two-phase field below the eutectic, as in Figure 8.5, are incorporated into these equations. In order to use this method, it is necessary to fit the relevant region of the phase diagram to a regular solution model, and then set the desired temperature and matrix composition. The numerical calculation is started with a nucleus of higher composition in the center of the computational field, and the program follows its growth. The program takes into account the curvature of the precipitate particle, as well as the

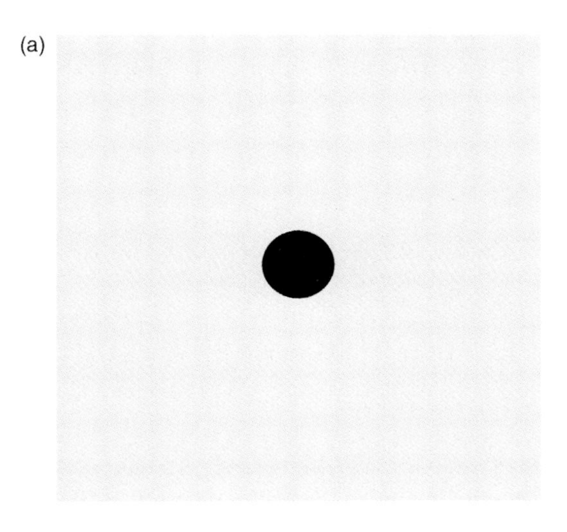

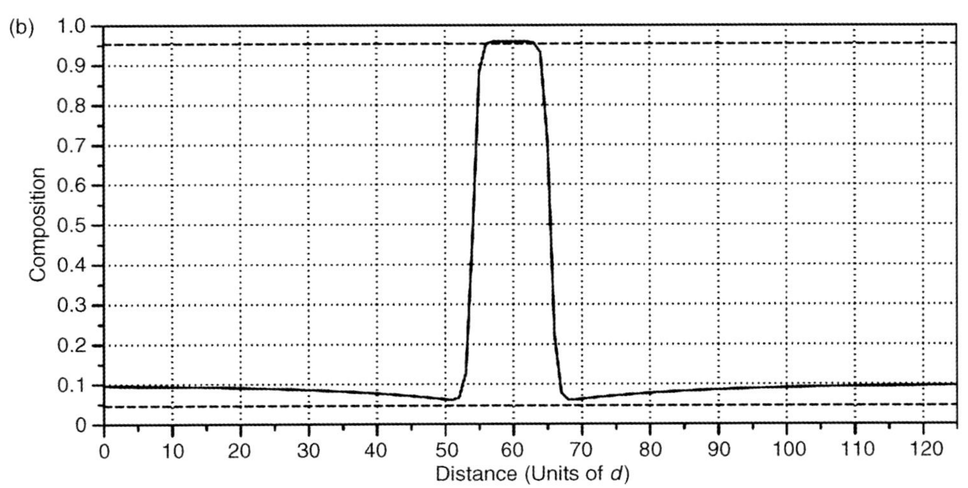

Figure 8.6 (a) Gray scale image of a precipitate growing into a matrix of composition 0.1 at $T/T_C = 0.6$ (b) Composition profile through the center of the precipitate in (a). The dashed lines are the phase boundary compositions.

boundary conditions on the moving boundary. An example of this method is shown in Figure 8.6 for the growth of a precipitate in two dimensions.

Figure 8.6a is an image of a precipitate growing at 0.6 of the critical temperature, T_C, into a composition of 0.1. The composition profile through the center of the precipitate is in Figure 8.6b. The equilibrium phase boundary compositions, and the boundary conditions at the moving surface of the precipiate are automatically included. The compositions both inside and outside the particle are higher than the equilibrium phase boundary compositions because the effect of the curvature of the surface of the precipitate is taken into account. The radius of the precipitate increases with the square root of time, as in Equation 8.32. This is a relatively simple method of obtaining an accurate solution to a Stefan problem.

References

1 Deal, D.E. and Grove, A.S. (1965) *J. Appl. Phys.*, **36**, 3770.
2 Plummer, J.D., Deal, M.D., and Griffin, P.B. (2000) *Silicon VLSI Technology*, Prentice-Hall, Upper Saddle River, NJ, p. 313.
3 Reiss, H., Patel, J.R., and Jackson, K.A. (1977) *J. Appl. Phys.*, **48**, 5274.
4 Ham, F.S. (1958) *J. Phys. Chem. Solids*, **6**, 335.

Problems

8.1. Calculate the thickness of the oxide layer on a silicon wafer after (a) 1 min, (b) 10 min, (c) 1 h, in a wet oxygen atmosphere at 1050 °C.

8.2. A 1 nm diameter particle of gold nucleates in a glass matrix containing 1% gold at 1000 °C. The precipitate particle is essentially pure gold, and the equilibrium concentration of gold in the glass at 1000 °C is 0.1%. Assuming that the growth of the particle is controlled by diffusion, and the diffusion coefficient of gold in glass at 1000 °C is $10^{-10}\,cm^2\,s^{-1}$, use the steady state approximation for diffusion to a spherical particle to calculate how big the particle will be after 1 h.

9
Phase Transformations

The rate at which a first order phase transformation proceeds depends on the rate of atomic processes at the interface, on how rapidly the transformation heat can be removed from the vicinity of the interface, and on diffusion to or from the interface region of the species involved in the growth process. There are instances when only one of these three dominates the growth process. There are other cases where two of these three control growth, and there are cases where all three are important. But all three are always present; it is their relative dominance in the growth process which varies.

In general, the rate at which a phase transformation proceeds depends on all three: the composition at the interface, the reaction kinetics, and the heat flow.

The slow process dominates and controls the transformation rate.

There is always a finite driving force for a phase transformation. The slow process dominates by taking up more of the available driving force, so leaving less for the other processes, which are thus slowed.

9.1
Transformation Rate Limited Growth

A first order phase transformation proceeds by the rearrangement of atoms at the interface between the two phases. These atomic processes depend only on the local conditions at the interface. The rate at which a transformation proceeds depends on the composition and temperature *at* the interface.

There are situations where the atomic processes dominate the transformation process. Transformation rate limited growth occurs, for example, during crystallization of glasses, where the mobility of the atoms or molecules which are crystallizing is very small. For example, the crystallization rate of silica, as shown in Figure 9.1, proceeds at a maximum rate of about 12 μm min^{-1}. Glass crystallization is usually studied by placing a sample of the glass in a furnace for minutes or hours, and then removing the sample to examine how far the crystals have grown. For a single component glass, chemical segregation is not an issue, and the growth proceeds sufficiently slowly that the sample is at the temperature of the furnace.

Kinetic Processes: Crystal Growth, Diffusion, and Phase Transitions in Materials. Kenneth A. Jackson

ISBN: 978-3-527-32736-2

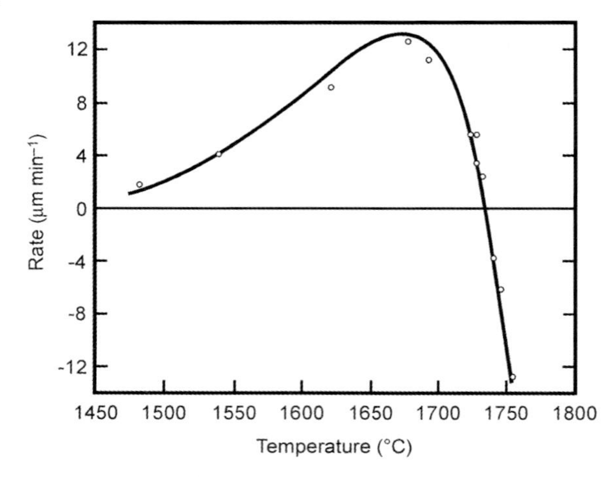

Figure 9.1 Crystallization rate of silica at various temperatures [1].

Transformation limited growth also occurs during growth from solution. Crystals are often grown from solution in times of the order of weeks or months. The growth rate is limited, not by the mobility of the molecules in the solution, but by the rate of formation of new layers of the crystal, which requires a finite amount of supersaturation (ΔC_K) or equivalently, supercooling (ΔT_K), for growth at a reasonable rate (see Figure 9.2).

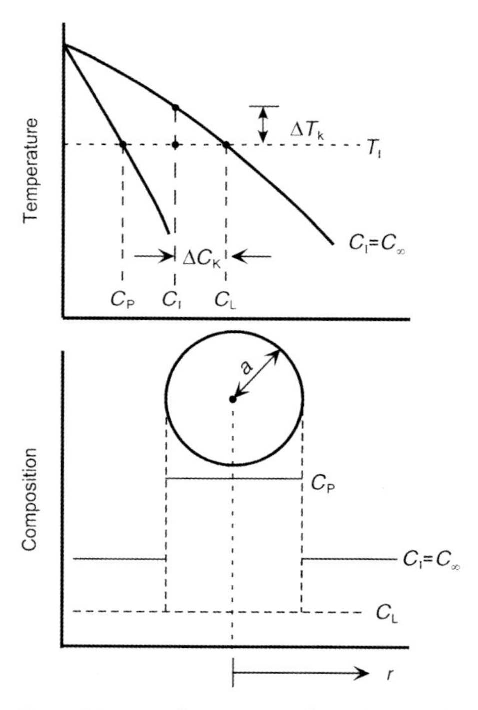

Figure 9.2 Transformation rate limited growth from a solution.

The growth is so slow that the solution is also isothermal. The crystals grow so slowly that the composition field is also uniform throughout the solution at the interface composition, C_I, although there can be a local diffusion limitation to the motion of steps across a surface. In this case, the temperature at the interface is the externally imposed temperature. The composition of the solution is uniform, although it changes with time as the crystal grows. The average interface composition is not at the equilibrium value, since there must be a finite supersaturation to promote the growth process.

9.2
Diffusion Limited Growth

Diffusion limited growth of an oxide film and of a precipitate were discussed in Chapter 8. Segregation during crystal growth will be discussed in more detail in Chapters 11 and 12.

9.3
Thermally Limited Growth

The precipitation process is usually exothermic, and the heat generated by the growth process changes the temperature at the growth front. For a slowly growing precipitate, the temperature rise may be very small, but it must be there, as illustrated schematically in Figure 9.3.

The thermal field around a growing precipitate or crystal has the same mathematical form as the concentration field around a growing crystal, but the diffusion of heat is much more rapid than the diffusion of matter.

The heat generated by the growth process plus the heat conducted to the interface through the crystal must be carried away into the liquid, so the boundary condition for heat flow in one dimension due to the evolution of heat by the growth process is:

$$K_L\left(\frac{dT}{dx}\right)_L + Lv = K_S\left(\frac{dT}{dx}\right)_S \tag{9.1}$$

The positive direction of x is the direction of motion of the growth front, K is the thermal conductivity, L is the latent heat, and v is the growth rate. The subscripts refer to solid and liquid phases. In two or three dimensions, the temperature gradients and the growth rates are those normal to the local interface. In Figure 9.3, there is no temperature gradient in the precipitate, and the temperature gradient in the liquid is negative. This temperature gradient must be large enough to carry away the latent heat. For faster growth, the temperature gradient must be steeper, and so the temperature of the growing crystal is likely to increase.

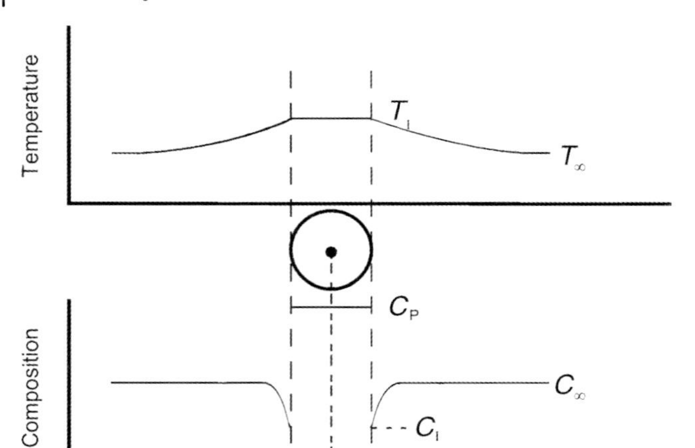

Figure 9.3 Thermal field and concentration field around a growing precipitate.

Most metals and semiconductors have very rapid transformation kinetics for solidification, and so the freezing rates of pure metals and semiconductors are limited by heat flow. The heat flow is used to control the growth process, as for example in Bridgman growth [2], or Czochralski growth [3], as illustrated in Figure 9.4.

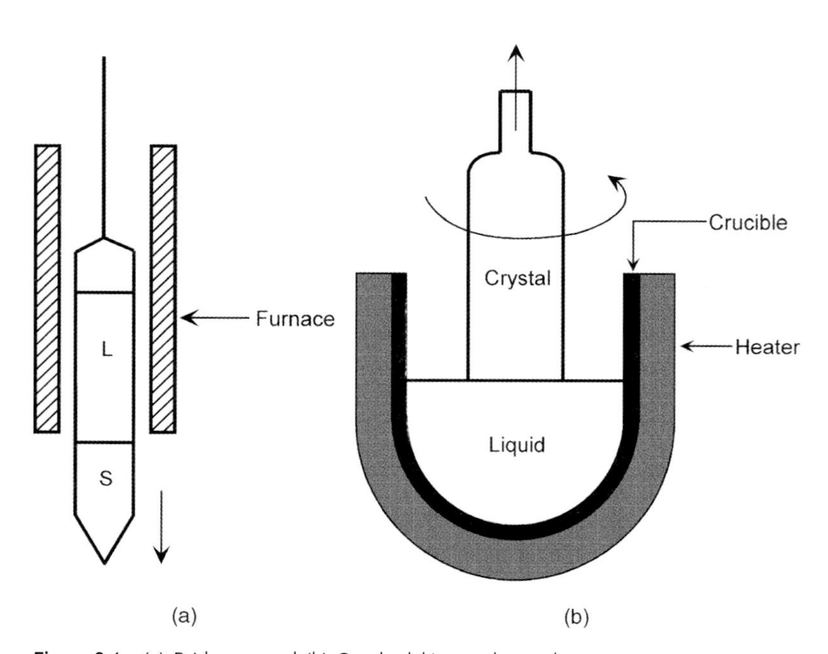

Figure 9.4 (a) Bridgman and (b) Czochralski crystal growth.

In both of these cases, the thermal conditions are established by heaters so that the liquid is hot and the crystal is cooler. In the Bridgman case, the sample is held in a container which is lowered out of a furnace at a controlled rate. In the Czochralski case, the crystal is pulled out of the liquid at the same rate that it is growing so that the interface stays near the surface of the melt. In both cases, heat flows through the interface from the liquid to the crystal.

The latent heat generated by the crystallization process and the heat flux to the interface due to the temperature gradient in the liquid must both be carried away by the temperature gradient in the crystal, as in Equation 9.1. Changing the growth rate in this case does not change the interface temperature due to heat flow. It changes the temperature gradients at the interface.

The interface position is stabilized by the temperature gradients. If the growth rate increases above the drive rate, the interface moves into hotter liquid, and the growth rate slows down. Conversely, if the growth rate slows, the interface moves into colder liquid, and the growth rate increases. Crystal growth processes are discussed in more detail in Chapter 10.

A similar process is used for continuous casting, which produces polycrystalline material. The solidified material is continuously lowered from the cooled mold, and more liquid is continuously added to the melt, as illustrated in Figure 9.5. Continuous processes are usually less expensive than batch processes, and the billet produced is smaller in diameter than a typical casting, and so is easier to roll and draw into a sheet or a wire.

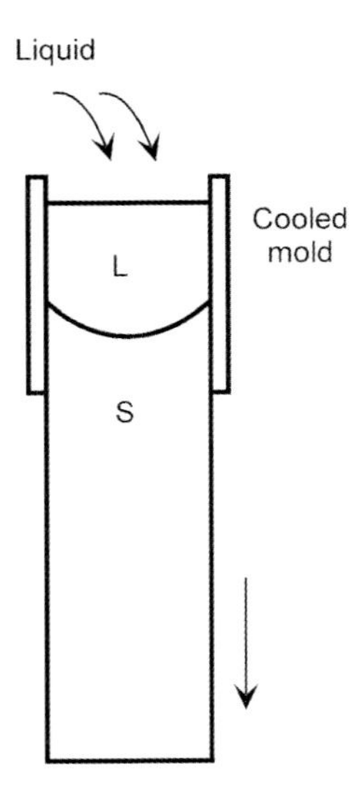

Figure 9.5 Continuous casting.

9.4
Casting of Metals

If the liquid of pure metal is supercooled and then the solid phase is nucleated, the growth will be dendritic. The heat of fusion is more than enough to heat the whole sample up to the melting point, so only some of the liquid will freeze until more heat is extracted. The growth process can reach the far corners of the container much faster than heat can diffuse there, and so the growing solid subdivides into a dendritic structure, leaving behind a region which is part liquid and part solid, but which has already been heated to close to the melting point. Dendritic growth is discussed in Chapter 26.

If a pure metal liquid is poured into a cold mold, many small dendritic crystallites will form initially, and then growth will proceed in from the wall as heat is extracted from the mold.

For an alloy, the situation is even more complicated. Dendrites can form as a result of chemical diffusion as well as heat flow. The scale of the dendrite structure depends on the diffusion coefficient, which is quite different for heat and for chemical diffusion. In an alloy dendrites will form even when the thermal field is carefully controlled. The structure of castings is discussed in more detail in Chapter 28.

9.5
Operating Point

Examples are given above for cases where chemical diffusion limits the rate of a phase transformation, where transformation kinetics are rate limiting, and where heat flow is rate limiting. There are cases where two of these factors are important and the third is not. In general, all three are present and are important to a greater or lesser degree. Figure 9.6 illustrates the case where all three are present during growth of a solid phase.

In Figure 9.6, T_M^A is the melting point of the pure material. C_∞ is the initial composition of the alloy or solution. The liquidus temperature for this composition is $T_L(C_\infty)$. The composition of the solid will lie on the solidus. As a result of the rejection of the second component from the growing phase (or the preferential incorporation of the primary component, depending on how you look at it), the concentration of the liquid at the interface has been increased, by an amount ΔC_D, to C_I. The liquidus temperature corresponding to this composition is $T_L(C_I)$. The temperature far from the growing phase is T_∞, and the liquidus composition at this temperature is $C_L(T_\infty)$. The interface temperature has increased above the far field temperature by an amount ΔT_H, to T_I, due to the latent heat generated by the growth process.

The liquidus composition at this temperature is $C_L(T_I)$, and the equilibrium composition of the solid phase at this temperature is $C_S(T_I)$.

The operating point is in the two phase field. It must be below the liquidus in order for the transformation to proceed. The composition of the solid phase which grows is, correspondingly, at C_P, which is below the equilibrium solidus.

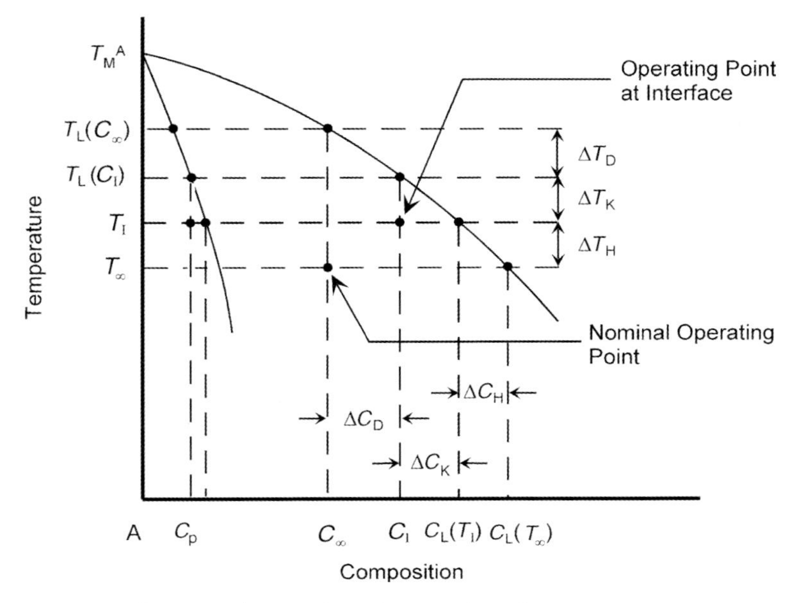

Figure 9.6 Illustrating the operating point on a phase diagram.

The distance below the liquidus represents the driving force for crystallization. It can be described either as an undercooling, $\Delta T_K = T_L(C_I) - T_I$ or as a compositional driving force, $\Delta C_K = C_L(T_I) - C_I$. These two representations are equivalent for the kinetic driving force for the transformation, which is formally defined as the difference between the free energy of the liquid which is transforming and of the solid which is forming.

Usually, an experimenter cannot control the interface composition and temperature directly. Only the initial concentration of the melt, and either the initial temperature or the rate of cooling can be controlled. As a result, the supercooling is usually defined as $T_L(C_\infty) - T_\infty$, and the supersaturation as $(C_L(T_\infty) - C_\infty)/(C_L(T_\infty) - C_P)$. These values of the supercooling and supersaturation define the *nominal* operating point. They include the driving forces necessary to drive the diffusion process, the heat flow, and the transformation kinetics.

The transformation rate is controlled by the local temperature and composition at the interface, These depend on the fluxes of heat and matter to or from the interface, as well as on the transformation rate.

In addition to the local temperature and composition, the motion of an interface is influenced by the local curvature of the interface. A positive local curvature (the center of curvature in the lower energy phase) lowers the local equilibrium temperature, and a negative curvature raises it. This effect plays a central role in nucleation, as discussed in Chapter 15.

The equilibrium phase diagram is applicable when the rate at which atoms or molecules at the interface move around rapidly compared to the rate at which the transformation proceeds. Each growth site at the interface then has a concentration which is, on average, the local concentration. Each component then acts independently, and the incorporation of species into the growing phase depends on their individual chemical potentials. When the growth rate becomes comparable to the rate at which atoms or molecules can move by diffusion, then the species perforce must transform between the phases in a cooperative manner, and so a kinetic version of the phase diagram is applicable, as discussed in Chapter 24.

References

1 Wagstaff, F.E. (1969) *J. Am. Ceram. Soc.*, **52**, 650.
2 Bridgman, P.W. (1925) *Proc. Am. Acad. Arts Sci.*, **60**, 303.
3 Czochralski, J. (1917) *Z. Phys. Chem.*, **92**, 219.

Problems

The answers to these questions are to be found by consulting reference sources.

9.1. What are the thermal and chemical diffusivities in water at room temperature?
9.2. What are the thermal and chemical diffusivities in liquid and solid copper at the melting point of copper?

10
Crystal Growth Methods

Crystals will grow from a melt much more rapidly than they will grow from the vapor phase or from a solution. This is simply because the density of material in the melt is comparable to that in the crystal, so the atoms or molecules are, essentially, there already to grow the crystal. For both vapor and solution growth, the density of atoms or molecules in the mother phase is much lower, and the growth rate depends on the rate at which they arrive at the surface of the crystal.

For the commercial growth of crystals, the faster that crystals of acceptable quality can be grown, the better. This is also true for the non-commercial growth of experimental crystals. So melt growth is the preferred method. There are various reasons why many crystals cannot be grown from the melt, but if a crystal can be grown from its melt, it will be.

10.1
Melt Growth

There are several different schemes for melt growth [1, 2]. In melt growth, the intrinsic growth process is usually so rapid that the growth is controlled by heat flow. The various schemes differ primarily in the configuration of the growth apparatus.

10.1.1
Czochralski Growth

This method is named after the person who is given credit for inventing it. It is a preferred method of growth because there is no container around the growing crystal. Most of the silicon used for microelectronic applications is grown by the Czochralski method [3]. Figure 10.1 shows a schematic drawing of a Czochralski growth apparatus.

It is a very simple method, in principle. The liquid is slightly undercooled so that the crystal grows down into the liquid. However, there is a drive mechanism which is adjusted to pull the crystal up as fast as it grows so the interface between the crystal

Kinetic Processes: Crystal Growth, Diffusion, and Phase Transitions in Materials. Kenneth A. Jackson

ISBN: 978-3-527-32736-2

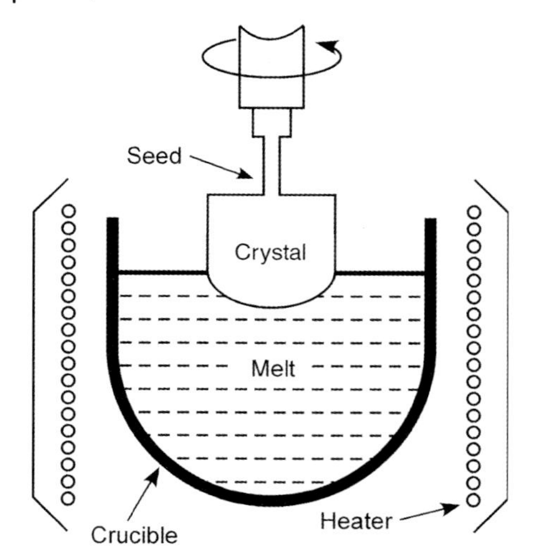

Figure 10.1 Schematic drawing of Czochralski crystal growth apparatus.

and the melt stays more or less stationary as the crystal grows. Of course, the liquid level drops as the crystal grows, and the crystal growers try to convert as much as possible of the liquid into good crystal.

The growth rate of the crystal is determined by the undercooling at the interface. The temperature of the interface depends on the temperature of the liquid in the crucible, which depends on the heat flow through the system. Heat flows from the heater which is around the crucible into the liquid, through the interface, and then up the crystal. The latent heat, which is generated by the freezing process, must also be conducted up the crystal. Heat is lost from the crystal to the surroundings by conduction, by convection and by radiation. Overall, the heat flow depends on the design of the crystal puller, on how much power is coupled into the melt, and on how fast heat is lost from the crystal into the upper part of the growth chamber. The temperature gradient at the interface tends to stabilize the position of the interface. If the crystal grows too rapidly, it grows into a region of higher temperature, and so it slows down, and vice versa, if it grows too slowly, the undercooling increases, and so the growth rate increases. The cross-section of the crystal, and the area of the interface are the same, so that the heat flow conditions are relatively independent of the diameter of the crystal.

As the crystal grows, the heat loss from the crystal changes because the surface area of the crystal increases as it gets longer, and the liquid level drops, so that the coupling from the heater into the melt changes. There must be continuous minor adjustments to maintain the balance of heat flow into the crucible and out through the crystal as growth proceeds. This can be done manually, by a crystal grower monitoring the process, or using automatic diameter control. The control system, whether manual or automatic, has only two variables to play with: the power to the heater, which changes the temperature of the melt, and the pull rate of the crystal. The first of these has a fairly long response time, because of the thermal mass of the crucible and the liquid. Rapid corrections are made by adjusting the pull rate.

Automatic diameter control is often accomplished by weighing either the crucible or the growing crystal. Modern weighing equipment is accurate enough to control the rate at which weight is gained by the crystal or lost from the crucible. This method works well for crystals which are denser than the liquid, but is not so good for silicon, which expands on freezing. If a silicon crystal grows down into the melt, the liquid level can rise, increasing the buoyancy on the crystal, and reducing its apparent weight. For silicon there is a bright ring, caused by reflections, at the meniscus where the liquid and crystal join. Diameter control is effected by imaging this bright ring onto an array of photocells, and using the output from the photocells for feedback control.

It is difficult to make the thermal field around the growing crystal precisely circularly symmetric, and so the crystal is rotated to smooth out any angular asymmetries in the thermal field. This often results in periodic variations in the local growth rate, as the crystal alternately rotates through the hotter and cooler zones. The crucible is sometimes rotated as well to modify the fluid flow patterns in the melt. Magnetic fields are sometimes used to reduce convective flows in the melt.

Dash [4] discovered that all of the dislocations in a silicon or germanium crystal will grow out, that is, they move laterally during growth so that they terminate on the sides of the crystal, leaving a perfect crystal, if the crystal is less that about 3 mm in diameter. When a seed crystal is first dipped into a hot melt, many dislocations are produced by the thermal stresses. These dislocations will propagate down the entire length of a large crystal, but they can be grown out of a small diameter crystal by the Dash procedure. So a small diameter seed crystal is grown a few inches, until the operator sees that the facets on the surface of the seed crystal have changed, which indicates that the seed is dislocation free. The seed is then grown another inch or so before it is "shouldered," that is, widened out to the desired final diameter, usually under computer control. Then the automatic diameter control system takes over. It is a tribute to the properties of silicon that a 300 mm diameter boule which is 1 m long can be suspended on a 3 mm diameter seed.

Figure 10.2 contains a sequence of photographs, looking down into the crucible at an angle, taken during the Czochraksi growth of a silicon crystal.

The process starts with lumps of high purity polysilicon in the crucible. The next two photos were taken as melting of the poly proceeded. At the left of the middle row, the seed is dipped into the melt. In the middle picture, the seed has been grown some distance, and the shouldering has started. In the next frame, the shouldering is continued. In the left picture in the bottom row, the shouldering is complete, and the growth at the final diameter has begun. In the last two pictures in the bottom row, growth proceeds at a constant diameter. The bright ring where the crystal leaves the liquid, which is used for diameter control, is clearly visible in the pictures in the bottom row.

The (111) face of silicon facets during crystal growth. The crystals can be grown either with a central facet or with a peripheral facet, as shown in Figure 10.3.

Figure 10.2 Photographs taken during Czochralski growth of a silicon crystal.

These configurations can be controlled by changing the temperature gradient. The temperature at the coolest part of the facet is about five degrees below the melting point so that new layers will nucleate rapidly enough for growth. The undercooling in the off facet directions is less than a milli-degree. These differences in growth rate are discussed in Chapter 21. The peripheral facet configuration tends to suppress the formation of twins at the edge of the interface.

The Czochralski method has the advantage that no container is required to hold the crystal but a crucible is required to contain the melt. For silicon, the crucible is made

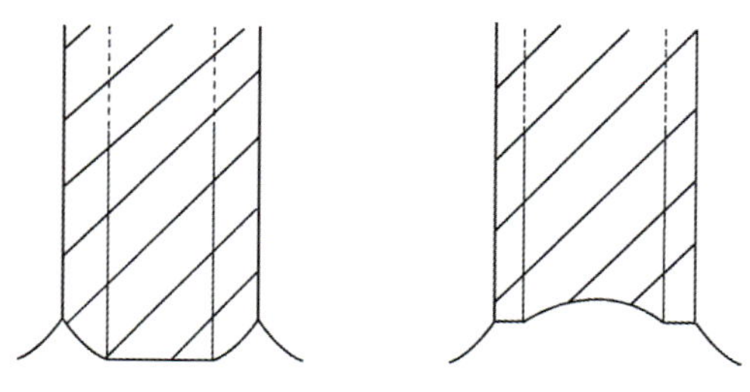

Figure 10.3 Growth profiles for (111) silicon/melt interfaces: (a) central facet, (b) peripheral facet.

from high purity "semiconductor grade" silica. Silica softens below the melting point of silicon, and so it must be supported, usually by shaped graphite. The graphite can also serve as a susceptor for radio frequency heating, or, more commonly, as a resistive heating element. Silica is slightly soluble in liquid silicon, which introduces oxygen into the melt. Some of this oxygen evaporates from the surface of the melt, and some of it is incorporated into the growing crystal. Czochralski silicon typically contains about 10^{18} oxygen atoms per cubic centimeter.

Single crystals of many materials are grown using the Czochralski method, because of its simplicity, and because it permits the seeding of the growth process, so that crystals of any desired orientation can be grown.

The components of some crystals, phosphorus in InP for example, have high vapor pressures at the melting point of the crystal. High pressure pullers and liquid encapsulated Czochralski (LEC) methods (Figure 10.4) have been developed for these materials. The chamber is typically filled with argon under pressure.

Growth is carried out in a pressurized chamber. Boron oxide, which is an inert, low melting point material, can be floated as a liquid layer on the top of the melt, and the crystal is pulled through this layer. The combination of the liquid encapsulant and the high pressure of argon above it work together to contain the volatile components in the melt.

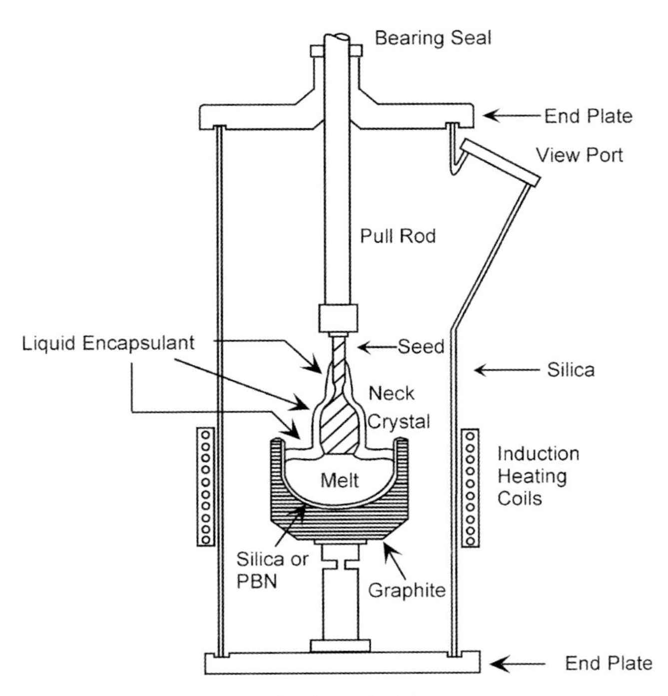

Figure 10.4 Liquid encapsulated Czochralski.

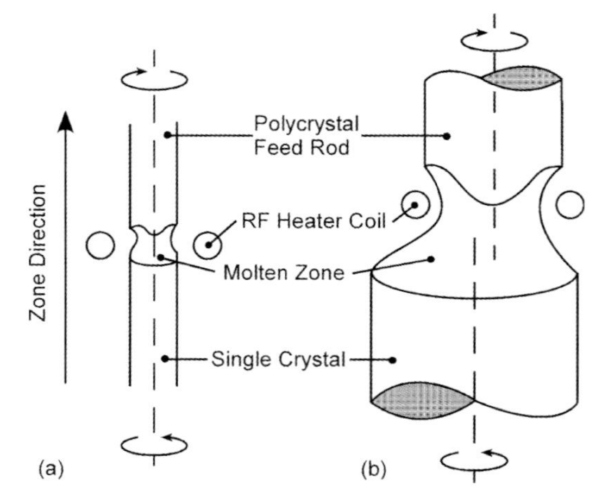

Figure 10.5 Floating zone crystal growth.

10.1.2
Floating Zone

The floating zone method [5] is used to make high resistivity silicon crystals, which are used to make power handling devices. The crystals do not contain oxygen, and so have resistivities as high as 10 000 Ω cm, instead of 10 Ω cm, which is typical for Czochralski silicon. This method does not require a container for the liquid. Instead, the liquid is held in place by surface tension, as illustrated in Figure 10.5. Growth proceeds by moving the heater coil, which moves the liquid zone, upwards.

On the left is axi-symmetric growth with the feed rod the same diameter as the final crystal. This method usually results in a single torroidal flow pattern in the liquid, and so the center of the liquid zone is not heated very efficiently. In the scheme shown on the right, which is the more usual configuration, the feed rod has a smaller diameter than the growing crystal. The smaller diameter feed rod is off-center, and this configuration results in a single horizontal convective roll in the melt, which heats the center of the liquid efficiently. The feed rod must be moved relative to the growing crystal at a different rate than that at which the heater coil is moving.

This method requires the preparation of a seed rod of appropriate diameter in a separate zone melting run, making this a more costly process than Czochraski growth. Semiconductor device people have learned to live with the oxygen in Cz silicon, and even turn it to advantage for internal gettering. For power transmission devices, the higher resistivity of floating zone silicon results in a higher breakdown voltage, which makes the extra cost worthwhile.

A variant on the floating zone method has been used to grow single crystal rods with small cross-sections of high melting point oxides. A high power laser or lasers are focused onto the rod to create a molten zone, and the molten zone is moved along the rod to achieve growth. This is usually done for high melting point oxide materials which are stable in air. The temperature gradients in the rod during growth are very large, because heat is lost rapidly by radiation.

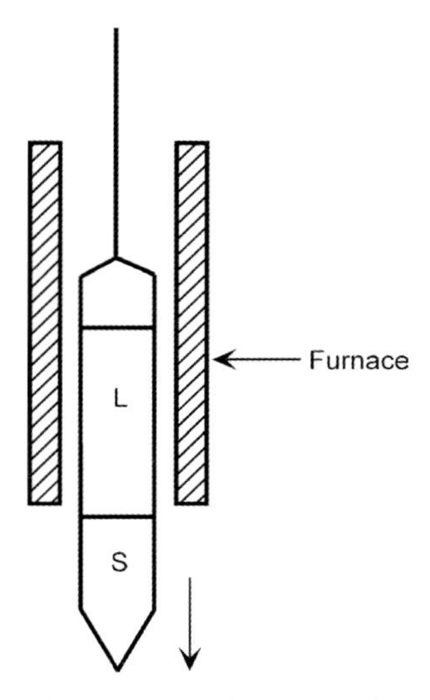

Figure 10.6 Bridgman crystal growth method.

10.1.3 Bridgman Method

In the simple version of Bridgman growth [6] the sample is contained in a tube, as illustrated schematically in Figure 10.6.

Initially the sample is heated in the furnace, and melted completely. It is then lowered slowly from the furnace. The crystal nucleates when the bottom end of the container gets cold enough. The bottom end of the container is usually tapered to a point to minimize the probability of forming many nuclei. In principle, a seed crystal can be used in the bottom end of the container, but, in practice, it is difficult to see the interface in the seed in order to determine when it is partly melted. If the starting material, which is usually polycrystalline, is not all melted, or if the seed melts completely, then the seeding process fails. Usually the time interval between these two events is short. So Bridgman growth is usually unseeded, and the orientation of the resultant crystal is random.

In more sophisticated set-ups, baffles or multi-zone heaters are used to control the temperature of the sample.

Special precautions are usually necessary to remove the crystal from the tube after growth.

10.1.4 Chalmers Method

The Chalmers method [7] is similar to the Bridgman method, except that the sample is held in an open, horizontal boat rather than in a vertical tube. Either the boat is withdrawn from a stationary furnace, or the furnace is withdrawn slowly from around

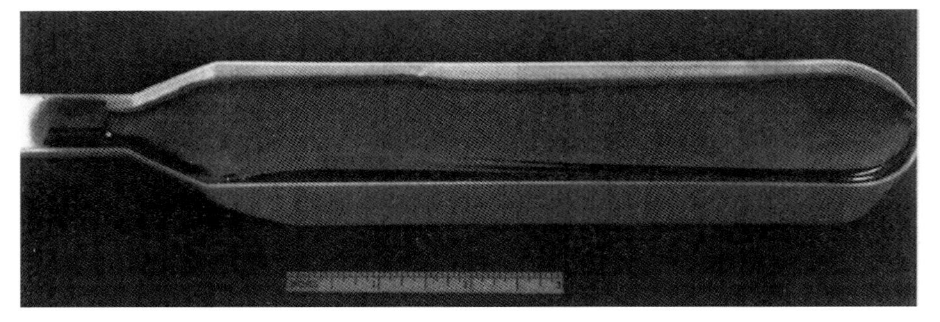

Figure 10.7 Horizontal boat used in the Chalmers method. The seed is visible on the left, and it is clear how much of the seed was melted.

the boat. In this configuration, the position of the interface can be observed on the top surface of the sample. This permits the use of a seed in one end of the boat, so that the partial melting of the seed can be controlled. The boat can be made with sloping sides to facilitate the removal of the crystal, as illustrated in Figure 10.7.

10.1.5 Horizontal Gradient Freeze

This method is similar to the Chalmers method, except that the sample is contained in a multi-zone furnace. The temperature gradient in the crystal is imposed by the various heating zones. The boat and furnace are not moved relative to each other, but rather the crystal is grown by lowering the temperature of each zone in the furnace, usually with computer control

There is also a method called vertical gradient freeze, which is a vertical version of this scheme, the Bridgman method without any moving parts.

10.2 Solution Growth

Crystals are grown from solution only if they cannot be grown from their melt. There are several reasons why certain crystals cannot be grown from their melt. Crystal growth from the melt requires that the molecules which make up the crystal should not decompose below the melting point. It requires that the material should not sublime below the melting point, although this can be circumvented for some materials with a high pressure growth apparatus. It requires that the material should melt congruently. It requires that there should not be a solid state phase transformation between the melting point and room temperature which will destroy the single crystal on cooling.

A variety of solvents are used for growing crystals from solution.

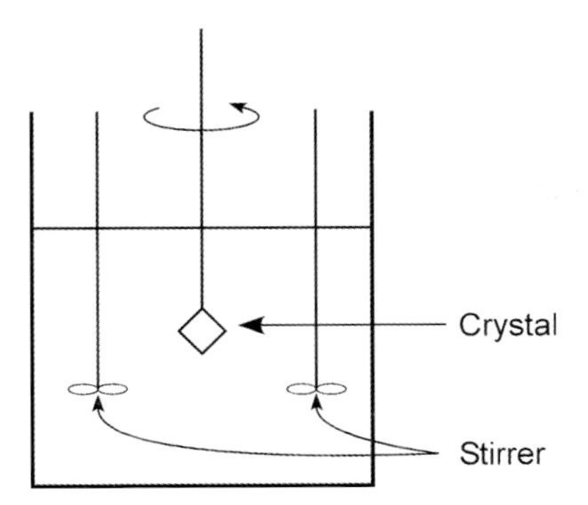

Figure 10.8 Apparatus for growth from an aqueous solution.

10.2.1 Growth from Aqueous Solutions

Water soluble crystals are grown from aqueous solutions, in an apparatus such as is illustrated in Figure 10.8.

The growth is isothermal, and usually several weeks of growth is required to obtain a crystal of reasonable size. The growth is started from a seed which is suspended from a rod. The crystal is rotated, and the solution is stirred. The temperature of the solution may be lowered progressively to compensate for the depletion of the solution.

10.2.2 Flux Growth

Flux growth is solution growth where the solvent is a low-melting oxide. Often the flux is a mixture of boron oxide and lead oxide, and it is used to grow crystals of oxides, such as garnets. Many garnets cannot be grown directly from the melt, but the components of the garnet are soluble in the flux. The growth temperatures are usually much higher than the boiling point of water, so crystals cannot be grown from aqueous solutions.

The growth crucible, which is usually platinum, is placed in a constant temperature furnace, without stirring. A cold finger is attached to the bottom of the crucible to promote localized nucleation. Growth of a crystal of reasonable size usually takes several weeks.

10.2.3 Hydrothermal Growth

Hydrothermal growth [1] uses superheated steam as the solvent. The crystals are grown in an autoclave, at a temperature and pressure above the critical point for water. Quartz crystals are grown hydrothermally because there is a transition from alpha to beta quartz at about 580 °C, which destroys the crystal, so a crystal grown from the melt at 1700 °C does not survive on cooling to room temperature. Superheated steam is used as the solvent because SiO_2 is minimally soluble in anything else.

There are usually two temperature zones in the autoclave, separated by a perforated baffle. Nutrient material is placed in the higher temperature zone, which is in the lower part of the chamber to promote convection, and seed crystals are suspended in the upper zone, at a lower temperature. The pressure is determined by the amount of water added to the chamber. When heated to the operating temperature, this turns into pressurized, superheated steam. Typical operating conditions for growing quartz are 350 °C in the dissolving zone, 400 °C in the growth zone, with a pressure of about 2000 bar. The growth of a crystal of reasonable size takes several weeks.

Natural quartz crystals and natural crystals of many other minerals are believed to have grown under conditions like this, well below the surface of the earth.

10.3 Vapor Phase Growth

Growth from the vapor phase is very slow, so that only a few bulk crystals are grown by this method. On the other hand, thin films are usually deposited from the vapor phase, where deposition rates of a fraction of a micron per minute are tolerable.

For growth from the vapor phase, the material should have high vapor pressure, and it should not decompose on vaporization if it is a compound. Alternatively, if the components have high vapor pressures, the desired compound can be assembled during deposition. One such scheme is illustrated in Figure 10.9, where two components are vaporized separately, and combine at the growing crystal.

10.4 Stoichiometry

There is always a range of compositions over which a compound crystal can grow, although this range may be very small. The composition, at which a liquid alloy

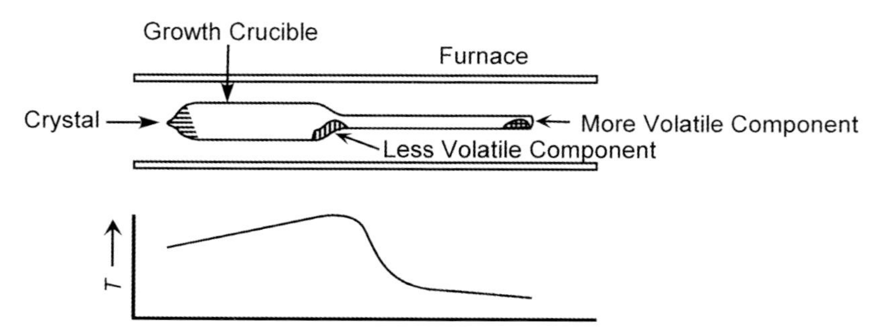

Figure 10.9 Vapor phase growth of a compound from its components.

transforms into a compound having the same composition as the liquid, is known as the congruent melting composition. On the phase diagram, this is the composition corresponding to the highest melting temperature of the compound, where the solidus and liquidus lines have a common, horizontal tangent. This is the preferred composition for growing crystals, because at this composition there is no compositional segregation of the major species during growth, and so growth rate fluctuations do not create composition fluctuations in the crystal, and instabilities due to constitutional supercooling do not occur.

Usually, the congruent melting composition is not at the stoichiometric composition, and so crystals grown at the congruent melting composition are not stoichiometric. As a result, there are vacancies or anti-site defects in the crystal, rather than more serious compositional variations which would result from interface instabilities. This is illustrated in Figure 10.10a, which shows a conventional phase diagram for Cd–Te alloys. The semiconducting compound CdTe is shown as a vertical line. Figure 10.10b is an enlargement of the central part of the phase diagram. The single phase region for CdTe has a distorted shape, and the congruent melting composition is displaced from the stoichiometric composition.

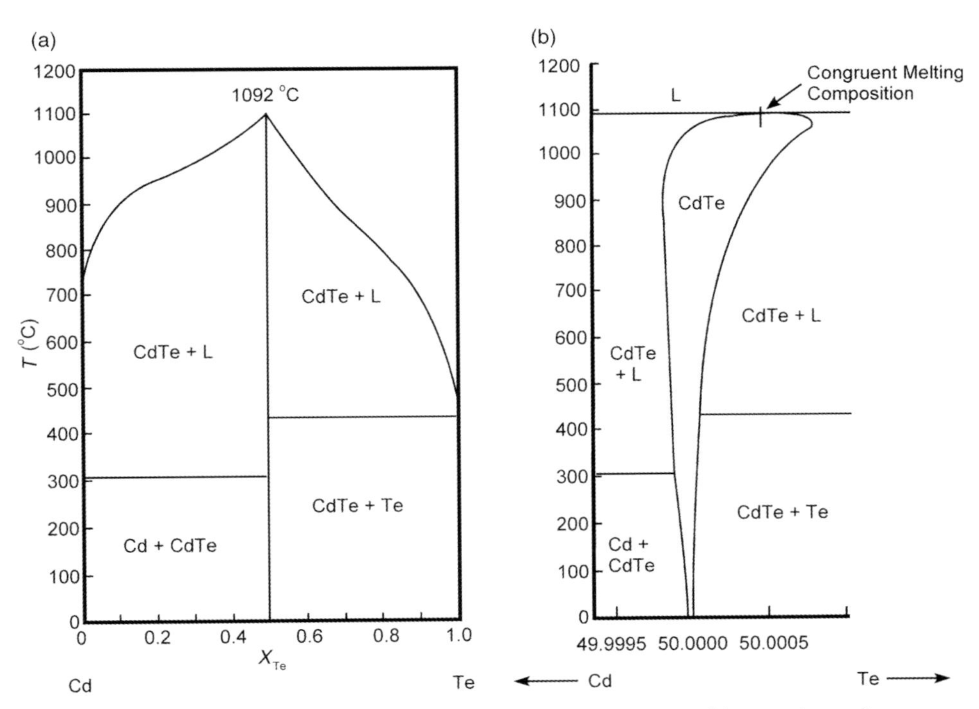

Figure 10.10 (a) A conventional phase diagram for Cd–Te; (b) an enlargement of the central part of the phase diagram.

References

1 Laudise, R.A. (1970) *The Growth of Single Crystals*, Prentice-Hall, Englewood Cliffs, NJ.
2 Wilkes, J.G., Benson, K.E., and Lin, W. (2000) *Handbook of Semiconductor Technology, Vol. 2, Processing of Semiconductors* (ed. K.A. Jackson), Wiley-VCH, Weinheim, p. 1.
3 Czochralski, J. (1917) *Z. Phys. Chem.*, **92**, 219.
4 Dash, W.C. (1958) *J. Appl. Phys.*, **29**, 739; (1959) **30**, 459; (1960) **31**, 736.
5 Pfann, W.G. (1958) *Zone Melting*, Wiley.
6 Bridgman, P.W. (1925) *Proc. Am. Acad. Arts Sci.*, **60**, 303.
7 Chalmers, B., (1964) *Principles of Solidification*, Wiley, New York.

Further Reading

Chernov, A.A. (1984) *Modern Crystallography III: Crystal Growth*, Springer-Verlag, Berlin.
Rosenberger, F. (1979) *Fundamentals of Crystal Growth*, Springer-Verlag, Berlin.
Tiller, W.A. (1991) *The Science of Crystallization*, Cambridge University Press, New York.

Problems

10.1. Discuss the difference between the Czochralski growth of silicon and of sapphire.

10.2. Discuss the relevant factors to be considered in selecting a growth medium for crystal growth.

11
Segregation

11.1
Segregation During a Phase Change

In this chapter, segregation during a phase change will be discussed. We will focus on these effects for the solidification process, which is the most common and most widely used phase transformation in materials processing. Segregation processes are similar for any phase transformation, but the details may differ due to differences in transformation rates, mobilities, diffusion rates, and so on of the species involved.

At equilibrium, the composition of one phase is, in general, different from that of the others. This difference in equilibrium concentrations is described by the phase diagram, as illustrated in Figure 11.1.

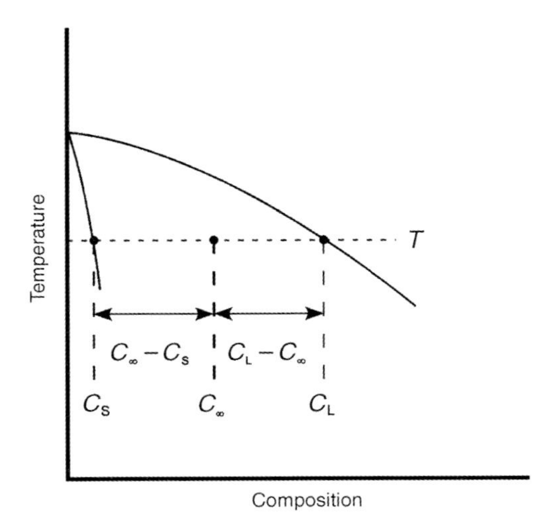

Figure 11.1 Phase diagram.

Kinetic Processes: Crystal Growth, Diffusion, and Phase Transitions in Materials. Kenneth A. Jackson

ISBN: 978-3-527-32736-2

At true equilibrium there are no composition gradients in any phase, and so the composition of each phase is uniform, at the equilibrium concentration for the appropriate temperature. In practice, a solid of varying composition can form as the growth conditions change during a phase transformation. Once such a solid has formed, it can take a very long time for its composition to become uniform, and so true equilibrium is seldom achieved. Below, we will discuss how the growth conditions can change the composition distribution in a solid while it is forming.

An important quantity in discussing segregation is the ratio of the compositions in the two phases. For solidification this is usually discussed in terms of the ratio of the concentration in the solid to that in the liquid, as illustrated in Figure 11.1. This ratio is known as the k-value.

$$k \equiv \frac{C_s}{C_L} \tag{11.1}$$

It is also commonly called the segregation coefficient.

11.2 Lever Rule

The simplest form of segregation is to assume that true equilibrium exists, so the compositions of both the phases are uniform. In this case, we can write that the sum of the compositions of a component in each of the phases must equal the total overall composition:

$$C_s V_s + C_L V_L = C_\infty V = C_\infty (V_s + V_L) \tag{11.2}$$

Where C_∞ is the overall average composition, C_S and C_L are the compositions of the solid and liquid respectively, V_S and V_L are the volumes of the solid and liquid present, and $V = V_S + V_L$ is the total volume. We can rewrite this equation in the form:

$$(C_\infty - C_S)V_S = (C_L - C_\infty)V_L \tag{11.3}$$

Figure 11.1 illustrates why this is called the lever rule. The closer the overall composition is to a phase boundary, the more of that phase is present.

Defining $g = V_S/V$ as the volume fraction of solid, the compositions of the solid and liquid at equilibrium can be written in terms of g and k as:

$$\begin{aligned} C_L &= \frac{C_\infty}{1 + g(k-1)} \\ C_S &= \frac{kC_\infty}{1 + g(k-1)} \end{aligned} \tag{11.4}$$

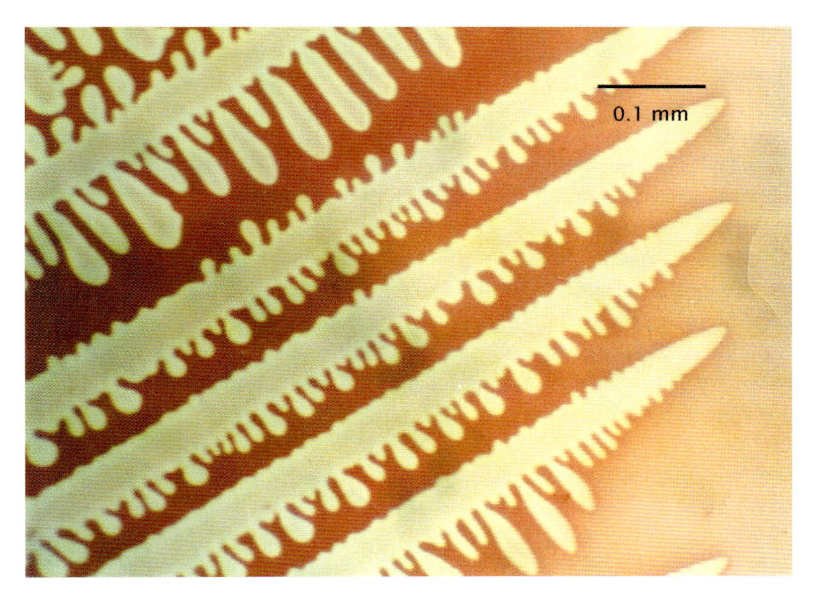

Figure 11.2 Dendritic growth with a red dye. There was a temperature gradient on the sample so that the left side was colder.

The lever rule is the classical expression for segregation of a component between two phases, but it seldom applies in practice because the composition of the solid as it forms is not uniform. It takes a very long time for the composition of a solid to become uniform if significant non-uniformities are grown into it during the growth process.

The lever rule is often used as an approximation for analyzing fluid flow and segregation during dendritic growth of alloys where the dendrites are slender and the liquid channels between them are narrow, as illustrated in Figure 11.2. The sample in Figure 11.2 was growing in a temperature gradient, cooler on the left, so that the volume fraction of solid increases towards the left, and the concentration of the red dye in the interdendritic liquid also increases towards the left.

There is a concentration gradient around the dendrite tip, but, in the liquid channels back from the tip, the concentration in the liquid is relatively uniform across the channels, because of the narrowness of the channels, and the relatively slow lateral growth rate. The concentration in the liquid is close to the local equilibrium temperature as given by the phase diagram, and so the concentration of dye in the liquid is higher at the lower temperatures. The volume fraction at each temperature is given approximately by the lever rule. However, there is a variation in the composition in the solid across the width of the dendrite stem. This can be ignored during the early stages of the process if the k-value is small enough that there is little of the second component in the solid. Ultimately, the liquid in the channel will freeze, and the composition of the last solid to freeze will be very high. This is described by the Scheil equation.

11.3 Scheil Equation

The Scheil equation [1] is also called the *normal freezing* equation. It is based on the assumption that the local composition of the solid does not change after it has formed. It also assumes that the concentration of the liquid is uniform during the solidification process, as illustrated in Figure 11.3. For $k < 1$, the first solid which forms contains less of the second component. The second component which is rejected by the solid stays in the liquid, and so the composition of the liquid increases as solidification proceeds. The composition of the liquid is uniform throughout, but increases with time. So the composition of the solid which forms, given by $C_S = kC_L$, also increases with time. The average composition of the final solid must be the same as the average composition of the starting liquid.

Defining g as the fraction solidified, we can write that the total amount of the second component in the solid and the liquid is equal to the total amount of the second component in the sample, as in Equation 11.3, except that the concentration in the solid must be expressed as an integral:

$$\int_0^g C_S \mathrm{d}x + (1-g)C_L = C_\infty \tag{11.5}$$

In this form, the equation looks difficult to solve, but it can be converted into an ordinary differential equation by differentiating with respect to g:

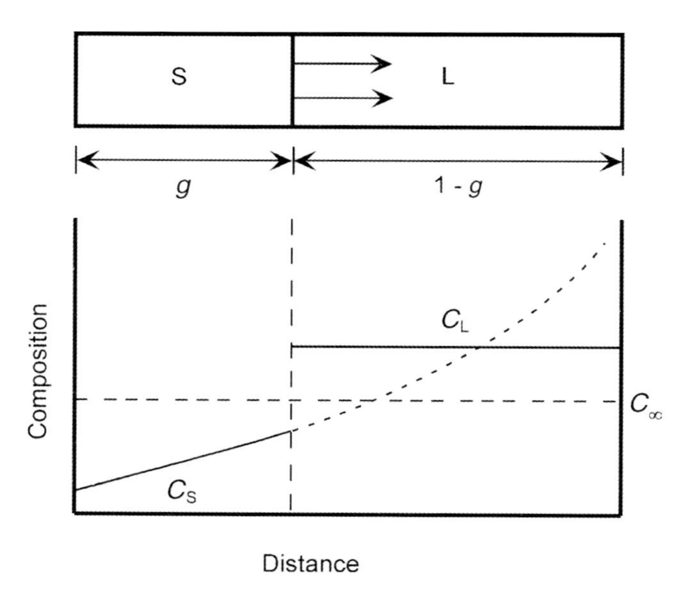

Figure 11.3 Composition distribution in the solid and the liquid for "normal freezing".

$$C_S + (1-g)\frac{dC_L}{dg} - C_L = 0 \tag{11.6}$$

Replacing C_S with kC_L and rearranging gives:

$$\frac{dC_L}{C_L} = \left(\frac{1-k}{1-g}\right)dg \tag{11.7}$$

which can be integrated to give:

$$\ln C_L = -(1-k)\ln(1-g) + \text{constant} \tag{11.8}$$

Applying the boundary condition that $C_L = C_\infty$ at $g = 0$ gives:

$$C_L = C_\infty(1-g)^{k-1} \tag{11.9}$$

The usual form of the Scheil equation is for the composition of the solid:

$$C_S = kC_\infty(1-g)^{k-1} \tag{11.10}$$

Notice that for $k < 1$, the concentrations of the liquid and solid go to infinity at $g = 1$.

The solid which forms can have any shape. The factor g is the volume fraction which is solid, but the fraction solidified is not given by the lever rule, since the composition of the solid is not uniform.

This expression describes the composition of the solid which forms as the interdendritic liquid freezes. It can only be applied where the composition of the liquid is uniform.

It does not apply at the dendrite tip, nor for unidirectional crystallization, where there is a boundary layer of the rejected components at the interface, which means that the composition of the liquid is not uniform. The analysis has been applied to unidirectional crystallization, but with an "effective" k-value which depends on the growth conditions, as discussed in more detail below.

11.4 Zone Refining

With normal freezing, the first part of the solid to freeze is purer in the second component than the last part to freeze. This effect can be used to purify a sample, by discarding the last part of the sample to freeze. Pfann [2] realized that the efficiency of the refining process could be improved by multiple freezing passes, but only if the mixing in liquid could be eliminated on successive passes. So he invented zone refining, where only a small zone of liquid is melted and passed from one end of the sample to the other. The next pass of a molten zone then acts on the distribution for the previous pass, and the purity of the first part of the sample to freeze can be improved.

For zone refining, on the first pass, as illustrated in Figure 11.4, the change in the composition of the liquid when an increment of the solid freezes can be written as:

$$l\,dC_L = (C_\infty - C_S)dx \tag{11.11}$$

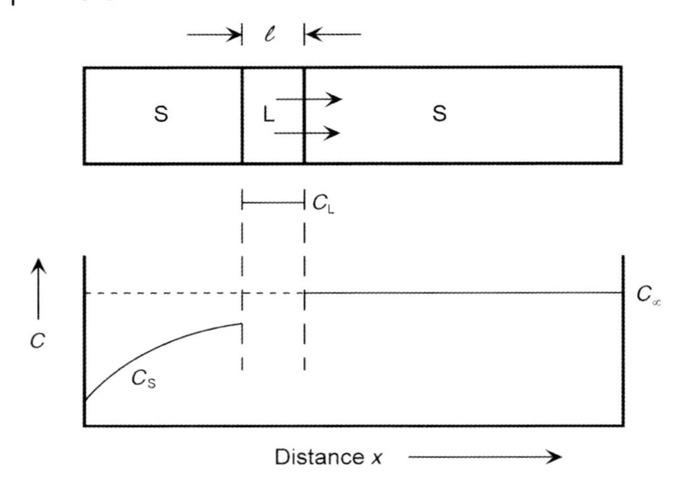

Figure 11.4 Zone refining.

Where l is the zone length. Using an analysis similar to that in Equations 11.5 to 11.10, the composition of the ingot after the first pass is:

$$C_S = C_\infty + C_\infty(k-1)\exp\left(-\frac{kx}{l}\right) \qquad (11.12)$$

This expression is only valid until the molten zone reaches the end of the sample. It does not describe the composition distribution resulting from the final freezing of the molten zone.

Repeated passes are very effective for purification, especially when the k-value is small, as is the case for many of the impurities found in semiconductors.

This process was used in the early days of semiconductors to purify silicon. The purified ingot was then used as starting material to grow single crystals by the Czochralski method. Today, all of the silicon used for semiconductors is purified by distilling silane or a chlorosilane. The silanes are then cracked at a high temperature in the Siemens process to deposit high purity polysilicon, which is then converted to single crystal silicon using Czochralski growth. Zone refining is still used to purify some of the elements used to make compound semiconductors. It is also used to purify organic compounds, but it can only be used for compounds which can be melted without decomposing.

11.5 Diffusion at a Moving Interface

11.5.1 Steady State Diffusion at a Moving Interface

In this section we will examine the distribution of a second component resulting from crystallization when there is no convective mixing in the melt [3]. The second

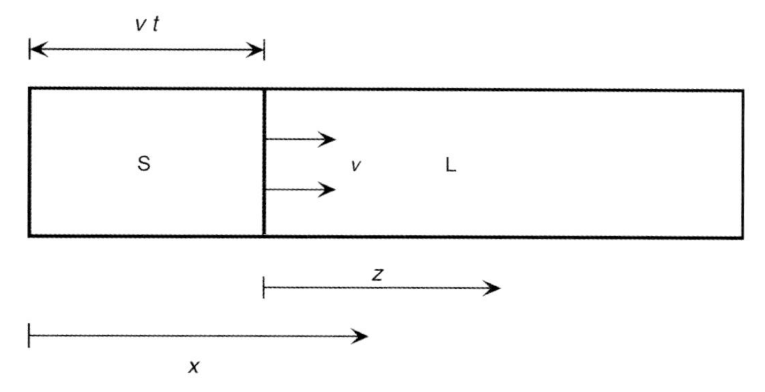

Figure 11.5 Unidirectional solidification at a constant velocity, v.

component will be assumed to redistribute in the liquid by diffusion only. It will be assumed that there is no diffusion in the solid. For the one-dimensional case, there is a steady state solution to the time-dependent diffusion equation. This steady state solution is not like the approximate steady state solutions discussed in Chapter 7. It is an exact solution in a coordinate system that moves with the interface.

The one dimensional time-dependent diffusion equation is:

$$D\frac{\partial^2 C}{\partial x^2} = \frac{\partial C}{\partial t} \tag{11.13}$$

The interface is assumed to be moving with a constant velocity, v, as illustrated in Figure 11.5. A new variable z, which moves with the interface, is defined as:

$$z = x - vt \tag{11.14}$$

The partial derivatives with respect to x and t can be expressed in terms of z:

$$\begin{aligned} \frac{\partial^2 C}{\partial x^2} &= \frac{\partial^2 C}{\partial z^2} \\ \frac{\partial C}{\partial t} &= -v\frac{\partial C}{\partial z} \end{aligned} \tag{11.15}$$

So that the diffusion equation becomes an ordinary differential equation which describes the steady state motion of the interface:

$$D\frac{\mathrm{d}^2 C}{\mathrm{d}z^2} + v\frac{\mathrm{d}C}{\mathrm{d}z} = 0 \tag{11.16}$$

which has the solution:

$$C = A\exp\left(-\frac{vz}{D}\right) + B \tag{11.17}$$

where A and B are constants. The composition far from the interface is the starting composition, C_∞, so that $B = C_\infty$. The amount of the second component which is rejected by the solid per unit time is given by the difference between the composition of the liquid and the solid at the interface, $(C_L - C_S)_I$, times the growth rate, v. The rejected second component stays in the liquid, and, at steady state, it must diffuse away from the interface at the same rate:

$$v(C_L - C_S)_I = -D\left(\frac{dC}{dz}\right)_I \tag{11.18}$$

From Equation 11.17 the concentration gradient at the interface is:

$$\left(\frac{dC}{dz}\right)_I = -A\frac{v}{D} \tag{11.19}$$

and so

$$A = C_L - C_S \tag{11.20}$$

For a steady state solution, the concentration at the interface must not change in time. As the interface moves, it is running into liquid of composition C_∞, and so, at steady state, it must be leaving behind solid of the same composition. If the composition of the solid forming is C_∞, then the composition of the liquid at the interface from which it is forming is C_∞/k. The composition distribution in the liquid is thus:

$$C = C_\infty + C_\infty\left(\frac{1}{k} - 1\right)\exp\left(-\frac{vz}{D}\right) \tag{11.21}$$

as illustrated by Figure 11.6.

The composition of the final solid is the same as the composition of the initial liquid, so this solution conserves the overall composition. The composition of the solid and liquid at the interface for steady state growth are indicated on the phase diagram in Figure 11.7.

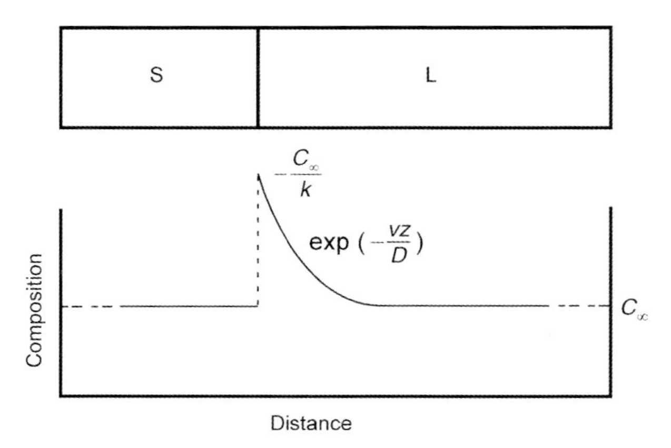

Figure 11.6 Concentration distribution in the liquid.

The concentrations at the interface are independent of the growth rate for steady state growth. The thickness of the boundary layer is given by the diffusion distance:

$$l_D = D/v \tag{11.22}$$

The thickness of the boundary layer depends on the growth rate. The solid rejects the second component at a rate that depends on the growth rate. For steady state growth, the rejected second component must be carried away from the interface by diffusion. For faster growth, the concentration gradient in the liquid must be steeper in order to do this.

If the growth rate changes, then the width of the boundary layer also changes. For example, if the growth rate increases, the second component is rejected faster, and the diffusion process cannot keep up because the old concentration gradient is too flat. So the concentration in the liquid at the interface increases, increasing the concentration in the solid. This dumps the excess of the second component into the solid, until the concentration gradient in the liquid becomes steep enough to remove the second component at the new rate. The composition of the liquid and the solid at the interface then return to the steady state values.

Thus, a fluctuation in the growth rate produces a fluctuation in the composition of the solid. When the diffusion boundary layer is thin, it is unlikely to be affected strongly by convection in the melt, because the shear velocity at the interface is small. But convection in the melt does result in temperature fluctuations, and these temperature fluctuations cause growth rate fluctuations, and the growth rate fluctuations cause fluctuations in the composition of the solid.

11.5.2
Initial and Final Transients

The initial composition of the liquid is C_∞, and so the composition of the first solid to form will be kC_∞, as indicated by the phase diagram in Figure 11.7. The composition distribution is illustrated in Figure 11.8.

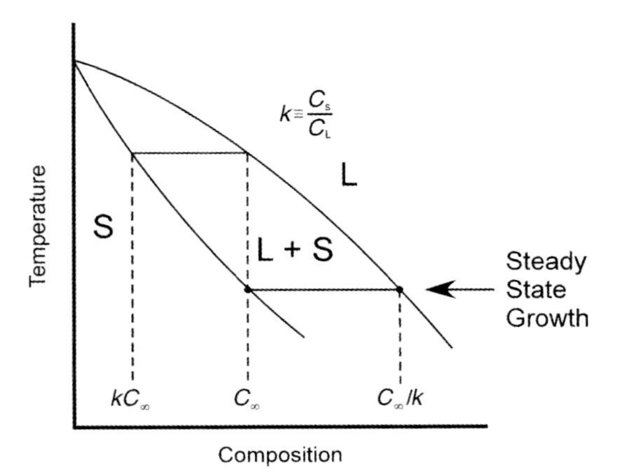

Figure 11.7 Phase diagram showing the concentrations at the interface.

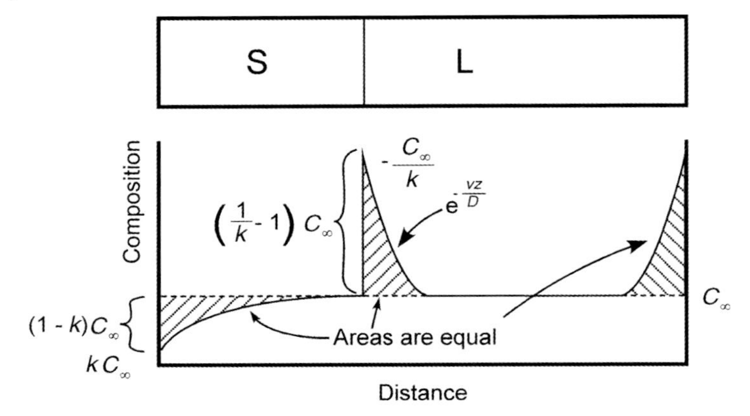

Figure 11.8 Illustrating the initial and final transients.

The steady state boundary layer at the interface has height C_∞/k and width D/v. The material in the boundary came from the first part of the crystal to freeze, and so the hatched areas in Figure 11.8 must be equal. Since the height of the missing material in the initial transient is k times that in the boundary layer in the liquid, the width of the initial transient should be about $1/k$ times the width of the boundary layer in the liquid. That is, its width should be about D/kv. A detailed analysis indicates that the initial transient is approximately exponential in shape, with an exponent $(-kvx/D)$. For small k, the length of the initial transient is much greater than the thickness of the boundary layer in the liquid.

All of the second component in the boundary layer gets dumped into the last part of the solid to freeze.

As with the Scheil equation, the formal analysis [4] predicts that the composition in the last bit of solid to freeze goes to infinity. In practice, the composition in the liquid is likely to reach the eutectic composition or some other phase boundary before that happens, as indicated in Figure 11.9.

11.6 Segregation in Three Dimensions

The steady state diffusion analysis above was done for one-dimensional solidification. There is no corresponding steady state solution for a cylinder or sphere. However, there is still a boundary layer present at the interface in those cases, and the thickness of the boundary layer is approximately the diffusion length, D/v. The first solid to form will have the composition kC_∞, and the second component in the boundary layer will be deposited in the last solid to freeze. This often results in the formation of a second phase at grain boundaries where the last bit of solid to form is between two growing grains, and the boundary layer is dumped there.

The stability of the growth fronts will be discussed in Chapter 12. These instabilities result in cellular and dendritic growth forms, as illustrated in Figure 11.2.

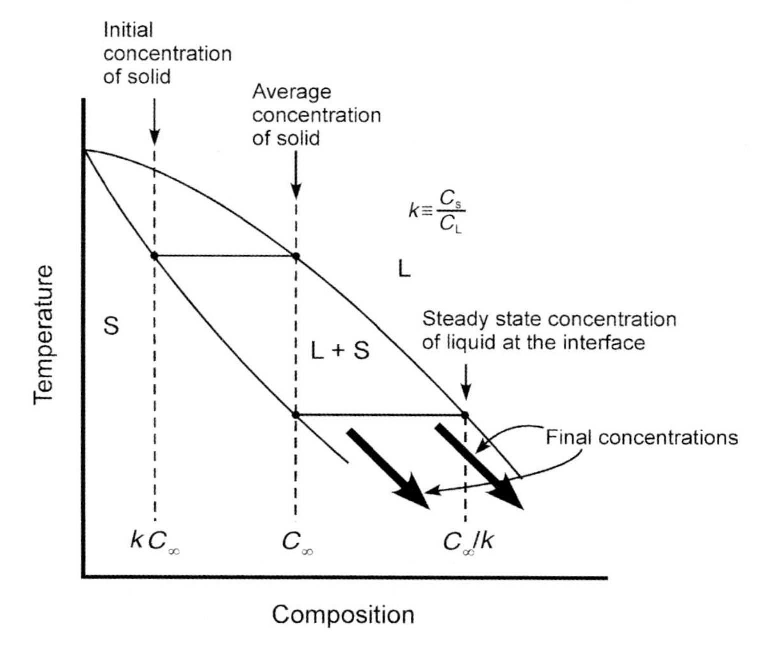

Figure 11.9 Concentration distributions in a finite sample, showing the steady state values, and the initial and final transients.

Eutectic can grow into the interdendritic liquid, as illustrated in Figure 27.20. With these growth morphologies, there is lateral segregation, parallel to the growth front, as is evident in Figure 11.2. The volume is subdivided by the instabilities into solid dendrites separated by regions of liquid. There is a thin boundary layer of the second component at the dendrite tip, but, behind the growth front, the width of the liquid channels is small enough so that the composition is uniform at the local equilibrium temperature in the interdendritic liquid. All the red dye which remains in the liquid will be concentrated in the interdendritic spaces which are the last to freeze.

11.7
Burton, Primm and Schlicter Analysis

The Burton, Primm and Schlicter (BPS) [5] analysis describes the segregation where there is both diffusion and convection in the melt, and so provides a bridge between the Scheil equation and the diffusion-only analysis. Their analysis provides some insight into the origin of the "effective" k-value which is used to fit experimental data to the Scheil equation and to the zone melting equations. BPS assumed that there is a boundary layer of width δ at the interface, and that the transport of the second component within this region was by diffusion only. They assumed that the composition of the liquid beyond this boundary layer was uniform, as illustrated in Figure 11.10.

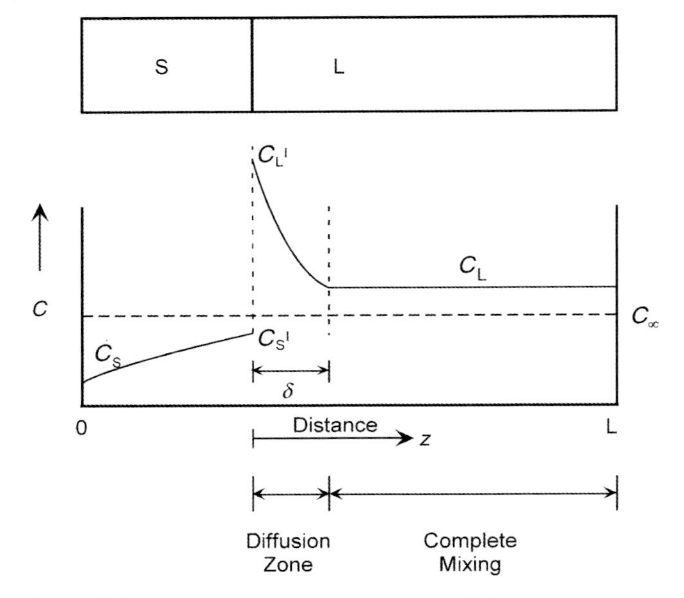

Figure 11.10 Concentration distributions in the BPS analysis.

Fluid dynamicists do not like this simplification of a boundary layer in which no convective flow occurs, but full scale modeling of this problem is a major computational effort. The model is presented here to provide some insight into what is happening in the liquid ahead of an advancing interface when there is some diffusion and some convection in the liquid. The liquid right at the interface does not flow along the interface by convection, and there is a region adjacent to the interface where the flow velocity is reduced. The BPS analysis is a simplification of this situation.

In the diffusion zone, $0 \leq z \leq \delta$, the steady state diffusion equation is applied,

$$D\frac{\mathrm{d}^2 C}{\mathrm{d}z^2} + v\frac{\mathrm{d}C}{\mathrm{d}z} = 0 \tag{11.23}$$

so that the composition distribution has the form of Equation 11.17

$$C = A + B\exp\left(-\frac{vz}{D}\right) \tag{11.24}$$

Applying the boundary condition for segregation at the interface, as in Equation 11.18,

$$v\left(C_L^I - C_S^I\right) = -D\left(\frac{\mathrm{d}C}{\mathrm{d}z}\right)_I = vB \tag{11.25}$$

gives

$$B = \left(C_L^I - C_S^I\right) \tag{11.26}$$

so that

$$C = A + (C_L^I - C_S^I)\exp\left(-\frac{vz}{D}\right) \tag{11.27}$$

At $z = \delta$, the composition in the diffusion field is:

$$C_L = A + (C_L^I - C_S^I)\exp\left(-\frac{v\delta}{D}\right) \tag{11.28}$$

This is assumed to be the composition throughout the liquid beyond δ. This composition changes as solidification proceeds. Inserting this value of A into the expression for C:

$$0 \le z \le \delta: \quad C(z) = C_L + (C_L^I - C_S^I)\left(\exp\left(-\frac{vz}{D}\right) - \exp\left(-\frac{v\delta}{D}\right)\right)$$

$$\delta \le z \le L: \quad C = C_L \tag{11.29}$$

At the interface,

$$C_L^I = C_L + (C_L^I - C_S^I)\left(1 - \exp\left(-\frac{v\delta}{D}\right)\right) \tag{11.30}$$

which can be rearranged to give:

$$1 = \frac{C_L}{C_S^I} - \left(\frac{C_L^I}{C_S^I} - 1\right)\exp\left(-\frac{v\delta}{D}\right) \tag{11.31}$$

Since

$$k_{\mathrm{eff}} = \frac{C_S^I}{C_L}, \quad \text{and} \quad k = \frac{C_S^I}{C_L^I} \tag{11.32}$$

Equation 11.31 can be written as:

$$k_{\mathrm{eff}} = \frac{1}{1 + \left(\frac{1}{k} - 1\right)\exp\left(-\frac{v\delta}{D}\right)} \tag{11.33}$$

For $\delta = 0$, $k_{\mathrm{eff}} = k$, which corresponds to the Scheil equation.

For $\delta = \infty$, $k_{\mathrm{eff}} = 1$, which is the diffusion-only solution.

Experimental data for the distribution of a dopant in a directionally grown crystal can often be fitted with the Scheil equation, using an effective k-value. Similarly, the distribution of a dopant after one-pass zone melting can be fitted with Equation 11.12, using an effective k-value. The BPS analysis suggests why this works, and what determines the effective k-value. The effective k-value in the BPS model depends strongly on δ, and, in practice, δ is difficult to estimate. It depends on the vigor of the convection in the melt, which in turn depends on the temperature gradients, the shape of the growth container, the growth rate, and so on. This means that there is no simple way to predict k_{eff}. But it is reasonably similar for successive growth runs in the same equipment.

The equilibrium k-value applies at the interface, and so there is always an enriched boundary layer there. The convective velocities in the melt are zero right at the interface, and so mixing there is primarily by diffusion. The contribution of convective flow to the mixing increases with distance from the interface. The liquid composition far from the interface is best described as uniform if there is convection in the melt, as is assumed in the derivation of the Scheil equation. Even though the composition profile of a solidified sample can be fitted to the Scheil equation with an effective k-value, the onset of interface instability in the same sample is often correctly predicted by the diffusion-only solution, using the equilibrium k-value. This apparently anomalous state of affairs is possible because using an effective k-value in the Scheil equation compensates for the presence of the boundary layer, and the composition gradient right at the interface is not affected significantly by the convective flow. The instability of the interface, which is discussed in Chapter 12, depends on the composition gradient right at the interface.

References

1 Scheil, E. (1942) *Z. fur Metall.*, **34**, 70.
2 Pfann, W.G. (1958) *Zone Melting*, John Wiley & Sons, New York.
3 Tiller, W.A., Jackson, K.A., Rutter, J.W., and Chalmers, B. (1953) *Acta Met.*, **1**, 428.
4 Smith, V.G., Tiller, W.A., and Rutter, J.W. (1955) *Can. J. Phys.*, **33**, 723.
5 Burton, J.A., Prim, R.C., and Schlichter, W.P. (1953) *J. Chem. Phys.*, **21**, 1987.

Further Reading

Chalmers, B. (1964) *Principles of Solidification*, John Wiley & Sons, New York.

Tiller, W.A. (1991) *The Science of Crystallization*, Cambridge University Press, New York.

Problems

11.1. Write a finite difference program and calculate the final concentration distribution of a dopant with an initial concentration of 1% after directional solidification in a sample 10 cm long. Assume a k-value of 0.1, a diffusion coefficient $5 \times 10^{-5}\,\text{cm}^2\,\text{s}^{-1}$, and a constant growth rate of $1\,\text{mm}\,\text{min}^{-1}$.

11.2. Discuss the final concentration distribution of a small amount of a second component in a sample after directional solidification.

11.3. Derive Equation 11.10.

11.4. A Czochralski silicon crystal which is about 1 m in length is grown in 8 h, so that a crystal can be grown by a worker in one shift. For a diffusion coefficient in the liquid, $D = 5 \times 10^{-5}\,\text{cm}^2\,\text{s}^{-1}$, what is the thickness of the diffusion boundary layer?

11.5. Some dopants in silicon have a k-value which is approximately 10^{-3}. What is the characteristic length of the initial transient (where the concentration reaches e times its initial value) for such a dopant in a Czochralski crystal?

11.6. Crystals which are grown from solution are grown much more slowly, at a rate of perhaps $1\,\mathrm{mm\,d^{-1}}$. For a liquid diffusivity, $D = 5 \times 10^{-5}\,\mathrm{cm^2\,s^{-1}}$, what is the diffusion length? What does this imply about the concentration in a growth vessel which is 20 cm in diameter?

11.7. The thermal diffusion length is the thermal diffusivity divided by the growth rate. For a thermal diffusivity of $0.1\,\mathrm{cm^2\,s^{-1}}$, what is the thermal diffusion length for the solution growth in Problem 3?

11.8. What is the diffusion distance for a solid state transformation which is proceeding at a rate of $1\,\mathrm{mm\,min^{-1}}$ in a solid where the diffusion coefficient is $10^{-10}\,\mathrm{cm^2\,s^{-1}}$?

11.9. Nickel dendrites have been observed to grow into supercooled molten nickel at $40\,\mathrm{m\,s^{-1}}$. For a diffusion coefficient, $D = 5 \times 10^{-5}\,\mathrm{cm^2\,s^{-1}}$, what is the diffusion distance at this growth rate?

11.10. For a thermal diffusivity of $0.1\,\mathrm{cm^2\,s^{-1}}$, what is the thermal diffusion distance ahead of the growing nickel dendrite?

12
Interface Instabilities

The steady state solution to the diffusion equation for a second component when a sample of constant cross-section is freezing from one end at a constant growth rate, v, is given by Equation 11.21. The steady state solution is described using a coordinate system that moves with the interface at the velocity, v. But that solution is not mathematically unique. Under some conditions, the interface is unstable to perturbations, and we will discuss these instabilities in this chapter.

There are many cases where there is a valid mathematical solution which is not stable. For example, if a cylinder is held so that it is not horizontal and a ball is started rolling down from the upper end, a perfectly valid solution is that the ball will roll straight down the upper edge of the cylinder. But that is unlikely to happen in practice because any slight perturbation in the surface of the cylinder will deflect the ball so that it rolls sideways off the cylinder. There are, in fact, multiple solutions for the path of the ball.

Another example is that of a flapping flag. There is a valid solution where the flag flies straight out in the breeze. But a small perturbation makes the flag look like an airfoil, generating a pressure difference which increases the size of the perturbation. But the perturbation cannot grow indefinitely, because the breeze will blow it back when it gets too big, and so the perturbation travels down the flag, and it flaps. This is complex time-dependent behavior, rather than the much simpler behavior of a non-flapping flag.

The motion of an interface can also present complex, time-dependent behavior, as is discussed below.

12.1
Constitutional Supercooling

The steady state solution which we derived for diffusion in the liquid becomes unstable when there is *constitutional supercooling* [1] in the liquid ahead of the interface.

Figure 12.1 shows the steady state concentration profile which we obtained.

Kinetic Processes: Crystal Growth, Diffusion, and Phase Transitions in Materials. Kenneth A. Jackson

ISBN: 978-3-527-32736-2

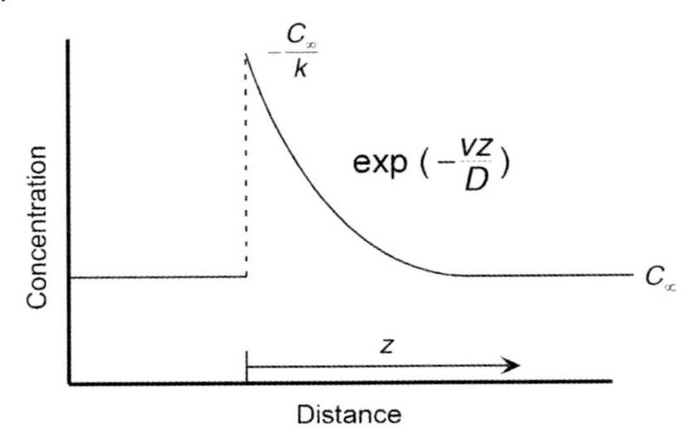

Figure 12.1 Steady state concentration distribution ahead of an advancing interface.

The concentration of the phase that forms is the same as the initial concentration, C_∞, and the concentration at the interface is much greater, C_∞/k. Equation 11.21, which we derived for the concentration in the liquid is:

$$C = C_\infty + C_\infty\left(\frac{1}{k}-1\right)\exp(-vz/D) \tag{12.1}$$

The growing phase always rejects those components which lower the melting point. So the melting point of the liquid at the interface is lower than that of the starting material.

On the phase diagram, Figure 12.2, the liquidus temperature for the starting concentration, $T_L(C_\infty)$, is much higher than the liquidus temperature for the concentration, $T_L(C_\infty/k)$, at the interface.

It is usual to make the approximation that the liquidus line is a straight line, so that there is a linear relationship between the composition and the corresponding liquidus temperature, given by:

$$T_L = T_m - mC \tag{12.2}$$

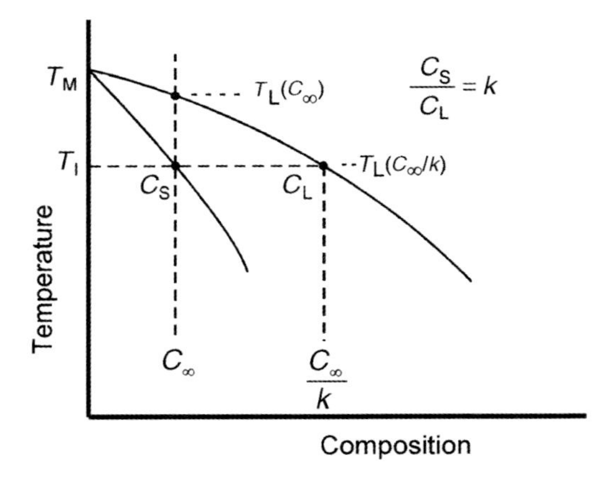

Figure 12.2 Phase diagram corresponding to Figure 12.1.

The melting point of the alloy increases as the composition of the second component decreases. The composition of the liquid ahead of the interface decreases with distance, so the melting point of the liquid increases with distance from the interface.

If the temperature of the sample is uniform at the melting point of the material at the interface where the concentration is C_∞/k, the liquid far from the interface, where the concentration is C_∞, will be below its melting point. It will be supercooled by an amount $T_L(C_\infty) - T_L(C_\infty/k)$.

Usually there is a positive temperature gradient at the interface, so that the liquid is hotter than the solid. But even under these conditions, when the liquid is all hotter than the solid, there can still be supercooled liquid, because the composition of the liquid also changes. This is known as *constitutional supercooling*. The boundary layer depresses the melting point at the interface and so, and even though the liquid is hotter, there can still be supercooled liquid ahead of the interface.

In Figure 12.3 the actual temperature of the sample has been plotted, assuming that the liquid is hotter than the solid, and that the temperature increases linearly with distance. The temperature at the interface is assumed to be the equilibrium temperature for the composition there. On the same plot is the liquidus temperature corresponding to the local composition. This curve can be constructed by noting the composition at some distance from the interface, as in Figure 12.1, going to the phase diagram, Figure 12.2, to determine the corresponding liquidus temperature, and then plotting this temperature at the corresponding distance from the interface. This gives the local melting point as a function of distance ahead of the interface. This curve can be calculated by inserting Equation 12.1 into Equation 12.2.

For a shallow temperature gradient, most of the liquid will be at a temperature below its local melting point. For a steeper temperature gradient, less of the liquid will be below its melting point. There is some critical temperature gradient above which there will be no constitutional supercooling.

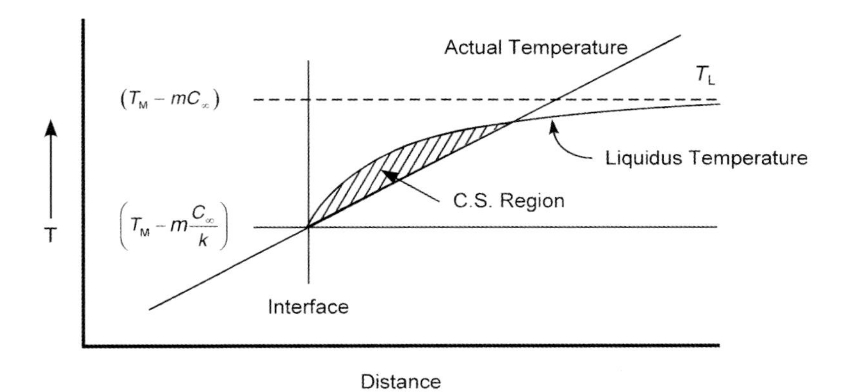

Figure 12.3 Actual temperature and liquidus temperature of the liquid.

This critical value can be calculated as follows. The composition gradient at the interface is obtained by evaluating the derivative of Equation 12.1 at the interface:

$$\left(\frac{dC}{dz}\right)_I = -\frac{v}{D}C_\infty\left(\frac{1}{k}-1\right) \tag{12.3}$$

The rate at which the liquidus temperature increases with distance ahead of the interface, as indicated in Figure 12.3, is obtained by multiplying the concentration gradient in Equation 12.3 by m, which is the slope of the liquidus line on the phase diagram. This slope is then compared with the actual temperature gradient, $G = dT/dx$, in the liquid. If the slope of T_L is greater than G, there is constitutional supercooling in the liquid.

For *constitutional supercooling*:

$$G < mC_\infty\left(\frac{1}{k}-1\right)\frac{v}{D} \tag{12.4}$$

There is no constitutional supercooling for

$$G > mC_\infty\left(\frac{1}{k}-1\right)\frac{v}{D} \tag{12.5}$$

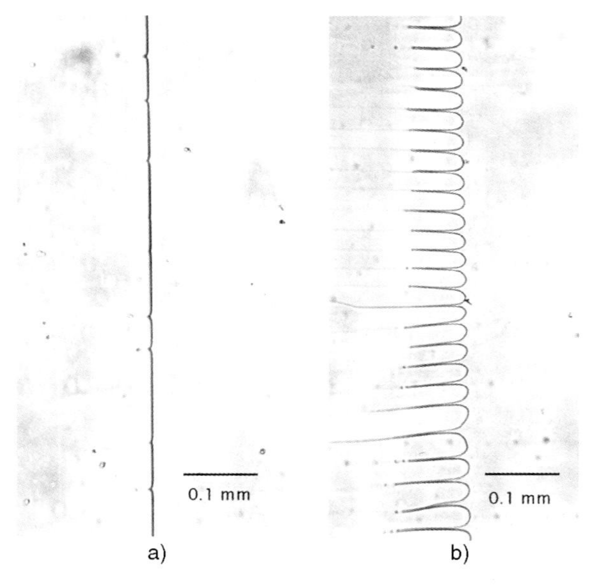

Figure 12.4 (a) Planar interface growth in a thin cell. The interface temperature is very close to the melting point isotherm. The small grooves are due to grain boundaries. (b) Cellular growth in the same growth cell.

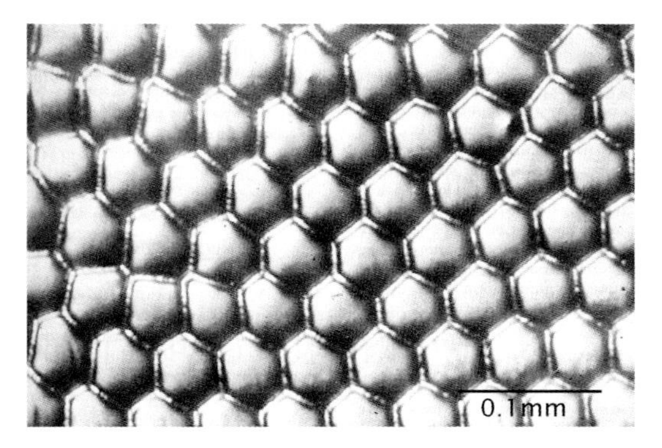

Figure 12.5 Cellular growth on a decanted interface.

For materials which have rapid growth kinetics, such as metals and semiconductors, where the growth rate is not limited by surface nucleation effects, the interface shape is determined by diffusion processes. When there is no constitutional supercooling the interface will follow an isotherm, as illustrated in Figure 12.4a.

When constitutional supercooling is present, the interface becomes unstable, and cells form, as shown in Figure 12.4b. The crystals in Figure 12.4 were growing in a thin transparent container, and are seen in profile. In a thicker sample, the cells form a fairly regular hexagonal array on the interface, as in Figure 12.5, which is why this is known as a cellular substructure.

More of the second component results in more constitutional supercooling, and the cells become more elongated and separated. Still more of the second component results in dendritic growth, as shown in Figure 11.2, where the cells have become so elongated and separated that they develop side branches. Dendritic growth is discussed in Chapter 26.

What is happening here is that, if a plane interface gets a bump on it, then the bump can get rid of the second component more readily than can the flat interface, as illustrated in Figure 12.6.

If the bump also sees a supercooled region ahead in the liquid, it will grow faster than the average interface, and so constitutional supercooling leads to interface instability. A stability analysis of the interface is presented in the next section.

There is a simple interpretation of the constitutional supercooling condition, which also provides a simple way to remember it. On the phase diagram, Figure 12.2, the liquidus temperature for the initial concentration of the liquid, $T_L(C_\infty)$, is given by $T_M - mC_\infty$. The liquidus temperature for the steady state liquid concentration at the interface, $T_L(C_\infty/k)$ is $T_M - mC_\infty/k$. The difference between these two temperatures is $mC_\infty(1/k - 1)$, which is the first part of the right-hand side of Equation 12.4. But this temperature difference is also the difference between the liquidus and solidus temperatures at the composition C_∞, which we will call $\Delta T(C_\infty)$. So:

$$mC_\infty(1/k-1) = T_L(C_\infty)-T_L(C_\infty/k) = \Delta T(C_\infty) \tag{12.6}$$

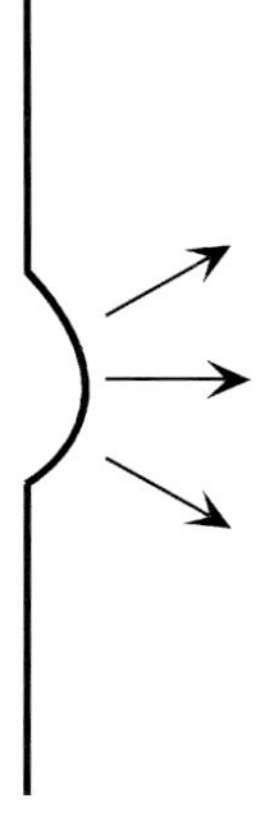

Figure 12.6 A second component will diffuse more rapidly away from a bump on the interface.

Recalling that D/v is the diffusion length, l_D, the constitutional supercooling condition, Equation 12.4 can be written as:

$$G < \frac{\Delta T(C_\infty)}{l_D} \tag{12.7}$$

So the constitutional supercooling condition compares the actual temperature gradient in the sample with the temperature gradient obtained by dividing the temperature difference between the liquidus and the solidus for the initial composition by the diffusion length.

Constitutional supercooling can be present for fairly small concentrations of a second component, as can be seen by inserting some typical numbers into Equation 12.4. A temperature gradient of about 100 °C cm^{-1} is fairly typical for melt growth. The heat flux necessary to maintain such a temperature gradient is given by multiplying this gradient by the thermal conductivity. For aluminum, the corresponding heat flux is 130 W cm^{-2}, for alumina, it is 20 W cm^{-2}, and for silicon it is 94 W cm^{-2}. For a 12 in diameter silicon boule, this amounts to a total of 50 kW. Using $k = 0.1$, a liquidus slope, $m = 1$ °C (%concentration)$^{-1}$, and a growth rate of 1 mm min^{-1} = 1/600 cm s^{-1}, the critical concentration above which there will be constitutional supercooling is:

$$C_\infty^{\text{C.S.}} = \frac{100 \times 5 \times 10^{-5} \times 600}{1 \times \left(\frac{1}{0.1} - 1\right)} \approx 0.3\% \tag{12.8}$$

So materials which are nominally pure (but not very pure) will exhibit constitutional supercooling under fairly typical growth conditions. For a second component which has $k = 0.001$, as do several dopants in silicon, the critical composition for the same growth conditions is 30 ppm. For a k-value close to 1, several % of the second component will not cause constitutional supercooling.

For solidifying metals, the presence of cells or even dendrites often does not have a deleterious effect on properties. In many cases, alloying elements are added to enhance the properties. However, in the growth of semiconductor or optical crystals, inhomogeneities in composition are bad, so interface instabilities are to be avoided. In addition, the presence of interface instabilities often leads to the creation of other crystalline defects, which can be even worse.

12.2 Mullins and Sekerka Linear Instability Analysis

We derived an expression for the diffusion field in the liquid ahead of an interface which was advancing at a constant rate, v. The analysis was carried out in a coordinate system which moved with the interface. The planar front solution which we found is a valid solution. The discussion of constitution supercooling indicated that this analysis, although valid, is probably not stable, and therefore not a unique solution. In this section, the stability analysis of an interface as it was first done by Mullins and Sekerka [2], will be outlined. The analysis concludes that an infinitesimal perturbation of the interface will grow if a condition which is very similar to the constitutional supercooling condition is obeyed.

We employ, as before, a coordinate system which moves with the interface at an average velocity V. z is the distance ahead of the moving interface.

We start with a sinusoidal perturbation of the interface which has a very small amplitude δ, and a wavelength $2\pi/\omega$, as illustrated in Figure 12.7.

$$Z_{\mathrm{I}} = \delta(t)\sin \omega x \tag{12.9}$$

We will ask the question: does the amplitude δ increase or decrease with time? We will do a linear stability analysis, which means that we assume that δ is small, so we will keep only terms which are linear in δ in our analysis, and throw out terms which are higher order in δ. Our solution will be valid only for small values of δ. It will indicate whether a perturbation will initially grow or shrink, but it does not describe how the interface shape evolves after it begins to grow.

The equilibrium temperature along the interface will depart from the melting point of the pure material because of the variations in the concentration and in the

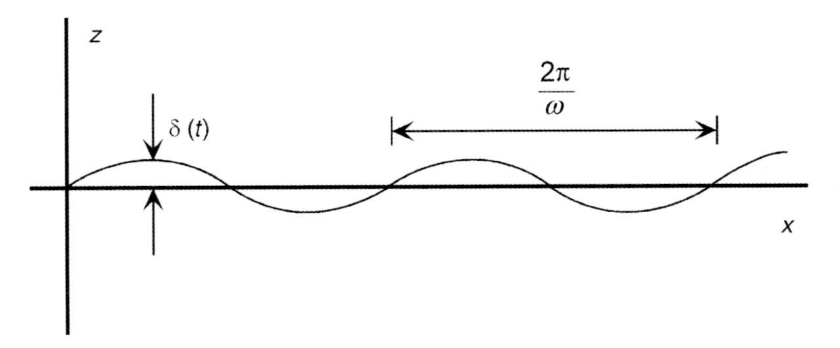

Figure 12.7 Sinusoidal perturbation of an interface.

curvature along the interface:

$$T_M - T_I(x,t) = mC_I(x) - \frac{T_M \sigma}{L r_I} \quad (12.10)$$

The first term on the right is due to variations in the concentration of the second component along the interface, and the second term is due to the variations in the curvature of the interface. This second term is due to the Gibbs–Thompson effect, which will be derived in Chapter 15 when we discuss nucleation. Here T_M is the melting point of the pure material, $C_I(x)$ is the concentration of the second component along the interface, and m is the slope of the liquidus line, so that $mC_I(x)$ is the local melting point depression. σ is the surface tension, L is the latent heat and r_I is the radius of curvature of the interface given approximately by:

$$\frac{1}{r_I} = \frac{d^2 Z_I}{dx^2} \quad (12.11)$$

So we can write Equation 12.10 as:

$$T_I(x,t) = T_M - mC_I(x) - \frac{T_M \sigma}{L} \delta(t) \omega^2 \sin(\omega x) \quad (12.12)$$

The local growth rate of the interface depends on the average rate of motion of the interface, plus a term which depends on the rate at which the amplitude of the perturbation is growing or shrinking, $\dot{\delta} = d\delta/dt$.

$$v(x) = V + \dot{\delta} \sin(\omega x) \quad (12.13)$$

As in the derivation of Equation 10.21, we will assume that the ratio of the solid concentration to the liquid concentration at the interface is k. The boundary condition at the interface is similar to Equation 10.18, except that it is applied at each point along the interface:

$$D\left(\frac{dC}{dz}\right)_I = -(1-k)C_I(x)v(x) \quad (12.14)$$

The thermal gradient in the solid must carry away both the heat conducted to the interface through the liquid, and the latent heat generated at the interface by the solidification process:

$$K_S G_S = vL + K_L G_L \quad (12.15)$$

Where K_S and K_L are the thermal conductivities of solid and liquid, respectively, and G_S and G_L are the corresponding thermal gradients.

We will assume that the thermal field is not affected by the perturbation of the interface, because the wavelength of the perturbation is small compared to the thermal diffusion distance. This means that we will not consider variations in temperature along the interface in the x direction. Since the perturbation of the interface is symmetric, the thermal fields in the two phases are effectively averaged.

We will replace the actual temperature gradients in the z direction in the liquid and solid with an average temperature gradient, $\overline{G}$:

$$T_I = T_0 + \overline{G}z \tag{12.16}$$

where

$$\overline{G} = \frac{K_S G_S + K_L G_L}{K_S + K_L} \tag{12.17}$$

In their original analysis, Mullins and Sekerka did not make this assumption, but their analysis indicates that it is valid, and it simplifies our analysis.

We will look for a solution which has a variation of the composition along the interface of the form:

$$C_I = C_0 + bZ_I = C_0 + b\delta \sin(\omega x) \tag{12.18}$$

Here C_0 is the average composition of the liquid at the interface, which is equal to C_∞/k, where C_∞ is the composition of the liquid far from the interface. Combining Equations 12.12,12.16 and 12.18, we have, at the interface:

$$T_0 + \overline{G}Z_I = T_M - mC_0 - mbZ_I - \frac{T_M\sigma}{L}\omega^2 Z_I \tag{12.19}$$

The terms without Z_I give:

$$T_0 = T_M - mC_0 \tag{12.20}$$

and the terms with Z_I give:

$$b = -\frac{1}{m}\left(\overline{G} + \frac{T_M\sigma\omega^2}{L}\right) \tag{12.21}$$

We now seek a solution to the steady state diffusion equation:

$$D\left(\frac{\partial^2 C}{\partial z^2} + \frac{\partial^2 C}{\partial x^2}\right) + v\frac{\partial C}{\partial z} = 0 \tag{12.22}$$

which has a composition at the interface given by Equation 12.18. Such a solution is:

$$C(x,z) = C_0 + \frac{G_C D}{V}\left[1 - \exp\left(-\frac{vz}{D}\right) + \delta(b - G_C)\sin(\omega x)\cdot\exp(-\omega^* z)\right] \tag{12.23}$$

where

$$G_C = \frac{V}{D}C_0(k-1) = \frac{V}{D}C_\infty\left(1 - \frac{1}{k}\right) \tag{12.24}$$

and

$$\omega^* = \frac{V}{2D} + \sqrt{\left(\frac{V}{2D}\right)^2 + \omega^2} \tag{12.25}$$

ω^* is a new wavenumber which depends on both the wavelength of the perturbation and on the diffusion distance.

From Equation 12.23, the concentration gradient at the interface is given by:

$$\left(\frac{dC}{dz}\right)_I = G_C + Z_I\left(G_C + \left(\omega^* - \frac{V}{D}\right) - b\omega^*\right) \qquad (12.26)$$

where only the terms which are linear in Z_I have been retained. Using Equation 12.14, we can now write:

$$V + \dot{\delta}\sin(\omega x) = \frac{D}{(k-1)C_0}\left\{G_C + Z_I\left[G_C\left(\omega^* - \frac{V}{D}\right) - b\omega^*\right] - \frac{bG_C}{C_o}Z_I\right\} \qquad (12.27)$$

The terms which do not contain $\dot{\delta}$ or Z_I give:

$$V = \frac{DG_C}{(k-1)C_0} \qquad (12.28)$$

as in Equation 12.24, and the terms which contain $\dot{\delta}$ or Z_I give:

$$\frac{\dot{\delta}}{\delta} = \frac{V}{mG_C}\left[mG_C\left(\omega^* - \frac{V}{D}\right) + \left(\omega^* - \frac{(1-k)V}{D}\right)\left(\overline{G} + \frac{T_M\sigma\omega^2}{L}\right)\right] \qquad (12.29)$$

For small k we can write this as:

$$\frac{\dot{\delta}}{\delta} = -\frac{V}{mG_C}\left(\omega^* - \frac{V}{D}\right)\left[-mG_C - \overline{G} - \frac{T_M\sigma\omega^2}{L}\right] \qquad (12.30)$$

The first term in the square brackets is due to the composition field, the second to the thermal field, and the third to the surface tension acting on the interface curvature.

When the right-hand side of this equation is positive, the perturbation will grow. When it is negative, it will shrink. G_C is negative for $k < 1$, and ω^* is always greater than V/D, so the factors outside the square bracket are positive. When the square bracket is positive, the perturbation will grow. The composition term mG_C is positive, because G_C is negative, so it tends to make the perturbation grow. The thermal and surface tension terms in the square bracket are negative, so these tend to make the perturbation decay, stabilizing the interface. The surface tension term is usually small, so that major competition is between the composition field and the thermal field.

The perturbation will grow when the composition term in the square brackets is larger than the thermal term. This happens when

$$\overline{G} < mC_\infty\left(\frac{1}{k} - 1\right)\frac{V}{D} \qquad (12.31)$$

which is similar to the constitutional supercooling condition, Equation 12.4.

12.3 Anisotropic Interface Kinetics

When crystals are grown from a solution growth, the whole solution is undercooled, and yet the interface of the growing crystal does not become unstable unless the

Figure 12.8 Rapid lateral growth tends to stabilize an interface.

undercooling is quite large. For anisotropic interface kinetics, the slowest growing faces form the external boundary of the crystal. These faces form facets which are relatively resistant to instabilities. A perturbation on these faces is quickly removed by rapid lateral growth, to reform the facet, as illustrated in Figure 12.8.

This situation is much more complex to analyze than the case of isotropic growth, because lateral growth into the depressions changes the shape of the perturbation, not its amplitude. For small anisotropies, it has been shown [3] that the anisotropy tends to stabilize the growth front:

$$\frac{\overline{G}}{V} < \left(\frac{mC_\infty}{D}\right)\left(\frac{1}{k}-1\right)-\theta \qquad (12.32)$$

where θ is a term due to the anisotropy of the growth rate. The stability of an interface where the anisotropy of the growth rate is large has not been solved for the general case.

Crystals grown in solutions are always grown under conditions where the interface would be unstable if it were not for the anisotropy in the growth rate. Experimentally, crystal growers like to grow their crystals as rapidly as possible, which means using an undercooling (supersaturation) which is as large as possible, because the growth rate increases with supersaturation. But if the supersaturation is too large, the interface will become unstable, resulting in a poor quality crystal. The fastest growth rate at which good crystals can be grown is usually determined empirically.

References

1 Tiller, W.A., Jackson, K.A., Rutter, J.W., and Chalmers, B. (1953) *Acta Met.*, **1**, 428.

2 Mullins, W.W. and Sekerka, R.F. (1963) *J. Appl. Phys.*, **34**, 323; (1963) *J. Appl. Phys.*, **34**, 444.

3 Chernov, A.A. (1974) *J. Crystal Growth*, **24** 11.

Further Reading

Chalmers, B. (1964) *Principles of Solidification*, John Wiley & Sons, New York.

Langer, J.S. (1986) *Phys. Rev. A*, **33**, 435.

Tiller, W.A. (1991) *The Science of Crystallization*, Cambridge University Press, New York.

Problems

12.1. Discuss the growth forms which result from interface instabilities during melt growth.

12.2. A crystal is growing at a rate of 10 mm min^{-1} from a liquid containing 0.1% of a dopant in a temperature gradient of 100 °C cm^{-1}. The *k*-value (distribution coefficient) for the dopant is 0.1, its diffusion coefficient in the liquid is 10^{-4} $\text{cm}^2\,\text{s}^{-1}$, and it depresses the melting point 1 °C for each percent in the liquid. Discuss the interface shape.

13
Chemical Reaction Rate Theory

Chemical reaction rate theory [1] is used to describe homogenous reactions, that is, reactions which occur in only one phase. The reaction can take place in the gas phase, in a liquid solution, or even a solid solution. The model applies when the interaction takes place between isolated atoms or molecules which come together and then react to form a product.

13.1
The Equilibrium Constant

A simple reaction between two species, A and B, to form a third species, C, is described by:

$$aA + bB \rightleftharpoons cC \tag{13.1}$$

where a, b, and c are the numbers of atoms or molecules of each species which are involved in the reaction. The equilibrium constant, K, for the reaction is given by:

$$K = \frac{[\mathrm{C}]^c}{[\mathrm{A}]^a[\mathrm{B}]^b} \tag{13.2}$$

where the square brackets indicate the number of atoms or molecules of the species per unit volume, for example:

$$[\mathrm{C}] = \frac{n_\mathrm{c}}{V} \tag{13.3}$$

The equilibrium constant, K, gives the ratio of the concentration of the product to the concentrations of the reactants.

Kinetic Processes: Crystal Growth, Diffusion, and Phase Transitions in Materials. Kenneth A. Jackson

ISBN: 978-3-527-32736-2

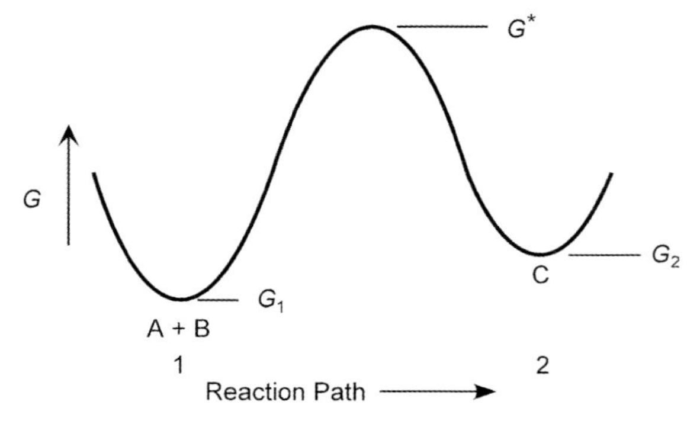

Figure 13.1 Free energy variation along the reaction path.

13.2
Reaction Rate Theory

The reaction path between the reactants and the product for the reaction:

$$A+B \rightleftharpoons C \tag{13.4}$$

is illustrated schematically in Figure 13.1.

G_1 and G_2 are the free energies in the initial and final states respectively, and G^* is the highest free energy along the reaction path. G^* depends on the reaction path as A and B come together, and can be different for different conditions, for example, when a catalyst is present. For simple cases, the energy along the reaction path can be calculated from quantum mechanics. The reaction rate for the reaction in Figure 13.1 is proportional to the number of A atoms per unit volume, times the probability that a B atom will be next to it, times the probability that they will have enough energy to get over the barrier to the reaction:

$$[A]\,[B]\,\exp[-(G^*-G_1)/kT] \tag{13.5}$$

where kT is Boltzmann's constant times the temperature. At equilibrium, this forward rate will be equal to the reverse rate:

$$[C]\,\exp[-(G^*-G_2)/kT)] \tag{13.6}$$

or

$$[C]=[A]\,[B]\exp\,[-(G_2-G_1)/kT]=[A]\,[B]\exp\left(-\frac{\Delta G}{kT}\right) \tag{13.7}$$

where $\Delta G = G_2 - G_1$.

The equilibrium constant is thus:

$$K=\exp\,(-\Delta G/kT) \tag{13.8}$$

It depends only on the free energies of the initial and final states. It does not depend on the reaction path, which determines the rates, or on how long it takes to reach equilibrium.

In a reaction starting with only reactants A and B present, the number of product species will increase and the number of reactants will decrease until the net rate is zero.

13.3 Reaction Rate Constant

For a reaction

$$aA + bB \rightarrow cC \tag{13.9}$$

with no back flux, the rate of production of C is:

$$\frac{1}{V}\frac{dn_c}{dt} = \frac{d}{dt}[C] = k^*[A]^\alpha[B]^\beta \tag{13.10}$$

where k^* is the reaction rate constant. α and β are not necessarily equal to a and b, but they are usually close. Valuable information about many reactions can be obtained from the order of the reaction, which is defined as follows:

The overall reaction is of order α + β.
The reaction is of order α in A.
The reaction is of order β in B.

13.4 Transition State Theory

For the reaction:

$$A + B \rightarrow C \tag{13.11}$$

There will be no back flux if the free energy in state 2 is very low, so that $G^* - G_2$ is large, as illustrated in Figure 13.2.

We assume that there is a transition state complex $(AB)^*$ which has free energy corresponding to the maximum along the reaction path, and that the concentration of this complex is in equilibrium with the reactants. The equilibrium constant for this transition state is:

$$K^*[A][B] = [(AB)^*] = [A][B]\exp[-(G^* - G_1)/kT] \tag{13.12}$$

The rate of formation of the product C is a frequency times the concentration of the transition state. The frequency is usually taken to be approximately the Debye frequency, kT/h.

$$\frac{d[C]}{dt} = k^*[A][B] = \frac{kT}{h}[(AB)^*] \tag{13.13}$$

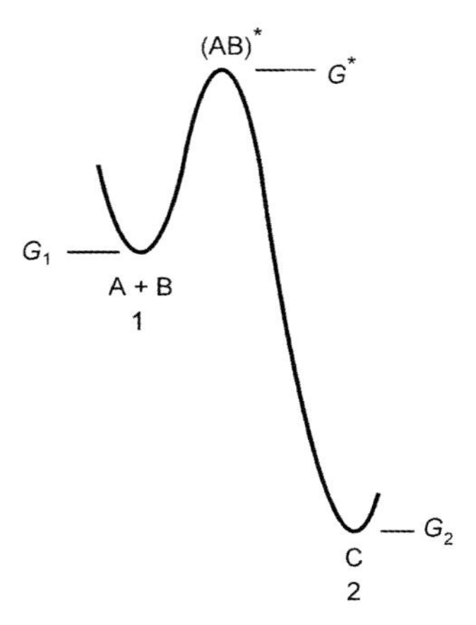

Figure 13.2 Free energy variation along the reaction path with no back flux.

So the reaction rate constant, k^*, is approximately given by:

$$k^* = \frac{kT}{h} K^* = \frac{kT}{h} \exp\left[-\Delta G^*/kT\right] \tag{13.14}$$

where $\Delta G^* = G^* - G_1$.

Writing $\Delta G^* = \Delta H^* - T\Delta S^*$, the reaction rate constant can be written as k_0^* times the Boltzmann factor:

$$k^* = k_0^* \exp\left(-\Delta H^*/kT\right) \tag{13.15}$$

where:

$$k_0^* = \frac{kT}{h} \exp\left(\Delta S^*/R\right) \tag{13.16}$$

The Debye frequency does not change very much over a limited temperature range, whereas the Boltzmann factor does.

13.5 Experimental Determination of the Order of a Reaction

First order reaction:

$$\text{A} \rightarrow \text{product} \tag{13.17}$$

Examples are nuclear decay, Newton's law of cooling, and outgassing of a solid where the rate of outgassing is limited by the evaporation rate not by diffusion in the solid.

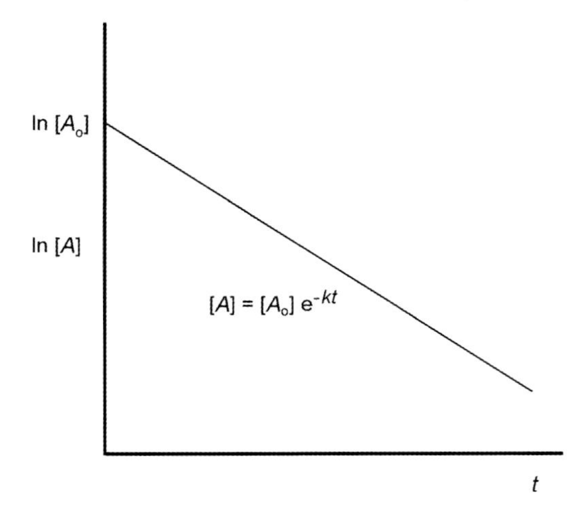

Figure 13.3 Plot for a first order reaction.

The time dependence of the concentration of A is:

$$\frac{d[A]}{dt} = -k^*[A] \tag{13.18}$$

which has a solution:

$$[A] = [A_0] \exp(-k^* t) \tag{13.19}$$

For first order reaction, plotting the logarithm of the concentration against time will give a straight line with a slope $-k^*$, and the intercept at $t = 0$ is $[A_0]$, as in Figure 13.3.

In a second order reaction, two entities come together:

$$\frac{d[A]}{dt} = -k^*[A]^2 \tag{13.20}$$

Examples are:

$$2A \rightarrow \text{product}$$

$$A + B \rightarrow \text{product}$$

$$N + N \rightarrow N_2$$

$$Na + Cl \rightarrow NaCl$$

$$\text{Interstitial} + \text{vacancy} \rightarrow 0$$

On solving Equation 13.20, the time dependence of the reaction is given by:

$$\frac{1}{[A]} - \frac{1}{[A_0]} = k^* t \tag{13.21}$$

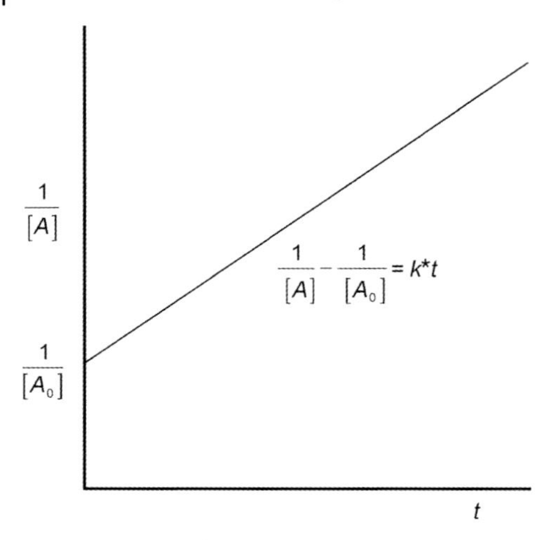

Figure 13.4 Plot for a second order reaction.

so that plotting $1/[A]$ against time will give a straight line, with a slope k^*, and an intercept $1/[A_0]$, as in Figure 13.4.

A reaction of order n involves n entities coming together:

$$\frac{d[A]}{dt} = -k^*[A]^n \tag{13.22}$$

The time dependence of the reaction is given by:

$$\frac{1}{n-1}\left[\frac{1}{[A]^{n-1}} - \frac{1}{[A_0]^{n-1}}\right] = k^* t \tag{13.23}$$

Defining the time rate of change of concentration of the reactants, $d[A]/dt \equiv -r$, Equation 13.22 can be rewritten as:

$$\ln r = \ln k^* + n \ln [A] \tag{13.24}$$

The order of the reaction can be determined by recording the time rate of change of concentration, as in Figure 13.5a, to determine r, and then plotting the logarithm of r against the logarithm of the concentration, [A], as in Figure 13.5b.

The slope of the line in Figure 13.5b gives n, the order of the reaction, which is the number of species involved in the reaction.

These methods work only for cases where there is no back reaction.

13.6
Net Rate of Reaction

When the reaction is reversible so that there is a back flux, for example, as illustrated above in Figure 13.1, the forward rate can be written:

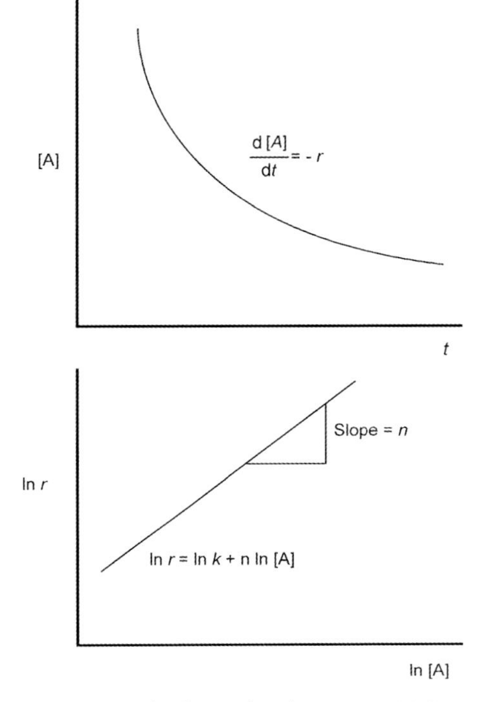

Figure 13.5 Plot for an *n*th order reaction. (a) Concentration versus time. (b) Logarithm of *r* versus logarithm of concentration.

$$\frac{kT}{h}[\mathrm{A}][\mathrm{B}]\exp\left[-(G^{*}-G_{1})/kT\right]=k^{*}[\mathrm{A}][\mathrm{B}] \tag{13.25}$$

And the reverse rate is:

$$\frac{kT}{h}[\mathrm{C}]\exp\left[-(G^{*}-G_{2})/kT)\right]=\frac{kT}{h}\frac{[\mathrm{C}]}{K}\exp\left[-(G^{*}-G_{1})/kT)\right]=k^{*}\frac{[\mathrm{C}]}{K} \tag{13.26}$$

The net rate is the difference between these two:

$$\frac{\mathrm{d}[\mathrm{C}]}{\mathrm{d}t}=k^{*}\left([\mathrm{A}]\,[\mathrm{B}]-\frac{[\mathrm{C}]}{K}\right) \tag{13.27}$$

Where k^* is the forward reaction rate constant and K is the equilibrium constant. The bracket is zero at equilibrium, and the rate of approach to equilibrium is proportional to the departure from equilibrium. Equation 13.27 can be written in expanded form as:

$$\frac{\mathrm{d}[\mathrm{C}]}{\mathrm{d}t}=\frac{kT}{h}\exp\left(\frac{\Delta S^{*}}{R}\right)\exp\left(-\frac{\Delta H^{*}}{kT}\right)\left[[\mathrm{A}][\mathrm{B}]-[\mathrm{C}]\exp\frac{\Delta G}{kT}\right] \tag{13.28}$$

In this form it is evident that the reaction rate consists of three factors. The first is a more-or-less constant pre-factor which is a frequency times a constant which depends on the entropy difference. The second is a Boltzmann factor containing the enthalpy difference. The third is a term which depends on how far the system is from equilibrium.

The time evolution of the concentrations is described by a differential equation. For this simple reaction, the differential equation can be solved readily. The time dependence of the concentration of the product can be calculated if the initial concentrations, the equilibrium constant and the reaction rate constant are known.

Often there are several or many reactions taking place at the same time, with the product from one reaction being a reactant in the next reaction. The coupled differential equations describing the overall reaction can be quite difficult to solve, even if all the equilibrium constants and reaction rate constants are known. There are computer programs which are designed to solve these coupled differential equations.

13.7 Catalysis

An effective catalyst reduces the maximum free energy, G^*, along the reaction path. It does not change the free energies of the initial or final states. It does not change the equilibrium condition. It reduces the time to get to equilibrium, or in a non-equilibrium system, it reduces the time for a reaction to take place.

There are homogeneous catalysts and heterogeneous catalysts.

Homogeneous catalysts are usually molecules which are dispersed homogeneously with the reactants. The reactants combine with the catalyst molecule, react, and then the product separates from the catalyst molecule.

Heterogeneous catalysis takes place on a surface. For a surface to be effective as a catalyst, the reactants must adsorb onto the surface, react there to form the product, and then the product must desorb, as illustrated in Figure 13.6. So the reactants and products must bind to the surface, but not too strongly. If the products do not desorb, the catalyst will become covered with the product, and will no longer be effective.

The most widely used heterogeneous catalyst is platinum. Platinum is relatively inert, it has a relatively high melting point, and the platinum atom can polarize readily

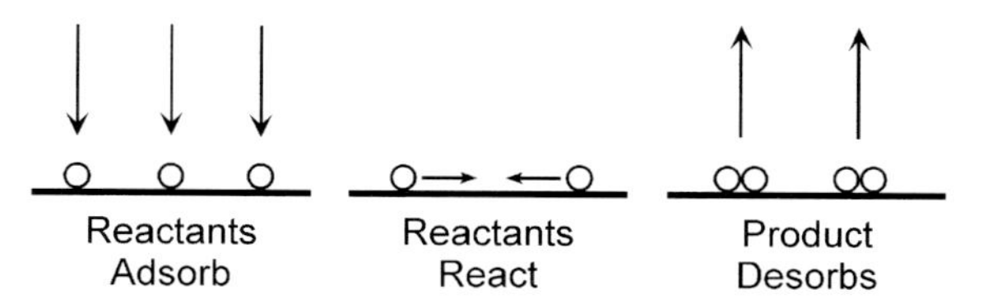

Figure 13.6 The reactants adsorb onto the surface of a heterogeneous catalyst, react, and then the product desorbs.

because it has many electrons. The polarizability means that the electronic charge on platinum can distort and so atoms or molecules will have van der Waals attraction to its surface. These are relatively weak bonds, so that the products can desorb readily. Most of the processing of bulk chemicals for fertilizers, plastics, and so on uses platinum as a catalyst.

The catalyst should have a large surface area, but a fine powder of pure platinum tends to sinter into bigger particles well below its melting point, which reduces the surface area. So platinum is usually coated onto a ceramic material which has a very large surface area. The ceramic is often alumina (Al_2O_3) or ceria (CeO_2) which has been prepared as a very fine powder, and then partially sintered. The very high melting points of these ceramic materials preclude further sintering at usual operating temperatures.

An example is the catalyst in your car. It is usually platinum on alumina. The catalytic converter on your car has a complex task. It is designed to oxidize hydrocarbons to make H_2O and CO, and to further oxidize the CO to CO_2, while at the same time reducing nitrous oxides, NO_x, to N_2. It turns out that a catalytic converter can do both reasonably well provided that the air to fuel ratio going through it is correct. 75–80% conversion for both the oxidation of CO and the reduction of NO_x can be achieved at an air to fuel ratio of 14.6, as illustrated in Figure 13.7.

13.8 Quasi-Equilibrium Model for the Rate of a First Order Phase Change

The rate at which an atom or molecule proceeds from state 1 to state 2, can be written in a form similar to Equation 13.25 as:

$$R_{1,2} = \frac{kT}{h} \exp\left(-\frac{(G^* - G_1)}{kT}\right) = R_{1,2}^0 \exp\left(\frac{G_1}{kT}\right) \tag{13.29}$$

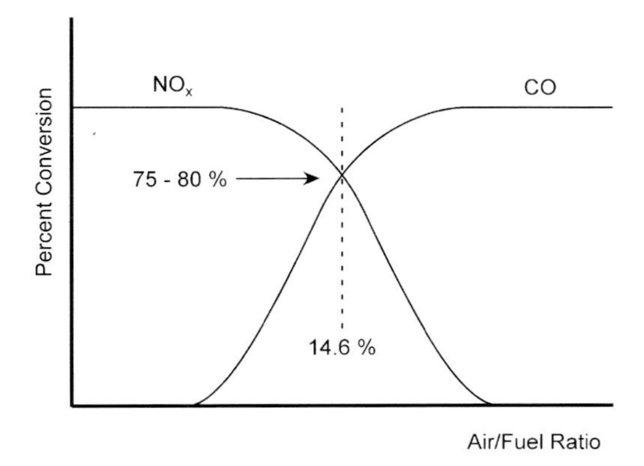

Figure 13.7 Conversion efficiency of an automobile catalytic converter.

where kT/h is the Debye frequency, which is approximately the atomic vibrational frequency, G_1 is the free energy of the system in state 1, and G^* is the free energy of the activated state, which is the state of highest free energy along the reaction path between states 1 and 2, as illustrated in Figure 13.1.

Similarly, the rate for the reverse reaction is given by:

$$R_{2,1} = \frac{kT}{h} \exp\left(-\frac{(G^* - G_2)}{kT}\right) = R_{2,1}^0 \exp\left(\frac{G_2}{kT}\right) \tag{13.30}$$

The rate of the reaction depends on the properties of the initial state.

The net reaction rate can be written:

$$R_{1,2} - R_{2,1} = \frac{kT}{h} \exp\left(-\frac{(G^* - G_2)}{kT}\right)\left[\exp\left(\frac{\Delta G}{kT}\right) - 1\right] \tag{13.31}$$

where $\Delta G = G_2 - G_1$. The term in the square bracket can be approximated for small $\Delta G/kT$:

$$\left[\exp\left(\frac{\Delta G}{kT}\right) - 1\right] \approx \frac{\Delta G}{kT} \tag{13.32}$$

For small departures from equilibrium, the reaction rate, Equation 13.31, can be written:

$$R_{1,2} - R_{2,1} \approx \frac{kT}{h} \exp\left(-\frac{(G^* - G_1)}{kT}\right)\left[\frac{\Delta G}{kT}\right] = R_0 \left[\frac{\Delta G}{kT}\right] \tag{13.33}$$

This formalism contains no details about the physical processes involved in a first order phase transformation, which is the subject of Chapters 20 and 21. It does predict that the transformation rate depends on the free energy difference between the two phases, and, for small departures from equilibrium, it predicts that the transformation rate is linearly proportional to the free energy difference, G. This conclusion is generally valid, but all the other details are buried in R_0.

Reference

1 Eyring, H., Lin, S.H., and Lin, S.M. (1980) *Basic Chemical Kinetics*, John Wiley & Sons, New York

Further Reading

Lupis, C.H.P. (1983) *Chemical Thermodynamics of Materials*, North-Holland, New York.

Ragone, D.V. (1995) *Thermodynamics of Materials*, John Wiley & Sons, New York

Problems

13.1. Discuss the difference between the temperature dependences of the equilibrium constant and the reaction rate constant.

14
Phase Equilibria

Phase changes can be first order, second order or higher order. These will be discussed below, together with the definitions and characteristics of each.

14.1
First Order Phase Changes

First order phase changes take place between two different states of matter: vapor and solid, liquid and solid, two different solid phases, and usually between vapor and liquid.

The essence of a first order phase change is that it is inhomogeneous in space. For example, part of the sample is liquid and part is solid. At the equilibrium temperature between the two phases, the interface separating them is stationary. The interface moves to create more or less of one of the two phases if it is above or below the equilibrium temperature. As the interface moves, the difference between the enthalpies of the two phases, also known as the latent heat, must be supplied to or removed from the interface region. This heat is generated (or absorbed) at essentially the same temperature as the interface moves through the sample. A large amount of heat is required to transform from all of one phase just below the equilibrium temperature to all of the other phase just above the equilibrium temperature. The input of heat does not change the temperature, it changes the relative amounts of the two phases. So the heat content of the material is discontinuous with temperature – which is the formal definition of a first order phase change.

Figure 14.1 illustrates the Gibbs free energy as a function of temperature for a solid and liquid. The free energies of the two phases are equal at the equilibrium temperature, which is referred to as the melting point. The equilibrium phase is the one with the lower free energy, and so there is a discontinuity in the slope of the equilibrium free energy curve at the melting point.

We can write the free energy of the solid as:

$$G_S = H_S - TS_S \tag{14.1}$$

Kinetic Processes: Crystal Growth, Diffusion, and Phase Transitions in Materials. Kenneth A. Jackson

ISBN: 978-3-527-32736-2

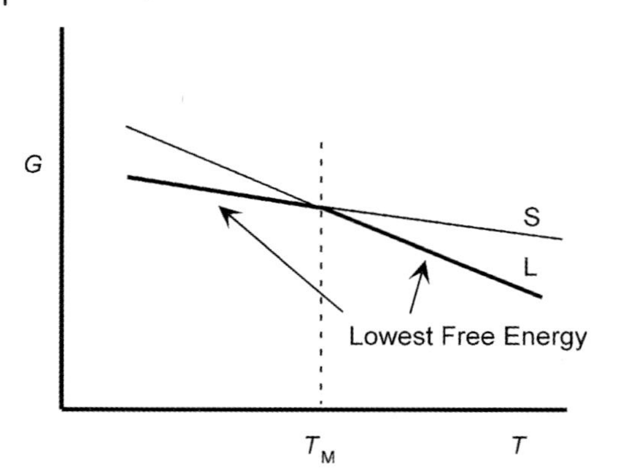

Figure 14.1 Gibbs free energy as a function of temperature for a solid and liquid.

and for the liquid:

$$G_L = H_L - TS_L \tag{14.2}$$

At equilibrium, above T_M

$$\frac{dG_S}{dT} = -S_S \tag{14.3}$$

and below T_M

$$\frac{dG_L}{dT} = -S_L \tag{14.4}$$

so that the entropy of the material is also discontinuous at the melting point.

At the melting point, $G_S = G_L$, so that

$$H_L - H_S = T_M (S_L - S_S) \tag{14.5}$$

which can be written as:

$$\Delta H = T_M \Delta S \tag{14.6}$$

There is a discontinuity in the enthalpy at the melting point, as illustrated in Figure 14.2. The enthalpy change is also known as the latent heat of transformation. It is the difference between the heat content of the two phases as measured with a calorimeter. The specific heat, which is the derivative of the enthalpy, is infinite at the melting point.

The melting point is the temperature of equilibrium between a solid and a liquid when both are present. The solid phase can exist above the melting point and the liquid phase can exist below the melting point. The Lindeman theory of melting suggested that melting occurs when the thermal vibrations of the atoms become large enough so that the solid is no longer stable. The Lindeman theory of melting is wrong.

There is a temperature above which a crystal structure is unstable, and this has been found in molecular dynamics (MD) simulations. In MD simulations it is

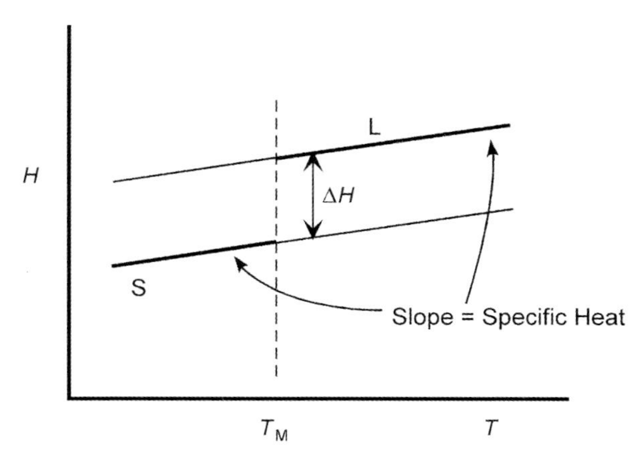

Figure 14.2 Enthalpy as a function of temperature for a solid and liquid.

possible to change the thermal energy of a sample very rapidly, between two computer time steps, for example, by stopping the simulation, doubling the velocities of all the atoms, and then continuing the simulation. Changing the temperature that far, that fast, in the real world is more difficult. In such computer simulations, where periodic boundary conditions are often used so that the crystals have no surface, the crystalline phase can be superheated to about 15–20% above the melting point before the crystal structure becomes unstable.

As discussed in Chapter 15, pure liquids can be supercooled to about 20% below their melting points, that is, to about 0.8 T_M, before small islands of solid form spontaneously in the liquid. These nuclei of solid then grow rapidly, transforming the liquid to crystal.

So the solid can be superheated about 20% above the melting point, and the liquid can be supercooled about 20% below the melting point. Nothing strange happens to the structure of the crystal or the liquid when they are superheated or supercooled. The properties of both continue uniformly through the melting point. The melting point is the temperature where the two phases are in equilibrium with each other.

There are many reports of pre-melting phenomena in the literature. These are reports that some property of a crystal changes as the melting point is approached. These effects are usually associated with re-melting of second components which segregated to surfaces or to grain boundaries during the solidification process.

It is well known that it is easier to supercool a liquid than to superheat a solid. The usual difficulty in superheating solids is the reason for the equilibrium temperature between a solid and a liquid being called the melting point. There is an intrinsic asymmetry between the initial formation of a solid in a liquid and the initial formation of a liquid in a solid which accounts for this. It depends on the relative values of the surface tensions of the solid and liquid. Only one of the relationships illustrated in Figure 14.3 can be valid for a particular material.

If $\sigma_{LS} + \sigma_{LV} < \sigma_{VS}$, then the free energy of the interface between the solid and vapor will be lowered if a layer of liquid forms. This will only happen near the melting

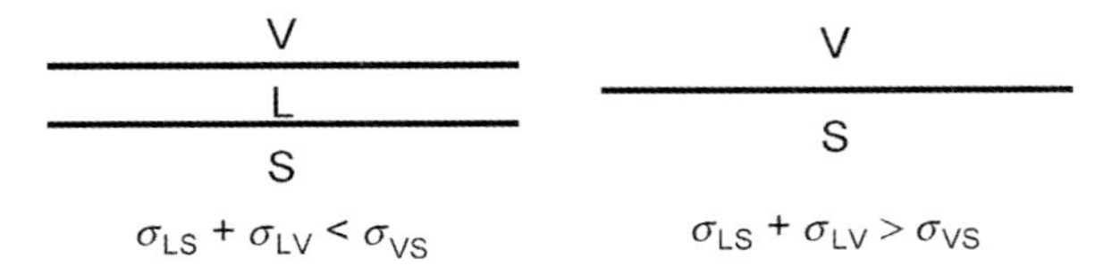

Figure 14.3 Surface tension relationships at a solid/vapor interface.

point, where the free energy of the liquid is close to that of the solid. On the other hand, if $\sigma_{LS} + \sigma_{LV} > \sigma_{VS}$, then the intermediate layer of liquid will not form.

Usually, the former relationship, illustrated on the left in Figure 14.3 applies:

$$\sigma_{LS} + \sigma_{LV} < \sigma_{LS} \tag{14.7}$$

If this relationship is valid, then a layer of liquid should form spontaneously at the surface of a solid at the melting point, as illustrated in Figure 14.3. Just below the melting point this layer is at most a few atom layers thick. This layer of liquid can grow in thickness above the melting point, without a nucleation barrier to the formation of the liquid phase. This is observed in MD simulations of crystal surfaces.

This asymmetry in the surface tensions has implication for nucleation. Liquids can be supercooled to their homogeneous nucleation temperature if they are clean and free of foreign particles. Usually, the nuclei of the solid phase do not form preferentially at the free surface, they form in the bulk of the liquid, which suggests that a layer of liquid separates the nucleating crystal from the vapor.

So the nucleation process is asymmetric. Both configurations in Figure 14.4 imply that Equation 14.7 is valid. Equation 14.7 is usually valid, so the liquid can be supercooled to the homogeneous nucleation temperature without nuclei forming at the surface, and melting occurs spontaneously at the surface of a crystal on heating.

Solids can be superheated by keeping the surface cool, and heating them internally with a high power, focused light source. This can be arranged so that the surface is below the melting point, while the interior of the sample is above the melting point. In this case, melting cannot proceed in from the surface, and so, if the liquid phase is to form, it must nucleate internally in the solid. This usually occurs first at grain boundaries or other defects. An example of this is dendritic melting of ice crystals to produce Tyndall figures.

By focusing sunlight into a block of ice with a magnifying glass, internal melting can occur in the form of liquid dendrites growing inside the ice. The dendrites are

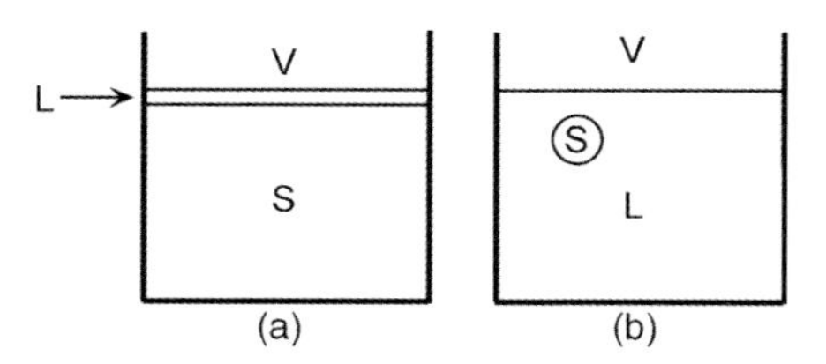

Figure 14.4 (a) Spontaneous formation of a layer of liquid on a solid, (b) Homogeneous nucleation of a solid in a liquid.

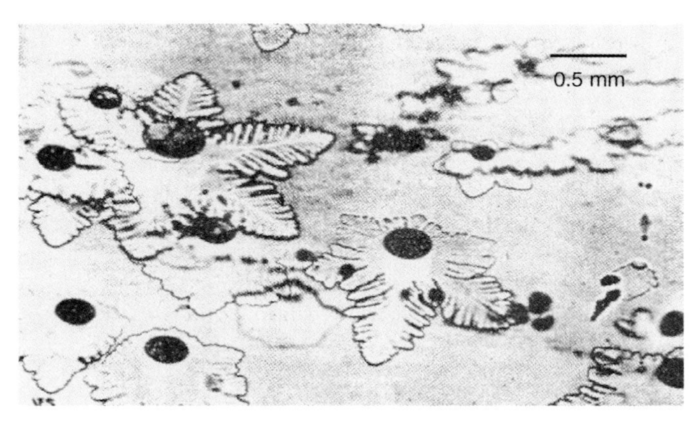

Figure 14.5 Tyndall figures: dendritic melting of ice.

planar and lie in the basal plane of the ice crystal. The morphology is similar to that of ice dendrites growing into supercooled water. The fact that Tyndall figures are dendritic in form implies that the ice was superheated, and that growth is limited by the diffusion of heat. Water is more dense than ice, so small vapor bubbles form in the liquid to take care of the volume change. These appear as dark circles in Figure 14.5. Glaciologists have used Tyndall figures to determine the orientation of the ice crystals in glaciers by focusing sunlight into the ice with a magnifying glass.

14.2 Second Order Phase Changes

Unlike first order phase changes, second order phase changes are homogeneous in space. There is a continuous change in structure as the critical point is approached from below. The prototypical example of a second order phase change is the ferromagnetic transition, where the critical temperature is known as the Curie temperature.

At a very low temperature, (0 K), all the spins in a ferromagnet align, as illustrated in Figure 14.6. There are magnetic domains in real magnetic materials so that the net

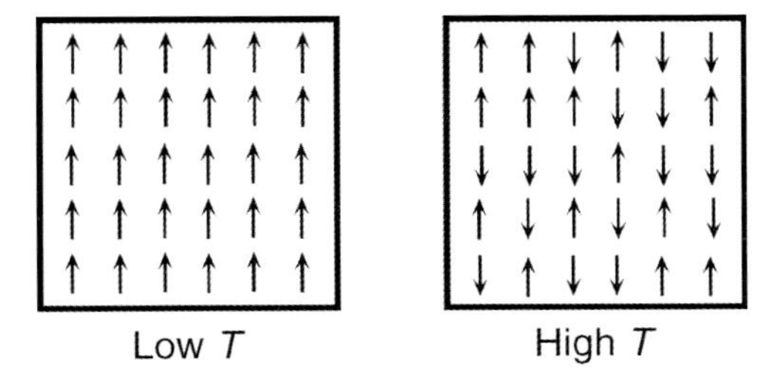

Figure 14.6 Spin alignment in a ferromagnet.

magnetization can be zero, but all the spins in any one domain are aligned. As the temperature is increased, some of the spins become misaligned. In the simple model which is used to describe this, it is assumed that there is a strong anisotropy to the alignment of spins, so that the spins can only point either up or down. As the temperature increases, more and more spins flip over. There is a critical point above which all of the spins are randomly up or down. This happens when the average thermal energy of the atoms is comparable to the increase in energy when a spin flips.

The enthalpy or energy content of the material changes as more spins flip, because the increased number of flipped spins represents a higher energy state. This is, of course, compensated for by a corresponding increase in entropy due to the increased randomness.

So the enthalpy increases continuously as the critical point is approached. Above the critical temperature, there is no more energy to be gained or lost, on average, by flipping spins, as shown in Figure 14.7.

The specific heat is the temperature derivative of enthalpy, as is shown schematically in Figure 14.8. The hatched area is the extra energy associated with flipping the spins, since the random configuration has a higher energy than the ordered state. There is a very large change in the specific heat at the critical point. This corresponds to a discontinuity in the second derivative of the free energy, which is why this is known formally as a second order phase transition.

However, the primary distinguishing characteristic of a second order phase change is that it is homogeneous in space. The spins flip randomly throughout the bulk of the sample. The transformation does not take place due to the motion of an interface.

In a second order phase transition, both the free energy and the temperature derivative of the free energy are continuous at the critical point. If you try to draw this, as in Figure 14.9, you will find that the curves do not cross, unless there is also an inflection point in the difference between the two curves, but that implies a third order phase transformation [1].

Some have argued that "second order phase transformation" is a misnomer, and they prefer to call this a lambda transition, since there is only one branch of the curve above the critical point, as illustrated in Figure 14.10.

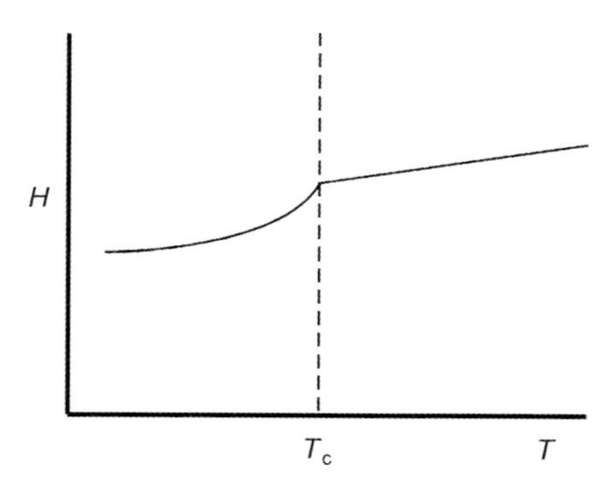

Figure 14.7 Enthalpy as a function of temperature for a second order phase change.

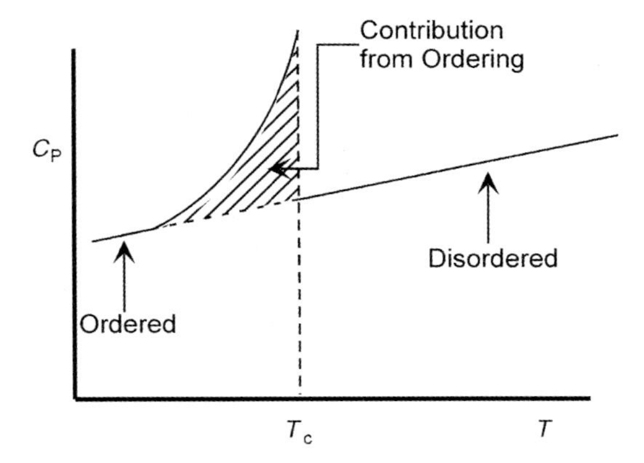

Figure 14.8 Specific heat as a function of temperature for a second order phase change.

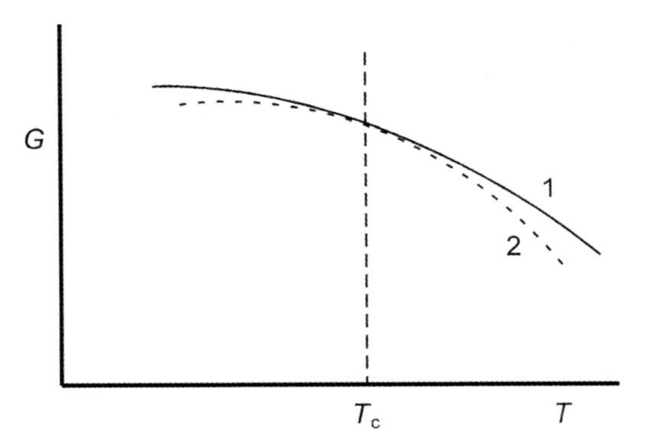

Figure 14.9 If both the free energy and the temperature derivative of the free energy are continuous at the critical point, the free energy curves do not cross.

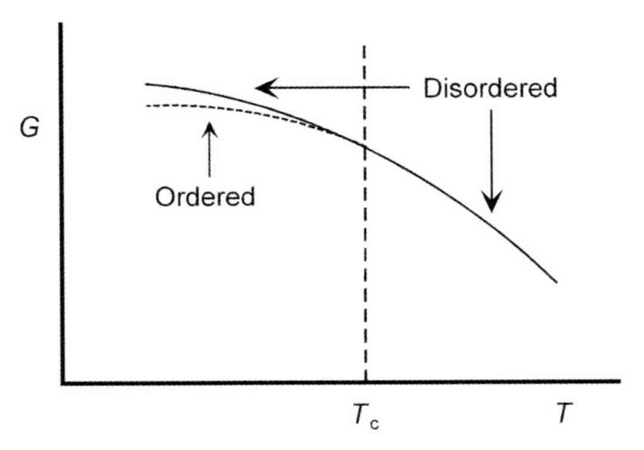

Figure 14.10 Free energy as a function of temperature for a second order phase change (lambda transition).

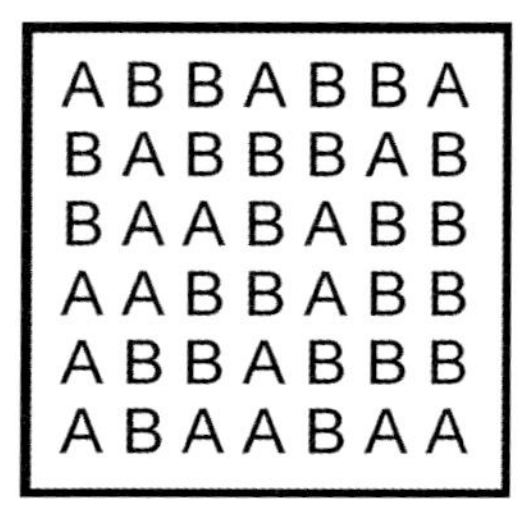

Figure 14.11 Disorder in an alloy that orders at low temperatures.

This is best illustrated by considering an alloy which has an order–disorder transition. In an ordering alloy, if the atoms are mobile enough at the critical point, then the transition from order to disorder will proceed in the vicinity of the critical point. However, the disordered high temperature phase, illustrated in Figure 14.11, can often be retained by rapid cooling from above the critical point to a temperature where the atoms are no longer mobile. So both the ordered and disordered phases can exist at low temperatures where there is not enough atomic mobility to enable a change in order but only the disordered phase exists above the critical temperature. The ordered phase cannot be maintained by rapid heating to a high temperature, because the mobility of the atoms increases with temperature. It is difficult to reproduce this effect in a ferromagnet because the spins can flip rapidly even at fairly low temperatures.

14.3
Critical Point Between Liquid and Vapor

There is a critical point between liquid and vapor, as illustrate in the phase diagram, Figure 14.12.

It is possible to go continuously from the liquid phase to the vapor phase by going around the critical point. Going through the critical point, for example by cooling, is a second order phase transformation. Below the critical point there is an area on the phase diagram which is bounded by the coexistence curve (the phase boundary). Within this area, a sample will consist of two distinct phases which are regions of greater and lesser density. As the critical point is approached from below, the densities and other properties of the two regions become increasingly similar, so that at the critical point they are indistinguishable. A transformation going directly through the critical point can proceed homogeneously in space, and so is second order. Elsewhere, however, there is a discontinuous density change when going from one phase to the other, and so a transformation proceeds by one phase growing at the expense of the other. Such a phase transformation is inhomogeneous in space, and so is first order.

People have looked for a critical point between a crystal and a liquid. At the limits of temperature and pressure available experimentally, either the density difference or

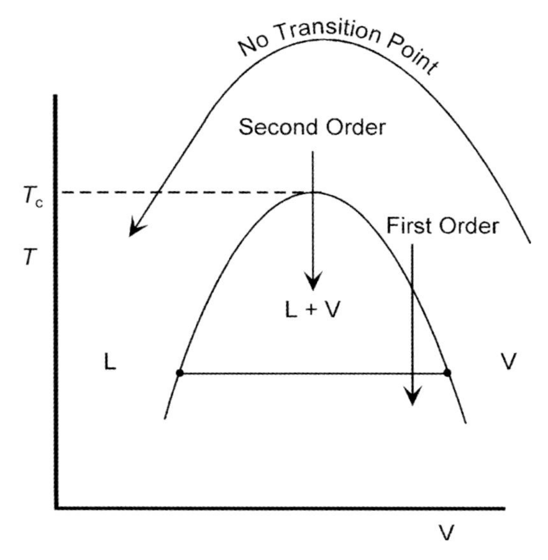

Figure 14.12 Phase diagram showing a critical point between liquid and vapor.

the entropy difference between the two phases is diverging. Some theorists state that the change in symmetry which is associated with the difference between the structure of a crystal and a liquid implies that this phase change must be first order, since it is impossible to change the symmetry gradually.

Reference

1 Pippard, A.B. (1964) *Elements of Classical Thermodynamics*, Cambridge University Press, Cambridge, UK.

Further Reading

Doremus, R.H. (1985) *Rates of Phase Transformations*, Academic Press, Orlando, FL.

Lupis, C.H.P. (1983) *Chemical Thermodynamics of Materials*, North-Holland, New York.

Kittel, C. and Kroemer, H. (1980) *Thermal Physics*, W. H. Freeman, San Francisco, CA.

Porter, D.A. and Easterling, K.E. (1992) *Phase Transformations in Metals and Alloys*, 2nd edn, Chapman and Hall, London, UK.

Problems

14.1. Discuss the difference between a first order and a second order phase transformation.

14.2. Why is there confusion about the difference between second and third order phase changes?

15
Nucleation

In the preceding chapter, the motion of an interface between two phases was discussed. This discussion assumed that both phases were present. In this chapter, the initial formation of one phase in another is discussed. There is usually a barrier to the formation of a new phase: the formation of a liquid in a gas, a solid in a gas, bubbles in a liquid, crystals in a liquid, precipitates in a solid, domains of reverse magnetization, and so on. The formation of a new phase begins with a small nucleus, which grows by atoms joining and leaving it, by the same processes discussed in the previous chapter. In this chapter, the barrier to the formation of a new phase is discussed. This barrier controls the rate of formation of the new phase.

In some instances, there is no barrier to the formation of a new phase, for example, in systems with a critical point, it is possible to go continuously from one phase to another. There is also a process known as phase separation where one phase spontaneously decomposes into two phases; this will be discussed in Chapter 23.

There is a barrier to the formation of new layers on the faces of some crystals, so that the growth of the crystal involves a nucleation process. We will return to the kinetics of crystal growth after discussing nucleation processes in general.

15.1
Homogeneous Nucleation

There are two classes of nucleation events, known as homogeneous and heterogeneous nucleation. Homogeneous nucleation involves the spontaneous formation and subsequent growth of small particles of the new phase. In heterogeneous nucleation, the new phase is initiated on a foreign material such as a particle or a surface layer. Homogeneous nucleation occurs when there are no heterogeneous nuclei present. A heterogeneous nucleating agent provides a lower barrier to the initial formation of the new phase. Most nucleation processes in the real world are heterogeneous, but the process depends on the nucleating agent involved, and so the details defy a generic description. The homogeneous nucleation process involves only the one material, and so it is intrinsic to the material. The conditions for homogeneous nucleation to occur represent a lower limit on the stability of the phase.

Kinetic Processes: Crystal Growth, Diffusion, and Phase Transitions in Materials. Kenneth A. Jackson

ISBN: 978-3-527-32736-2

It can be analyzed more readily than heterogeneous nucleation which involves a foreign, often unknown, material.

15.1.1 Volmer Analysis

Volmer [1] invented nucleation theory to explain the strange melting and freezing behavior of materials. Most materials melt at their melting points. Theories of phase transformation kinetics were available, but some liquids could be undercooled a lot, and others could be undercooled only a little, or not at all. The behavior of various materials was different, and even the behavior of one material was erratic and could vary from one sample to another. Volmer addressed the question: Why can liquids be undercooled into the metastable region indicated in Figure 15.1.

Volmer started from the fact that small particles are less stable than the bulk phase because of their surface tension, and so it is difficult to form a small particle, and this creates a barrier to the formation of a new phase. He suggested that the formation of small particles depends on fluctuations.

A supercooled liquid is metastable. It is stable in the absence of the solid phase, but it is not the lowest free energy state. Volmer considered the question: What is the probability of a fluctuation which is big enough to make a stable bit of the new phase?

He assumed that the change in free energy when a cluster of atoms of the new phase formed could be described by two contributions, one from the decreased free energy associated with the formation of the new phase, and the other from the surface tension of the small cluster. Assuming that the cluster of atoms of the new phase is a sphere of radius r, the change in free energy when the cluster forms can be written in terms of these two contributions:

$$\Delta G_r = -\Delta G_V \frac{4}{3}\pi r^3 + \sigma 4\pi r^2 \tag{15.1}$$

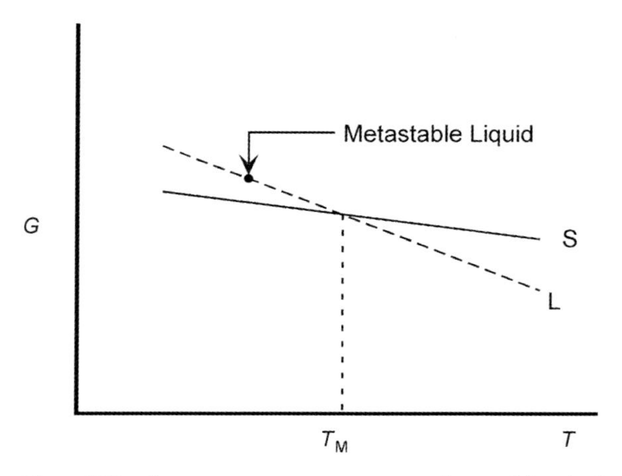

Figure 15.1 Free energy versus temperature for a solid and a liquid.

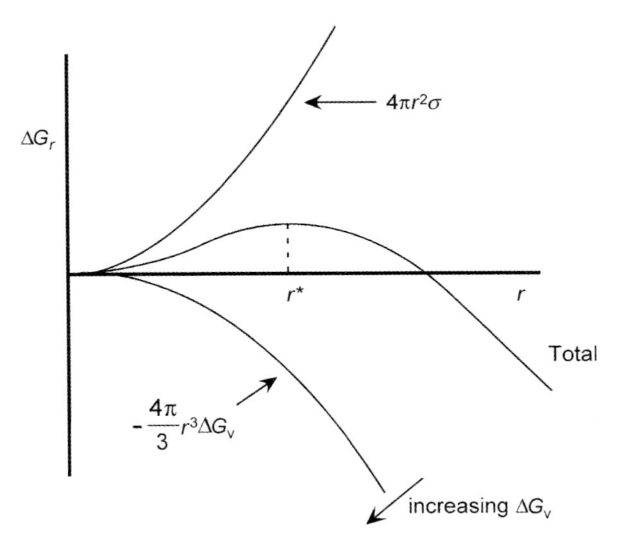

Figure 15.2 Free energy of a cluster of atoms for various radii of the cluster.

Here ΔG_V is the change in free energy per unit volume associated with the transformation, and σ is the surface tension, or specific surface free energy. For a liquid which is supercooled below its melting point, we can write $\Delta G_V \approx L\Delta T/T_M$, where ΔT is the undercooling, and L is the latent heat, as in Equation 20.7.

Figure 15.2 is a plot of the total free energy of the cluster, ΔG_r, for various sizes of cluster. The total surface area of the cluster increases as r^2, and the volume free energy term decreases as r^3. The magnitude of the volume term depends on the undercooling. Below the equilibrium temperature, the volume free energy term will dominate for large radius: a large enough crystal will grow. Very small crystals can reduce their free energy by shrinking. The transition between these two regions is at the maximum in the total free energy curve, which occurs at the critical radius, r^*. The value of r^*, which is the radius corresponding to the maximum in the free energy, is given by:

$$\frac{\Delta G_r}{dr} = 0 = -4\pi r^2 \Delta G_V + 8\pi r\sigma \tag{15.2}$$

so that the critical radius is:

$$r^* = \frac{2\sigma}{\Delta G_V} = \frac{2\sigma T_M}{L\Delta T} \tag{15.3}$$

Volmer made use of the fact that the probability of finding a fluctuation of energy, W, is given by a Boltzmann factor:

$$\exp(-W/kT) \tag{15.4}$$

The probability of finding a cluster of size r^* should then be given by Equation 15.4 with W given by the free energy required to form the cluster. The free energy to form a cluster of size r^* is obtained by substituting the value for r^* from Equation 15.3 into Equation 15.1:

$$\Delta G_{r^*} = \frac{16}{3} \frac{\pi \alpha \sigma^3}{(\Delta G_V)^2} \tag{15.5}$$

and so the probability of finding a cluster of size r^* among N atoms is:

$$\frac{N_{r^*}}{N} = \exp\left(-\frac{\Delta G_{r^*}}{kT}\right) = \exp\left(-\frac{16\pi\sigma^3}{3kT\,(\Delta G_V)^2}\right) \tag{15.6}$$

In general, the total free energy of a critical cluster, ΔG_{r^*}, includes strain energy, magnetic energy, electrical energy, and so on, in addition to the volume free energy and surface free energy terms. If they are relevant, these contributions should be added to obtain the total free energy of the critical cluster. The free energy in Equation 15.6 was calculated for a spherical cluster, but the energy in the exponent should always be simply the total free energy to form the critical nucleus.

For homogeneous nucleation of a solid in an undercooled liquid, there are typically about 300 atoms in a cluster of critical size. The total free energy to form a cluster is the total change in free energy for all 300 atoms. So the number in the exponential can be very large. For small undercooling, ΔG_V is small, so the exponent is large and negative, implying very few critical nuclei. For larger undercooling, the volume free energy becomes comparable to the surface free energy. When this happens, the exponent switches from being large to being small in a small temperature interval. The probability of finding a nucleus of critical size increases abruptly in a small temperature interval, as illustrated in Figure 15.3.

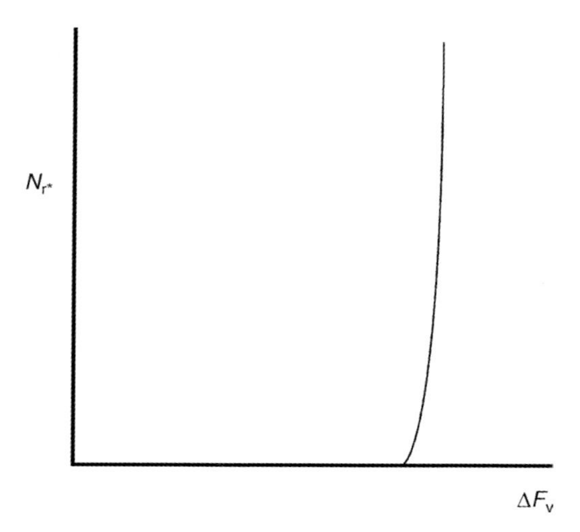

Figure 15.3 Number of critical nuclei as a function of undercooling.

For undercooled metals, the mobility of the atoms in the liquid is large, so the distribution of solid-like clusters in the liquid can change rapidly. Nucleation occurs very rapidly once the critical supercooling is reached. Supercritical clusters grow very rapidly, and their growth rate will be limited by heat flow. The latent heat of a typical metal is enough to raise its temperature by an amount which is about one third of its melting point. For most materials, homogeneous nucleation occurs at a supercooling of about 20%, that is, at a temperature that is about 0.8 T_M. The latent heat released by freezing of a sample at 0.8 T_M is more than enough to heat the sample to the melting point. So the nucleation process is terminated after a few nucleation events by the rapid growth and the associated heating of the sample.

In a glass, the situation is quite different. The rate at which the cluster distributions can change is slow, and the growth rate of critical nuclei is also slow. The growth rate is so slow that heat flow is not an issue. So a nucleus can form at one place and the supercooling is still maintained in another part of the glass sample, and so the nucleation rate can be measured. Nucleation of a new solid phase in a solid is usually similar to this. There is usually a difficulty in fitting the nucleation data to the expected nucleation rate for glasses. There are often defects in solids, and irregularities in the structure of glasses. It is also difficult to measure surface tensions and to determine strain energy contributions in order to estimate the work required to form the nucleus.

For precipitation of oxygen in silicon, the oxygen moves interstitially through the lattice. The growth rate of the precipitate depends on the rate of motion of oxygen through the lattice. The oxygen precipitates as crystoballite, a form of SiO_2. An SiO_2 molecule occupies about the same volume as two silicon atoms in a silicon crystal. There is a significant stress generated at the precipitate, and the energy in the stress field is a major component of the total energy of the precipitate. Silicon atoms are forced into an interstitial position, and the precipitate blows out dislocation loops to relieve the stress. The stresses generated and the effects of the stress relief mechanisms all contribute to the total work to form the precipitate.

Detailed experiments have been carried out to study the nucleation of liquid droplets in clean vapor. The surface tension of the liquid can be measured readily, and the clusters of liquid in the vapor phase are likely to be spherical. The experimental data on these systems verify the validity of the nucleation equations.

Continuous nucleation of particles can occur in a stream of hot atoms or molecules coming from a high temperature source, or expanding and cooling in a jet coming from a high pressure source. There can be a steady state nucleation process where particles nucleate continuously at some distance down stream from the source [2].

15.1.2
Turnbull's Droplet Experiment

Most liquids contain heterogeneous nucleating particles, and so it is difficult to observe homogeneous nucleation of crystals in liquids. Turnbull [3] reasoned that if he subdivided the liquid into small enough droplets, that most of the droplets would

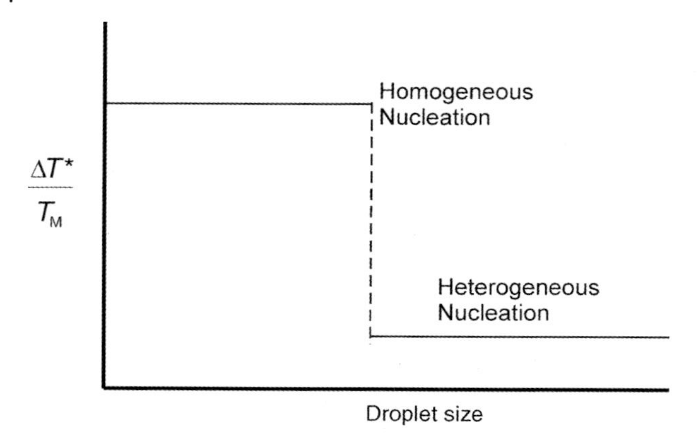

Figure 15.4 Observed undercooling versus droplet size.

not contain a nucleating particle. In order for this experiment to work, nucleation should occur in the bulk of the liquid, not at the surface or at the substrate on which the droplet is sitting. Turnbull chose mercury for the experiment, which can be readily purified by distillation. The results of the experiment are presented schematically in Figure 15.4.

For large droplets, the undercooling which could be achieved was small. Very small droplets could be undercooled to about 0.8 T_M. The transition occurred for a droplet diameter of about 60 μm. The volume of such a droplet is about 10^{-7} cm^3, which suggests that there were about 10^7 heterogeneous nuclei in a cubic centimeter of the liquid. This is not too different from the number of micron size particles in the air we breathe if the air is reasonably clean.

This experiment has been repeated hundreds of times by many experimenters using a variety of materials. The maximum undercooling is usually about 0.8 T_M. The free energy of solid–liquid interfaces is difficult to measure, and so the interfacial free energy of solid–liquid interfaces has been calculated for many materials from Equation 15.6 using the observed critical undercooling. For example, for the experiments illustrated in Figure 15.4, a droplet which is 60 μm in diameter contains about 3×10^{15} atoms, and so the probability of finding a critical nucleus is about 3×10^{-16}. The natural logarithm of 3×10^{-16} is -38, and so we have:

$$38 = \frac{16}{3} \frac{\pi\sigma^3}{kT\,(L\Delta T^*/T_M)^2} \tag{15.7}$$

At the nucleating temperature $kT \approx L\Omega$, where Ω is the atomic volume, and $L\Omega$ is the latent heat per atom. Using $\Delta T^*/T_M \approx 0.2$, we have approximately:

$$38 \approx \frac{16}{(0.2)^2} \frac{\sigma^3}{L^3\Omega} \tag{15.8}$$

so that:

$$\frac{\sigma}{a\,L} \approx \frac{1}{2} \tag{15.9}$$

where $a = \Omega^{1/3}$. This is known as Turnbull's Rule. The surface energy per atom is equal to about one-half of the latent heat per atom. For all the materials which have been measured, σ/aL is between 1/2 and 1/3, with 1/2 being more common for metals, and 1/3 being more common for organic compounds.

There are tables of solid–liquid surface tensions which have been determined by this method [5]. The game is that if someone repeats an earlier experiment, and supercools the liquid to a lower temperature, he wins, because he can say that the earlier experiments were obviously not as clean as his, and were subject to heterogeneous nucleation. So a new, higher, value of the surface tension is obtained.

Using Turnbull's rule, we can rewrite Equation 15.3 for the critical radius:

$$r^* = \frac{2\sigma}{\Delta F_V^*} = \frac{2\sigma T_M}{L\Delta T^*} \approx a\frac{T_M}{\Delta T^*} \tag{15.10}$$

or

$$\frac{a}{r^*} \approx \frac{\Delta T^*}{T_M} \tag{15.11}$$

So we have a simple expression for an approximate value for the amount by which surface curvature changes the melting point, which is known as the Gibbs–Thompson effect. This approximate value is accurate to within about a factor of two for all the materials which have been measured.

This equation says that there is a significant change in the melting point only if the radius of curvature of the surface is very small. For example, for a droplet with a 1 μm radius, $a/r^* \approx 3 \times 10^{-4}$. For a material with a melting point of 1000 K, the equilibrium temperature would be lowered by 0.3 K. For a sample 1 cm in radius, the effect of surface curvature on the melting point is negligible.

15.1.3 Surface Free Energy

The surface free energy of a clean surface is primarily due to missing bonds at the surface. There are also effects due to surface relaxation, surface stresses, surface reconstruction, and so on, and these modify the energy associated with the missing bonds but the basic effect is due to the missing bonds. To illustrate this, we will look at a simple case, in two dimensions. We will evaluate the total free energy of a group of atoms in order to determine the free energy per atom in the group. This must be done on a closed figure, since the contribution of the surface free energy to the total free energy depends on the relationship between the surface area and the volume of the cluster.

The square clusters in Figure 15.5 all have the same shape, and so there is a simple analytical relationship between the length of their peripheries and their areas. The

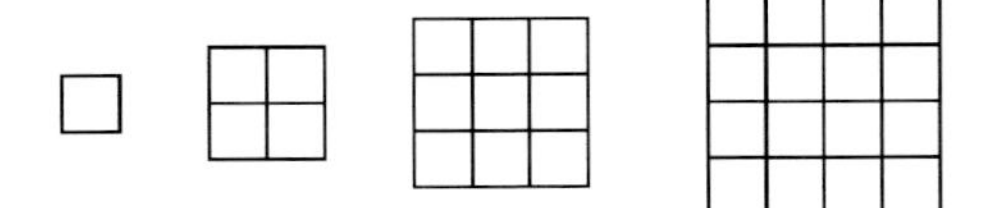

Figure 15.5 Square clusters in two dimensions.

table lists the number of atoms (squares) and the number of bonds in each cluster. It is assumed that there is a bond between any two adjacent squares.

Number of atoms, n	1	4	9	16
Number of bonds, N	0	4	12	24

The number of bonds, N, in each cluster is given in terms of n, the number of atoms in the cluster, by:

$$N = 2n - 2\sqrt{n} \tag{15.12}$$

The first term is proportional to the number of atoms in the cluster and the second term is proportional to the length of the periphery of the cluster. The total energy decrease on forming the cluster from isolated atoms is the number of bonds times the energy per bond, φ:

$$N\varphi = 2n\varphi - 2\sqrt{n}\varphi \tag{15.13}$$

The total free energy to form each cluster can also be written in terms of the latent heat, L, and the surface tension, σ. The area of each cluster is na^2, and the length of the periphery of each cluster is $4a\sqrt{n}$, so the energy of formation of a cluster is:

$$\Delta E = a^2 nL - 4a\sqrt{n}\sigma \tag{15.14}$$

Comparing terms in Equations 15.12 and 15.14 gives the latent heat and the surface tension in terms of the bond energy:

$$L = \frac{2\varphi}{a^2}; \quad \sigma = \frac{\varphi}{2a} \tag{15.15}$$

so that

$$\frac{\sigma}{aL} = \frac{1}{4} \tag{15.16}$$

If this is done in three dimensions, the result is:

$$\frac{\sigma}{aL} = \frac{1}{2} \tag{15.17}$$

which is Turnbull's rule for the relationship between the surface tension per atom and the latent heat per atom.

If we had not used squares for the geometry of all our clusters as in Figure 15.6, then we would not have found a simple relationship between the number of atoms and the number of bonds in the clusters.

Figure 15.6 Small clusters in two dimensions.

The relationship between the number of atoms and the number of bonds also depends on the lattice structure which is assumed for the clusters.

The geometry of the clusters averages out for large clusters, provided that they are reasonably compact. Compact clusters have the lowest total energy, and are the most likely configuration. This suggests why it is reasonable to use the ratio of the surface area to the volume of a sphere as an approximation to obtain the total energy of formation of a cluster.

15.1.4
Becker–Döring Analysis

Becker and Döring [4] derived the equilibrium cluster distribution, Equation 15.6, which Volmer simply assumed. Their analysis provides insight into the nucleation process, and into the kinetics of how the cluster distribution changes in time, as well as providing an expression for the rate of nucleation. They assumed, as Volmer did, that the clusters of atoms of the new phase are spheres. We will generalize that by assuming that there is a fixed relationship between the number of atoms, n, in a cluster, and its surface area, which is proportional to $n^{2/3}$, so the energy to form a cluster of n atoms can be written:

$$E_n = -nL + b\sigma n^{2/3} \tag{15.18}$$

where b is a geometrical constant. For a sphere:

$$b = (3\Omega)^{2/3}(4\pi)^{1/3} \tag{15.19}$$

where Ω is the atomic volume.

The amount by which the energy of a cluster of $n-1$ atoms changes when an atom is removed is:

$$\begin{aligned} \Delta E_n &= E_n - E_{n-1} \\ &= (-nL + b\sigma n^{2/3}) - (-(n-1)\,L + b\sigma\,(n-1)^{2/3}) \\ &\approx -L + \frac{2}{3} n^{-\frac{1}{3}} b\sigma = \frac{\mathrm{d}E_n}{\mathrm{d}n} \end{aligned} \tag{15.20}$$

From Equation 15.18, the energy per atom in a cluster of n atoms is:

$$\frac{E_n}{n} = -L + b\sigma n^{-\frac{1}{3}} \tag{15.21}$$

The energy per atom in a small cluster is higher than the energy per atom in the bulk of the nucleating phase, and increases as the number of atoms in the cluster decreases. This is illustrated in Figure 15.7.

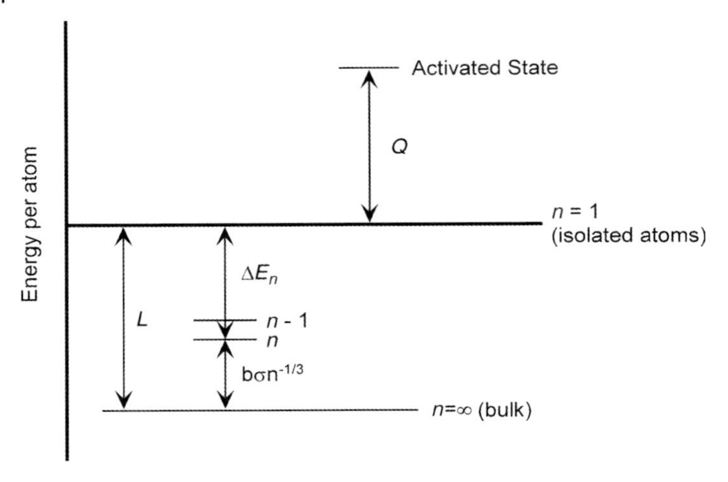

Figure 15.7 Energy per atom for clusters containing *n* atoms.

Assuming that the rate at which an atom joins a crystal is given by Equation 13.29, the rate at which atoms join a cluster containing n atoms can be written:

$$R_n^+ = R_0^+ A_n \exp\left(-\frac{Q}{kT}\right) \tag{15.22}$$

Q may be the same as the activation energy for diffusion, as in Equation 20.25, or it may be zero, as in Equation 20.26.

Here A_n is the capture cross-section (surface area) of the cluster. The rate at which atoms leave the cluster depends on the same activation energy, plus the amount by which the total energy of the cluster changes when the atom leaves, which can be written:

$$R_n^- = R_0^- A_n \exp\left[-\frac{(Q-\Delta E_n)}{kT}\right] \tag{15.23}$$

Note that ΔE_n is defined as a negative quantity in Equation 15.20, so that $Q_n - \Delta E_n$ in this equation is actually the sum of the two terms. This relationship is similar to the energy per atom in a cluster, which is given by Equation 15.21, and illustrated in Figure 15.7. However, in Equation 15.23 we want the change in total energy of the cluster when an atom leaves it, which is given by Equation 15.20.

The relationship between the R_0 terms is established by the equilibrium condition for the bulk phases, as in Equation 20.20:

$$R_0^+ = R_0^- \exp\left(-\frac{L}{kT_e}\right) \tag{15.24}$$

so that Equation 15.22 can be written:

$$R_n^+ = R_0^- A_n \exp\left(-\frac{L}{kT_e} - \frac{Q}{kT}\right) \tag{15.25}$$

Combining Equations 15.25 and 15.23

$$\begin{aligned} \frac{R_n^+}{R_n^-} &= \exp\left(-\frac{L}{kT_e} - \frac{\Delta E_n}{kT}\right) \\ &= \exp\left(\frac{L\Delta T}{kT_e T} - \frac{2}{3}\frac{b\sigma}{kT} n^{-\frac{1}{3}}\right) \end{aligned} \tag{15.26}$$

Becker and Döring assumed that most of the atoms are not in clusters, but are present as single atoms. They assumed that there are not very many clusters, so that the clusters do not interact or coalesce. They assumed that the clusters grow or shrink only by the addition or subtraction of single atoms, as in Figure 15.8.

With these assumptions, a cluster of size n is created only by adding an atom to a cluster of size $n-1$ or by an atom leaving a cluster of size $n+1$. A cluster of size n is destroyed if it either gains or loses an atom. So the rate of change of N_n, the number of clusters of size n is:

$$\frac{dN_n}{dt} = N_{n-1}R_{n-1}^+ + N_{n+1}R_{n+1}^- - N_n R_n^+ - N_n R_n^- \tag{15.27}$$

There is an equation like this for each cluster size, and these equations are all coupled. In general, they are not solvable. If dimers are permitted to join or leave the clusters, the equations become even more difficult to solve. To compare with Volmer's equation, we will look for the equilibrium distribution, which requires not only that $dN_n/dt = 0$, but also that there is no net flow from $n-1$ to n to $n+1$ clusters, which requires:

$$N_{n-1}R_{n-1}^+ = N_n R_n^- \tag{15.28}$$

or

$$\frac{N_n}{N_{n-1}} = \frac{R_{n-1}^+}{R_n} \tag{15.29}$$

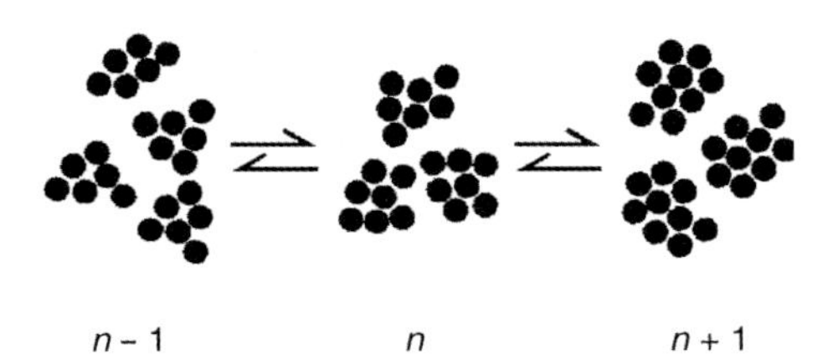

Figure 15.8 Equilibrium cluster distribution.

for each value of n.

We can write:

$$\begin{aligned}\frac{N_n}{N_1} &= \frac{N_n}{N_{n-1}} \cdot \frac{N_{n-1}}{N_{n-2}} \cdot \frac{N_{n-2}}{N_{n-3}} \cdots \frac{N_2}{N_1} \\ &= \frac{R_{n-1}^+}{R_n^-} \cdot \frac{R_{n-2}^+}{R_{n-1}^-} \cdots \frac{R_2^+}{R_3^-} \cdot \frac{R_1^+}{R_2^-} \\ &= \frac{R_1^+}{R_n^-} \cdot \prod_{i=2}^{n-1} \frac{R_i^+}{R_i^-}\end{aligned} \tag{15.30}$$

Using Equation 15.26, this becomes:

$$\begin{aligned}\frac{N_n}{N_1} &= \frac{R_1^+}{R_n^-} \cdot \prod_{i=2}^{n-1} \exp\left(\frac{L\Delta T}{kT_e T} - \frac{2}{3}\frac{b\sigma}{kT} i^{-\frac{1}{3}}\right) \\ &= \frac{R_1^+}{R_n^-} \exp\left[(n-2)\frac{L\Delta T}{kT_e T} - \frac{2}{3}\frac{b\sigma}{kT}\sum_{i-2}^{n-1} i^{-\frac{1}{3}}\right]\end{aligned} \tag{15.31}$$

Replacing the sum with an integral, and for large n, we have approximately:

$$\frac{N_n}{N} \approx \exp\left(\frac{nL\Delta T}{kT_e T} - \frac{b\sigma n^{2/3}}{kT}\right) = \exp\left(-\frac{\Delta G_n}{kT}\right) \tag{15.32}$$

as Volmer assumed.

Figure 15.9a is a plot of the free energy, ΔG_n, as in Figure 15.2, but with the abscissa n, the number of atoms in the cluster, rather than the cluster radius, r. In this plot, the volume free energy is linear in the number of atoms in the cluster, and the surface term increases as $n^{2/3}$. The maxima in the total free energy curves are at the critical cluster size.

Figure 15.9b is the corresponding plot of the logarithm of the number of clusters of size N_n as given by Equation 15.32, plotted against cluster size, n. It is a mirrored version of Figure 15.9a. The minima in the cluster distributions correspond to the maxima in the free energy.

If the vertical axis on this plot were linear in N_n, then the hatched area under each curve would be the total number of atoms in the sample. The horizontal lines at the bottom of the hatched areas are drawn where $N_n = 1$, which corresponds to the largest cluster which is likely to be present among the N atoms in the sample. As the undercooling is increased, the size of the largest cluster which is likely to be present increases but the size of the largest cluster does not increase as much as the size of the critical cluster decreases. The size of the critical cluster decreases from infinity at the melting point to a few hundred atoms at about 80% of the melting point. Nucleation takes place when the critical cluster size becomes small enough so that one is likely to spontaneously exist in the liquid.

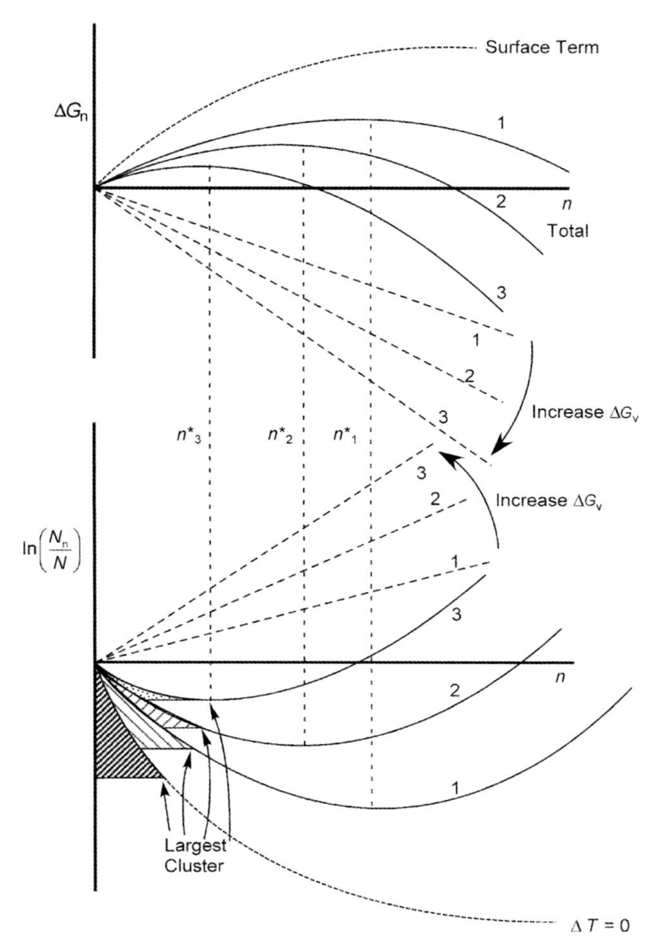

Figure 15.9 (a) Free energy to form a cluster containing n atoms, sketched for three different undercoolings. (b) The number of clusters containing n atoms, sketched for three undercoolings.

15.1.5
Nucleation Rate

Since a nucleus of critical size is equally likely to grow or shrink, a simple expression for the rate of nucleation, I, is obtained by multiplying the number of critical nuclei by half the rate at which atoms join a critical nucleus:

$$I = \frac{1}{2} R_{n^*}^{+} N_{n^*} \qquad (15.33)$$

The temperature dependence of the nucleation rate is dominated by the density of critical nuclei, and so it is similar to the curve in Figure 15.3.

There are more complicated versions than this which estimate the net rate at which the critical size clusters grow, notably one by Zeldovich, which is expressed as a correction to the above equation. However, the importance of this and other corrections to the simple expression above for the nucleation rate are overwhelmed by the sensitivity of the nucleation rate to the value of the surface tension. The logarithm of the nucleation rate depends on the surface tension cubed. For example, in an experiment where one critical nucleus in 10^{22} atoms was observed, the surface tension can be calculated from:

$$\ln 10^{-22} = -50 = \frac{16\pi\sigma^3}{3\,(L\Delta T/T_{\mathrm{M}})^2 kT} \tag{15.34}$$

But, if instead, only one critical nucleus in 10^{16} atoms was observed, the calculated surface tension would be 10% smaller. Turning this around, if the surface tension is known to an accuracy of $\pm 10\%$, then there is an uncertainty of six orders of magnitude in the nucleation rate. So if the Zeldovich factor changes the nucleation rate by a factor of 10, and it is left out of the calculation, that corresponds to a change in the surface tension of about 3%. That is within the typical experimental error for direct measurements of the surface tension. However, in principle, the correction should be included.

15.1.6 Limitations of the Becker–Döring Analysis

There is a statistical probability for the formation of each of the clusters in Figure 15.8. In reality, any possible shape of cluster can form, with some of the shapes being a lot more probable than others. The probability of a cluster being present at equilibrium depends on the free energy to form it in the matrix. There are a variety of configurations and corresponding free energies for various clusters containing n atoms. The average free energy to form a cluster of size n at equilibrium is a statistical average over all these clusters: (the free energy of the clusters of each possible shape) $\times$ (the probability of occurrence of each). Assuming that all the clusters are spheres is clearly a gross over-simplification. Attempting to improve the expression for the nucleation rate by refining the expression for the free energy to form a spherical cluster is silly.

The coupled equations above for the rate of change of the number of clusters of size n can be used, in principle, to assess how quickly the cluster distribution can respond, for example, to a change in temperature. But the kinetics of the nucleation process depend on the size and shape of the clusters which are present and on the details of how atoms can join and leave each of those clusters, and on the reaction paths linking the various clusters. There are many reaction paths joining clusters of various configurations. Many of these are relatively equivalent, and some are unimportant. So it is possible to think of an effective reaction path, and to approximate the response of the cluster distribution to a change in temperature by assuming that the most probable reaction path is that described by a sequence of spheres of increasing size. This is the spirit in which the Becker–Doring analysis should be viewed.

For nucleation of a solid in a liquid, the clusters involved in the nucleation process are relatively small, and the mobility of the atoms in the liquid is relatively large, so it is reasonable to assume that the cluster distribution reaches equilibrium very quickly. In a glass, however, the rate at which the atoms can rearrange is slow, so that there can be a large time lag for the distribution to change to the equilibrium distribution, and also a time lag associated with clusters which have become super-critical growing to a size where they can be observed. The same is true for many solid state transformations. Liquid droplet formation in a gas will depend on the density of the gas and how rapidly the appropriate atoms or molecules can diffuse.

It is surprising that the simple nucleation theory of Volmer, and the derivation of the cluster distribution provided by Becker and Döring work so well.

15.1.7 Assumptions in the Classical Nucleation Theory

The classical nucleation theory which is outlined above assumes that the critical nucleus is a sphere, but this is not a major deficiency, since the important factor is how the total free energy to form an average critical nucleus depends on its size, whatever its shape. The analysis leaves out contributions due to other sources, such as stress, to the total free energy of the cluster. These can also be added more or less readily if they are relevant.

The analysis is not applicable if the cluster distribution is not at equilibrium, for example if the temperature has changed too rapidly. The rate equations for non-equilibrium conditions are a large set of coupled differential equations, which are difficult to solve.

The Becker–Doring analysis assumes that most of the atoms are present as single atoms, so that the clusters grow and shrink only by the addition or removal of single atoms. In general, an analytical description of the nucleation process cannot be developed without this assumption, although there are various approximations.

The Volmer–Weber model assumed that the clusters were isolated spheres, but the expression they wrote down for the probability of finding a cluster of critical size is valid for any shape of cluster. It is also valid even if clusters interact and coalesce, but it is more difficult to define what is meant by a critical cluster in that case.

Formation of a new phase near a critical point violates the assumption that the clusters are separate, and do not interact or merge. The assumption is also violated in most processes of vapor deposition. It is also often violated for the nucleation of new layers during crystal growth. The cluster distributions in these cases are impossible to obtain analytically, but can be found readily using Monte Carlo computer simulations. This will be discussed in more detail below.

15.1.8 Nucleation of a Precipitate Particle

Nucleation of a precipitate particle can be described by the same equations which were derived above, by inserting the appropriate value for the free energy to form a

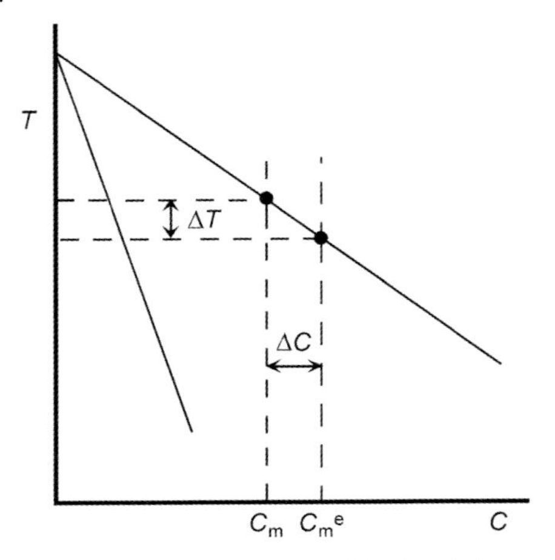

Figure 15.10 The instability of the liquid phase can be described as either a supersaturation or a supercooling.

nucleus. When the composition of the precipitate is different from the concentration of the matrix, then the chemical potential change required to form the nuclei can be expressed either as a supersaturation or as a supercooling, as illustrated in Figure 15.10.

If the concentration of the solvent component in the precipitate is small, the chemical potential change to form a cluster of atoms of the precipitating phase depends primarily on the properties of the precipitating component, and can be written as:

$$\mu_n = \sum_{i=1}^{n} \Delta\mu_i = -n\Delta\mu + b\sigma n^{2/3} \qquad (15.35)$$

where:

$$\Delta\mu = \Delta\mu_0 + kT \ln\left(\frac{C_P}{C_m}\right) \qquad (15.36)$$

C_m is the composition of the matrix and C_P is the concentration in the precipitate. The nucleation rate can be written as:

$$I = A_{n^*} R_0^+ \exp(-Q/kT) \exp\left[(n^*\Delta\mu - b\sigma n^{*2/3})/kT\right] \qquad (15.37)$$

For a precipitate where there is a significant concentration of both the solvent and the precipitating component in the precipitate, the contributions of both species to the free energy difference between the precipitate and the matrix must be included.

$\Delta\mu$ can be expressed either in terms of the supersaturation, or in terms of the undercooling.

For a precipitate nucleating in a vapor, the composition can be replaced by the partial pressure of the nucleating component. For nucleation of vacancies in a crystal, the concentration is the vacancy concentration, and so on.

Using this simple form of the nucleation theory assumes that the concentration of the precipitating component is given by the average composition or partial pressure, but there is often a depletion of the precipitating species around the nucleus. The clusters are formed by a local fluctuation in concentration, and it is reasonable to suppose that many of the atoms that were in some volume around the precipitate came together to make the nucleus. This would mean that the composition in the matrix around the nucleus is depleted. For nucleation in a vapor phase, the local depletion of atoms can result in a pressure release wave going out from the nucleus at the speed of sound. How much the matrix is depleted, and how this depletion will influence the growth of the nucleus depends on the time it takes to form a nucleus, compared to how rapidly the species can move around to change the environment in the vicinity of the nucleus.

15.2 Heterogeneous Nucleation

Most nucleation processes are not homogeneous. Nucleation occurs on particles, or at surfaces, or wherever. Turnbull's droplet experiment suggests that there are 10^7 or so nucleating particles in a cubic centimeter of material. Water can be supercooled to $-40\,^\circ$C by triple distillation using carefully cleaned containers, and by being careful not to expose the water to air at any time during or after the distillation process. Ordinary tap water can be supercooled to $-5\,^\circ$C.

15.2.1 Heterogeneous Nucleation Theory

The standard theory for heterogeneous nucleation assumes that clusters of the new phase form as droplets on a foreign substrate. The shape of a droplet on the substrate depends on the contact angle, θ, which is also know as the wetting angle. θ depends on the balance of surface tension forces at the edge of the droplet. θ will likely be small if the surface tension between the droplet and the substrate is small. For complete wetting, $\theta = 0$. When θ is small, as illustrated in Figure 15.11, the radius of curvature of the surface of the droplet is much larger that the radius of a sphere which contains the same number of atoms, and so this configuration will nucleate the second phase at a much smaller undercooling.

The illustration shows a liquid droplet on a solid surface in a vapor, but it could equally well be a solid droplet on a substrate in a liquid.

The critical condition for the growth of a cluster depends on the radius of its surface.

The volume free energy term in the expression for the free energy of the cluster depends on the number of atoms in the cluster. The surface free energy depends on the area of the vapor/substrate interface which was replaced with the liquid/substrate interface, plus the contribution from the liquid interface, which depends on the contact angle.

Heterogeneous nucleation is often observed at very small undercoolings, implying that the contact angle is small, or even that the substrate is completely wetted.

Heterogeneous nucleation usually takes place on small particles. For example, if the foreign particle is one micron in radius, and if the surface is completely wetted with a monolayer of the nucleating phase, then, from Turnbull's Rule, Equation 15.11, the critical undercooling for nucleation would be $10^{-4}T_M$, which is about 0.3 K for a T_M about 1000 K. This is the way most heterogeneous nucleating agents work. The nucleating phase adsorbs on a heterogeneous particle to form the nucleus. A plane substrate is a poor approximation for a small particle.

The heterogeneous nucleation theory provides an extra parameter, θ, to fit experimental data. For any observed nucleation temperature, the contact angle, θ can be calculated. This process often results in silly numbers for r^* and θ. This is quite different from measuring the contact angle directly. This can be done readily for liquid on a flat substrate with air or a vapor above, but it is a very difficult measurement to make for a crystal on a substrate surrounded by its melt.

There has been a lot of work done by surface scientists examining the interactions between deposits and surfaces. Some of this work will be discussed later in the context of vapor deposition. Surface scientists distinguish between physical and chemical adsorption onto surfaces. In physical adsorption the atoms or molecules are bound to the surface, but they can readily desorb. In chemical adsorption the adsorbed atoms or molecules react chemically with the surface, and so the surface chemistry is quite different from that of the adsorbing species or the substrate. If a droplet on a surface, as illustrated in Figure 15.11, is of a species which reacts with the surface atoms, then the surface between the droplet and the substrate will be covered with a layer which is the product of this reaction. As the droplet spreads across the surface, a chemical reaction occurs, forming the product layer. The energy associated with this reaction contributes to the free energy change associated with the spreading of the layer. This situation is not described by simple equations involving surface tensions.

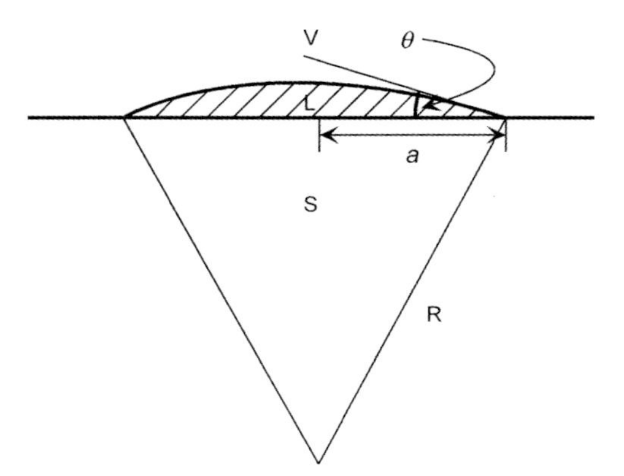

Figure 15.11 Heterogeneous nucleus.

In general, heterogeneous nucleation depends on the specific nature of the materials involved. There is no generally applicable description of this process.

15.2.2
Nucleation Lore

As mentioned above, ordinary tap water, or even water from a puddle in the street, can be readily supercooled to −5 °C. It can be maintained at that temperature for a long period of time. Ice will nucleate at about −6 °C. Very careful experiments have been performed to study the homogeneous nucleation of ice in water. By triple distillation from one still into the next, without exposure to air, water can be supercooled to about −40 °C, and maintained there for long periods of time (years). Exposing the purified liquid to air immediately returns it to a condition where it can only be supercooled to −6 °C. There is evidently a relatively powerful nucleant for ice which is common in the air.

It is commonly thought that supercooled liquids, such as a beaker of water at −5 °C, are very susceptible to vibration, but this is not so. This misconception is the result of sloppy experiments.

An experiment to measure the undercooling is often performed as illustrated in Figure 15.12, but the wall of the inner container above the liquid cools much more rapidly than the liquid. A little jiggling will move some water into contact with the cold wall. So the nucleation process is very sensitive to vibration, but the thermometer is not measuring the temperature where nucleation occurs. Keeping the level of the liquid above the level of the cooling liquid eliminates both problems.

There is an apocryphal story told about Joel Hildebrand, an outstanding physical chemist, who went to Berkeley from the University of Chicago. He wanted to study

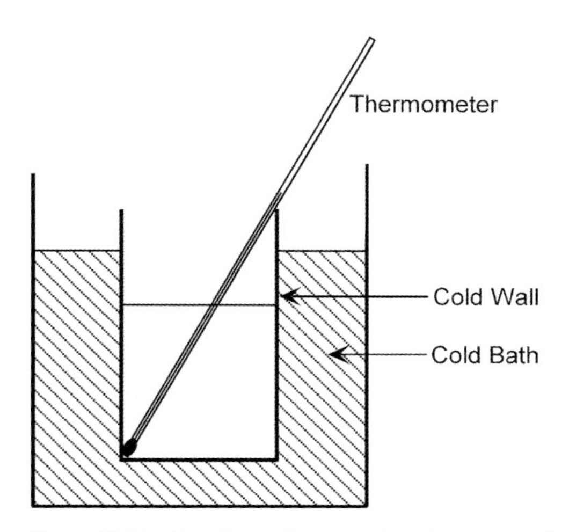

Figure 15.12 Experiment for measuring the supercooling of a liquid.

the crystallization of glycerin but, try as he might, he could not get the glycerin to crystallize at Berkley. It always supercooled into the glassy state. He hammered on it, shot bullets at it, but nothing worked. Finally, he contacted his former colleagues in Chicago, and asked them to send him some crystalline glycerin. They did, and after it arrived, it is said that they could no longer supercool liquid glycerin at Berkeley.

It was supposedly common practice for chemists in the good old days to comb their hair over their supersaturated solutions in order to induce precipitation.

It is well known that scratching the bottom of the beaker, which is filled with a supersaturated solution or a supercooled liquid, with a glass rod, will cause nucleation. This can be readily observed using Salol, which is a favorite material for demonstrations of crystal growth. It melts at 41.6 °C, and the liquid can be undercooled readily to room temperature. It can also be easily quenched to a glassy state. It has a maximum crystallization rate of 3.6 mm min^{-1} at about room temperature (see Figure 20.8).

The crystalline phase can be nucleated by scratching the bottom of the glass container with the broken end of a glass rod. This does not work if the end of the rod has been flame polished. The nucleation is not a result of the creation of a fresh glass surface by the scratching, since breaking a glass rod in the supercooled liquid does not nucleate crystals. It is probable that the scratching induces cavitation in the liquid, which results in nucleation.

15.2.3 Cavitation

Cavitation occurs when there is a negative pressure in the liquid. This can be induced by an ultrasonic sound generator, such as an ultrasonic cleaning bath, which produces standing waves in the liquid. Small cavities open up in the liquid in the regions of negative pressure, and collapse again when the pressure becomes positive. A relatively small negative pressure will create cavities, but a very high local pressure is generated when a cavity collapses. The pressure at the center of the collapse is increased by R / a, where R is the maximum diameter of the cavity, usually of the order of microns, and a is the smallest radius of the cavity when it collapses. If there is a gas component in the liquid, such as CO_2 in water, then this evaporates into the cavity and limits the collapsed diameter, a. If there are no gaseous components in the liquid, then a is of atomic dimensions, and the collapse pressure can be very high. After the collapse, which is accompanied by a compressive stress, there is a rebound which is accompanied by tensile stresses of a similar magnitude. The collapse pressure can be so high that light is emitted, an effect known as sonoluminescence. There is a sound, like a hiss, associated with the collapse of cavities in an ultrasonic field.

A standard test to see whether a liquid is cavitating is to dip a piece of aluminum foil into the liquid. Cavitation will make holes in it. Ultrasonic cleaners are used to clean jewelry, semiconductor wafers, and so on. The propellers on ships are carefully designed to eliminate cavitation, which will tear the propeller apart, over time.

Cavitation will induce nucleation in an undercooled liquid. For example, ice will nucleate in a beaker containing water which has been supercooled to $-4\,^{\circ}\mathrm{C}$ or so when the beaker is put into an ultrasonic cleaning bath so that cavitation occurs in the supercooled liquid. Nucleation can also be induced in undercooled water by rapidly separating two flat surfaces. This is believed to occur by generating a negative pressure which makes a cavity, which nucleates ice when the cavity collapses. Ice expands on freezing, and so a compressive stress should not nucleate ordinary ice, but there are more dense phases of ice which occur at high pressures. Alternatively, the nucleation could take place during the negative pressure release wave which follows the collapse of the cavity. It is not known which of these effects results in nucleation of ice.

This effect can be demonstrated dramatically with a sealed U-tube which is partially filled with de-gassed water. The water can be de-gassed by extensive boiling in a reflux condenser, which prohibits the access of air to the water. The water in the U-tube can be undercooled a few degrees by immersing it in a cooled bath. When the tube containing supercooled water is tipped so that the water runs into the end of the tube and collapses the vapor column, then ice will nucleate in the water. The pressure spike from the collapse of the column results in nucleation of ice.

15.2.4
Re-entrant Cavities

It is often found that the temperature to which a liquid can be supercooled depends on how high it was superheated before it was cooled. It is rather difficult to believe that a slowly cooled bulk liquid retains a memory of how high it was heated. This effect has been explained by postulating the presence of small volumes of the crystal which are retained in cavities in the wall of the container. When the proper relationship exists between the relevant surface tensions, a small crystal can exist in a microscopic cavity to a temperature far above the melting point. The surface tension can stabilize the crystal against melting, as illustrated in Figure 15.13.

When the liquid is supercooled, the crystal grows out into the bulk. Usually heating the liquid far enough above the melting point will eliminate this effect.

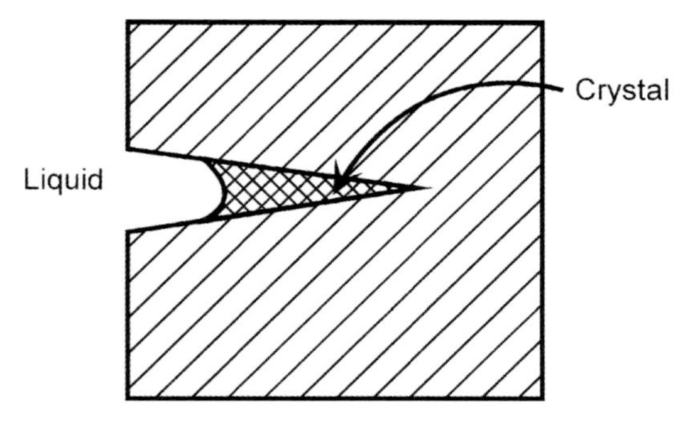

Figure 15.13 Solid retained in a re-entrant cavity.

15.2.5
Cloud Seeding

Cloud seeding to induce rainfall on parched fields is not as popular as it once was. It was sometimes done with carbon particles, which were said to work by radiative cooling. The commonest method made use of silver iodide smoke as a nucleating agent. The silver iodide smoke was generated in an airplane over the area where the farmers wanted rain. The idea was that the silver iodide would nucleate ice crystals which would then result in rain. Small particles of silver iodide will nucleate ice only below $-10\,^{\circ}\mathrm{C}$. There is no hope of nucleating ice under very dry atmospheric conditions. So the cloud seeders would wait until there was a likely day, one on which there was some probability of rain, and then do their thing to make it rain locally. If it rained, they would take the credit. But tests of this method were performed by choosing a likely day to do cloud seeding, and then randomly seeding or not seeding. On average, the rainfall turned out to be independent of the cloud seeding. Recently, claims have been made that the tests were not all that conclusive.

15.2.6
Industrial Crystallization

Many bulk chemicals are produced in crystalline form. Not only sugar and salt, but fertilizers, swimming pool chemicals, and so on, are produced by a crystallization process. The crystallization of sugar and salt is carefully controlled to produce the desired size of crystal. This is usually done in a continuous process, where the crystals are harvested continuously, and more nutrient is added to the solution to keep it supersaturated. The individual crystals continue to nucleate and grow. It seems unlikely that there is a large enough supply of heterogeneous nuclei around to keep the process going, especially when the crystallization is carried out in a closed chamber. Where do these nuclei come from? Some of the new crystals may form by fracture of existing crystals when they bump into each other or into the walls of the container. The fracture of a brittle material does not produce a simple fracture surface. For example, breaking a piece of blackboard chalk produces a shower of fine particles. Similarly, when a thin rod of ice is immersed in supercooled water and then broken, a shower of small ice crystals goes out into the liquid from the fracture. The ice crystals are initially very small, and if the liquid is not supercooled, they will melt and not be detected. However, in the supercooled water, they grow rapidly and became observable crystallites. This could be happening in industrial crystallization.

It is also well known that adding a crystal to a supercooled melt or solution to nucleate it often results in not one, but rather many growing crystals. This suggests that many small crystals are adhering to the surface of the crystal used as a seed. Often the multiple crystallization can be eliminated by washing the seed crystal first.

Salol crystals growing in a glass dish into supercooled liquid are usually observed to fracture when they reach a size of several millimeters. When the fracture occurs, many other small crystals are soon observed nearby.

Experiments have been carried out to attempt to identify the source of the nuclei in industrial crystallizers, and some of them quite clearly suggest that it is magic.

15.2.7
Grain Refiners

In metal castings, controlling the grain size of a casting is often desirable. Grain refiners are used to make castings with smaller grains. It should be simple to design a grain refiner. Just choose a material with a higher melting point which is wet by the crystal, and add it to the melt. But, unfortunately, materials which wet each other are also usually mutually soluble, and a material will not work as a grain refiner if it dissolves in the liquid.

The grain refiner must not dissolve in the liquid. And so grain refiners are usually some insoluble material which forms a compound or a surface layer in the melt which acts as the nucleus. There seem to be no simple rules to identify a good grain refiner.

In alloy castings new grains can form by partial remelting of dendrites, a process which is discussed in Chapter 28.

15.2.8
Residuals

When a droplet of water dries, it leaves behind a residue of material that was soluble in the water. This residue is likely to be hydroscopic. So, for example, when you wipe the mirror in your bathroom in the morning to clear off the water droplets so you can see your image, the pattern of wiping can often be seen in the water droplets which form on the mirror the next morning.

If there have been water droplets on a surface, and they have dried, then water droplets will form again on that surface much more readily than on a pristine surface.

A related process provides a sophisticated method for secret writing. A solution is made containing a low concentration of an appropriate chemical. When a droplet of the solution dries on a suitable substrate, it leaves behind a sub-microscopic residue. This residue can act as a nucleating agent. For example, large and clearly visible precipitates of a dye such as indigo can be grown by immersing such a substrate in an appropriate supersaturated solution. Unfortunately, this method can be compromised. Some of the residue can be transferred onto a piece of glass without destroying the pattern on the original, and then the pattern on the glass can be developed by immersing it in a supersaturated solution to reveal the pattern. So the receivers of the message are unaware that it has been intercepted.

15.3
Johnson–Mehl–Avrami Equation

For many solid state transformations, including glass crystallization, the rate of transformation is sufficiently slow that it can be followed in time. The progress of these transformations can be described by the Johnson–Mehl–Avrami equations.

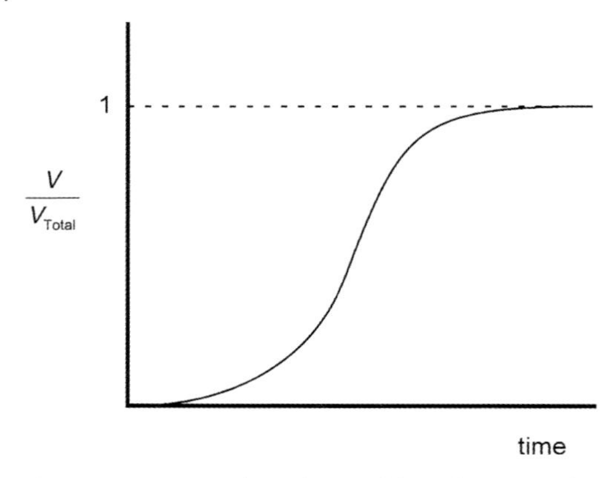

Figure 15.14 Time dependence of the volume transformed.

15.3.1 Johnson–Mehl Equation

For many transformations, when the volume transformed is plotted as a function of time, a plot such as shown in Figure 15.14 is obtained.

The volume transformed can be measured as the latent heat released using a calorimeter, or using differential thermal analysis (DTA). It can be measured as a change in volume with a dilatometer; or the formation of the new phase can be followed with X-rays. The spatial distribution of the transformed volume is irrelevant, as in Figure 15.15. The process involves the nucleation of the new phase, followed by the growth of the new phase. The growth proceeds until the whole sample has transformed. The nucleation and growth processes must be sufficiently slow that their progress can be monitored in time.

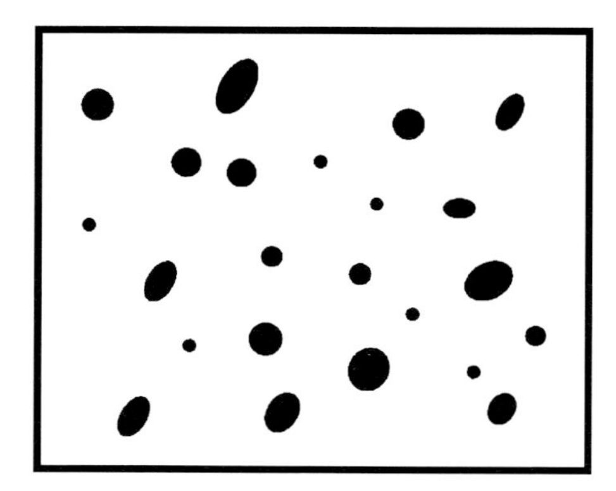

Figure 15.15 The transformation proceeds by nucleation, followed by growth of the nuclei.

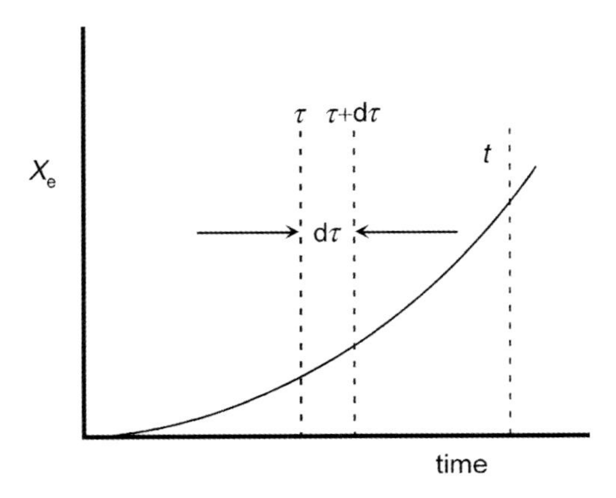

Figure 15.16 X_e as a function of time.

The Johnson–Mehl [6] equation describes the shape of the curve in Figure 15.14 for the case where the nucleation rate is constant in time, and the radial growth rate of the transformed particles is constant.

To derive the Johnson–Mehl equation we define X_e as the fraction of the volume which has transformed after a time t, as illustrated in Figure 15.16.

> dX_e is then the volume fraction which has transformed after time t due to particles which nucleated between times τ and $\tau + d\tau$.
> $dX_e =$ (number of particles which nucleated between τ and $\tau + d\tau$) $\times$ (volume, at time t, of a particle which nucleated between τ and $\tau + d\tau$).

Assuming that the nucleation rate, I, is constant in time, and that the radius of a particle increases linearly with time at a constant velocity v, so that its radius is $R = (t - \tau)\, v$,

$$dX_e = \frac{4}{3}\pi\,[v(t-\tau)]^3 I\, d\tau \tag{15.38}$$

so that:

$$X_e = \frac{4}{3}\pi v^3 I \int_0^t (t-\tau)^3 d\tau \tag{15.39}$$

Now,

$$\int_0^t (t-\tau)^3\, d\tau = -\frac{1}{4}(t-\tau)^4\Big|_0^t = \frac{1}{4}t^4 \tag{15.40}$$

resulting in:

$$X_e = \frac{\pi}{3} v^3 I\, t^4 \tag{15.41}$$

In this formulation, the volume fraction transformed becomes infinite with time. The growth of the new phase can only proceed into the untransformed volume. In order to normalize the volume transformed, we can write the transformation rate as the product of the untransformed volume fraction times the transformation rate:

$$\mathrm{d}X = (1-X)\,\mathrm{d}X_{\mathrm{e}} \tag{15.42}$$

So that:

$$X = 1-\exp\left(-X_{\mathrm{e}}\right) \tag{15.43}$$

Inserting the value of X_e from Equation 15.41 gives the volume fraction transformed as a function of time:

$$X = 1-\exp\left(-\frac{\pi}{3}v^3\, It^4\right) \tag{15.44}$$

The volume fraction transformed increases as t^4 for short times and saturates at $X=1$ at long times, as illustrated schematically in Figure 15.14.

15.3.2
Johnson–Mehl–Avrami

Avrami [7] modified the Johnson–Mehl formulation for the case where there are initially a fixed number, N_0, of nucleation sites per unit volume. These are used up as time proceeds, so that the nucleation rate decreases with time:

$$N = N_0 \exp\left(-\upsilon t\right) \tag{15.45}$$

where υ is a time constant. The nucleation rate, I, is:

$$I = -\frac{\mathrm{d}N}{\mathrm{d}t} = N_0\upsilon \exp\left(-\upsilon t\right) = N\upsilon \tag{15.46}$$

Putting the nucleation rate at time τ into the expression for the rate of transformation:

$$-\mathrm{d}(1-X) = \frac{4}{3}\pi v^3 \int_0^t I(\tau)\,(t-\tau)^3\,\mathrm{d}\tau \tag{15.47}$$

This can be integrated by parts to give:

$$X = 1-\exp\left[-\frac{8\pi N_0 v^3}{\upsilon^3}\left\{\exp\left(-\upsilon t\right)-1+\upsilon t-\frac{\upsilon^2 t^2}{2}+\frac{\upsilon^3 t^3}{6}\right\}\right] \tag{15.48}$$

For small υ, the number of nucleation events is given approximately by:

$$N \approx N_0\,(1-\upsilon t) \tag{15.49}$$

so the nucleation rate is:

$$I = -\frac{dN}{dt} = N_0 \upsilon \tag{15.50}$$

which is constant, as was assumed by Johnson and Mehl.

For small υ, the exponential in Equation 15.48 can be expanded as:

$$\exp(-\upsilon t) = 1-\upsilon t + \frac{\upsilon^2 t^2}{2} - \frac{\upsilon^3 t^3}{6} + \frac{\upsilon^4 t^4}{24} - \cdots \tag{15.51}$$

The first four terms in this expansion cancel out other terms in Equation 15.48, so the first remaining term in the brackets is in t^4:

$$\begin{aligned} X &= 1-\exp\left[-\frac{8\pi N_0 v^3}{\upsilon^3}\left(\frac{\upsilon^4 t^4}{24}\right)\right] \\ &= 1-\exp\left(-\frac{\pi}{3} I v^3 t^4\right) \end{aligned} \tag{15.52}$$

which is the Johnson–Mehl result, Equation 15.44.

For large υ, the number of nucleation sites decreases rapidly, so the t^3 term in the brackets dominates:

$$X = 1-\exp\left(-\frac{4}{3}\pi N_0 \upsilon^3 t^3\right) \tag{15.53}$$

And for intermediate values of υ, the exponent will be between 3 and 4.

In general, an expression of the form:

$$X = 1-\exp\left[-(kt)^n\right] \tag{15.54}$$

can be fitted to experimental data. n is known as the Avrami exponent. The time constant k depends on both the nucleation rate and the growth rate.

Analyses have been carried out for polymorphic phase changes which take place with various nucleation rates, geometries and growth rates. The corresponding Avrami exponents are presented below.

	Avrami exponent
Constant nucleation rate	4
Decreasing nucleation rate	3–4
Nucleation on grain boundaries	1
Nucleation on dislocations	2/3
Constant nucleation rate with diffusion controlled growth	2.5

Some researchers have obtained an Avrami exponent by fitting their data, and then used the value to identify how nucleation occurred in their samples. Better information about where and how nucleation occurred can be obtained by examining the sample under a microscope.

References

1 Volmer, M. (1934) *Z. Phys. Chem. Leipzig*, **170**, 273;(1939) *Kinetik der Phasenbildung*, Theodor Steinkopff, Dresden.
2 Katz, J.L. (1992) *Pure Appl. Chem.*, **64**, 1661.
3 Turnbull, D., (1950) *Science*, **112**, 448; (1956) *Solid State Phys.*, **3**, 225.
4 Becker, R., and Doering, W. (1935) *Ann. Physik.*, **24**, 719.
5 Jackson, K.A. (1965) *Indust. Eng. Chem.*, **57**, 28.
6 Johnson, W.A., and Mehl, R. (1939) *Trans. AIMMPE*, **135**, 410.
7 Avrami, M. (1941) *J. Chem. Phys.*, **9**, 177.

Problems

15.1. Using $\sigma = La/2$, $L\Omega = kT_M$, $n = (r/a)^3$, and the atomic volume, $\Omega = (4/3)\pi a^3$

(a) Plot log (N_n/N) vs. n for $\Delta T/T_M = 0, 0.1, 0.2, 0.3$.

(b) For $N = 10^{22}$ atoms, plot the size of cluster, n, for which $N_n = 1$ as a function of $\Delta T/T_M$. Plot the critical nucleus size, n^*, on the same plot.

(c) Plot the nucleation rate: $I/\nu^+ = \exp(-\Delta F_{n^*}/kT)$ versus the undercooling, $\Delta T/T_M$, in the temperature range where I/ν^+ is about 10^{-23}, that is where one nucleus forms in a cubic centimeter of material. ν^+ is the rate at which atoms join the critical nucleus.

15.2. A material (which is a lot like lead) melts at 327 °C. Its latent heat of fusion is 6 cal gm^{-1}, its density is 10 gm cm^{-3}, its molecular weight is 200, and its atomic diameter is 3 Å. What is the entropy of melting in units of R (= 1.98 cal mol^{-1} °C^{-1})? Assuming that the surface free energy is given by Turnbull's rule ($\sigma^a aL/2$), what is σ in erg cm^{-2}? Sketch the free energy as a function of radius for a spherical cluster of the solid in the liquid at 227 °C. Sketch the nucleation rate as a function of temperature, using $I = 5 \times 10^{22}$ $\exp(-\Delta F^*/kT)$, where I is the nucleation rate, ΔF^* is the free energy of a cluster of critical size, and $k = 1.381 \times 10^{-16}$ erg °C^{-1}.

16
Surface Layers

16.1
Langmuir Adsorption

The Langmuir adsorption isotherm [1] describes the equilibrium density of adsorbed atoms on a surface. It is assumed in the model that the adsorbed atoms interact only with the substrate, not with other adsorbed atoms. The rate at which atoms leave the surface, k^-, is the same for each adsorbed atom, and does not depend on the presence of neighbors. For a surface which has N sites per unit area, N_1 of which are occupied, the desorption flux is:

$$J^- = k^- N_1 \tag{16.1}$$

Assuming that arriving atoms can land only on empty sites, the rate of arrival of atoms onto the surface can be written as:

$$J^+ = k^+ P(N - N_1) \tag{16.2}$$

where P is the vapor pressure in the gas phase of the arriving atoms. At equilibrium the arrival and desorption fluxes are equal, and so

$$k^- \theta = k^+ P(1 - \theta) \tag{16.3}$$

where θ is the fraction of the surface sites which are occupied,

$$\theta = \frac{N_1}{N} \tag{16.4}$$

Equation 16.3 can be written:

$$\frac{\theta}{1-\theta} = \frac{k^+}{k^-} P = KP \tag{16.5}$$

where $K = k^+/k^-$

The surface coverage, θ, is given by:

$$\theta = \frac{KP}{1 + KP} \tag{16.6}$$

Equation 16.6 is illustrated in Figure 16.1.

Kinetic Processes: Crystal Growth, Diffusion, and Phase Transitions in Materials. Kenneth A. Jackson

ISBN: 978-3-527-32736-2

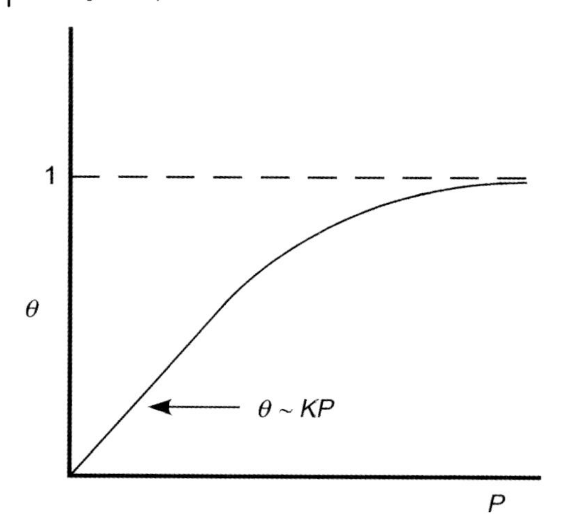

Figure 16.1 Langmuir adsorption isotherm.

The surface coverage increases linearly with pressure for low pressures, and then saturates for large KP at $\theta = 1$.

The rate of desorption of atoms from a surface usually has a temperature dependence given by:

$$k^- = \nu \exp\left(-\frac{Q_{ad}}{kT}\right) \tag{16.7}$$

where ν is a frequency and Q_{ad} is the adsorption energy. This is similar to the desorption rate which establishes the equilibrium vapor pressure as given by Equation 20.14. Using Equation 16.7, the surface coverage can be written as:

$$\theta = \frac{\dfrac{k^+ P}{\nu}\exp\left(\dfrac{Q_{ad}}{kT}\right)}{1 + \dfrac{k^+ P}{\nu}\exp\left(\dfrac{Q_{ad}}{kT}\right)} \tag{16.8}$$

So the surface coverage increases with pressure in the gas phase at constant temperature, because the rate of arrival of atoms at the surface increases with pressure, and the surface coverage decreases with increasing temperature at constant pressure, because the desorption rate increases as the temperature increases.

16.2 CVD Growth by a Surface Decomposition Reaction

The deposition of silicon onto a substrate, for example, the epitaxial growth of silicon on silicon, takes place by the decomposition of silane which has adsorbed onto the surface.

The deposition rate, R, depends on the surface coverage, θ, and on the surface decomposition rate:

$$R = k_{\mathrm{rxn}}\theta \tag{16.9}$$

where k_{rxn} is the decomposition rate, which usually has a temperature dependence given by:

$$k_{\mathrm{rxn}} = k^0_{\mathrm{rxn}} \exp\left(\frac{-Q_{\mathrm{rxn}}}{kT}\right) \tag{16.10}$$

The deposition rate is given by combining Equations 16.6, 16.8, 16.9, and 16.10:

$$R = \frac{k^0_{\mathrm{rxn}} K\, P \exp\left(\frac{Q_{\mathrm{ad}} - Q_{\mathrm{rxn}}}{kT}\right)}{1 + K\, P \exp\left(\frac{Q_{\mathrm{ad}}}{kT}\right)} \tag{16.11}$$

which for small surface coverage reduces to:

$$R = k^0_{\mathrm{rxn}} K\, P \exp\left(\frac{Q_{\mathrm{ad}} - Q_{\mathrm{rxn}}}{kT}\right) \tag{16.12}$$

16.3 Langmuir–Hinschelwood Reaction

When several non-interacting species are adsorbed on the surface, the coverage of any one species, i, will be given by:

$$\theta_i = \frac{K_i P_i}{1 + \sum_j K_j P_j} \tag{16.13}$$

where it is assumed that each species independently occupies surface sites.

When two of the adsorbed species react to form a deposit, as is the case for example for the MOCVD deposition of GaAs from tri-methyl gallium $(CH_3)_3Ga$ and arsene (AsH_3), the deposition rate can be written as:

$$R = k_{\mathrm{rxn}}\theta_1\theta_2 \tag{16.14}$$

where k_{rxn} is the rate at which the tri-methyl gallium and arsene react to form GaAs on the substrate. θ_1 and θ_2 are the fractional coverages of the two adsorbed species. The deposition rate can be written:

$$R = \frac{k_{\mathrm{rxn}} K_1 K_2 P_1 P_2}{(1 + K_1 P_1 + K_2 P_2)^2} \tag{16.15}$$

This is known as a Langmuir–Hinshelwood reaction.

16.4
Surface Nucleation

The Knudsen expression for deposition from a vapor, Equation 20.18, is:

$$J = J^{+} - J^{-} = \frac{(P - P_e)}{\sqrt{2\pi mkT}} \tag{16.16}$$

Increasing the pressure above P_e increases the rate of deposition; decreasing the pressure below P_e increases the rate of evaporation. The two fluxes are illustrated in Figure 16.2. At equilibrium they are equal.

This works well for a liquid surface, for example, evaporation from the surface of a liquid. It assumes reversible deposition and evaporation below and above $P = P_e$. Growth or evaporation are assumed to be independent of the configuration of the surface. This will be valid when there is no nucleation barrier to the formation of new layers.

The Langmuir model, Equation 16.6, on the other hand, does not suggest continuous growth with increased pressure. Increasing pressure merely increases the surface coverage. Increasing the adpopulation increases the flux of atoms from the surface until it matches the incident flux. Thereafter, there is no further net deposition.

The Knudsen and Langmuir models clearly predict different behavior. One predicts continuous growth in a supersaturated vapor, and the other does not.

The Langmuir model assumed that there is no interaction between the depositing atoms. This is valid for the deposition of a rare gas on a substrate, for example, but it is not true in general. It is especially not true when the material being deposited is the same material as the substrate, where there is always an interaction between the surface atoms, even if there is a barrier to the formation of new layers. If the adpopulation is very small, it may be difficult to nucleate new layers. The nucleation of new layers is discussed in more detail in the next section.

16.4.1
Nucleation on a Surface During Vapor Deposition

A model for nucleation on a surface can be made by combining the Langmuir model for surface coverage with the Becker–Doring model for nucleation [2, 3]. The model

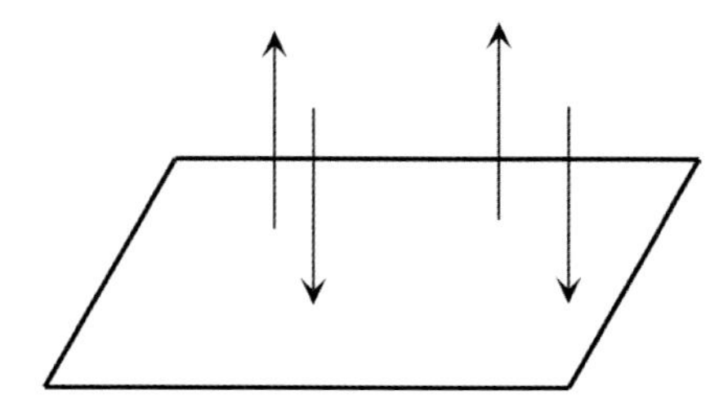

Figure 16.2 Vapor deposition.

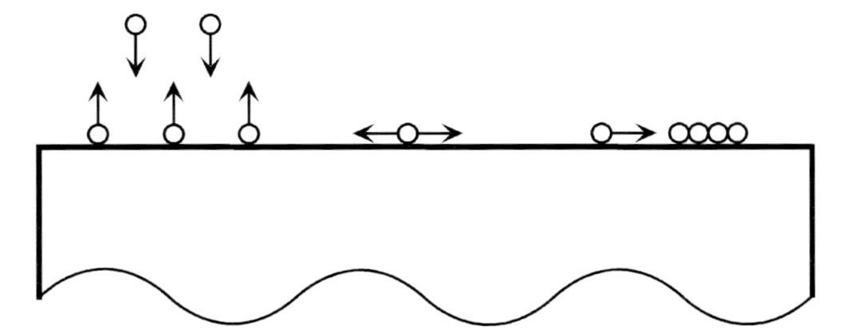

Figure 16.3 Atoms arrive from and leave to the vapor phase, and they move around on the surface to form clusters.

assumes that atoms arrive at the surface from the vapor phase, and leave the surface to go into the vapor phase, and they move around on the surface to form clusters, as illustrated in Figure 16.3. We will assume that most of the atoms on the surface are present as single atoms, which implies that the total density of adatoms is small.

The adpopulation density given by the Langmuir model as in Equation 16.8 is:

$$\theta = \frac{N_1}{N} = \frac{J^+}{N\upsilon} \exp(E_{\mathrm{ad}}/kT) \tag{16.17}$$

where N_1 is the number of adatoms per unit area, and N is the total number of surface sites per unit area. This is the approximate form of Equation 16.8 for small adatom density, $N_1 \ll N$. The incident flux J^+ has been retained explicitly, as in Equation 16.2, since the incident flux might be derived from a vapor source rather than from the local vapor pressure.

16.4.2 Cluster Formation

The rate at which atoms join a single layer cluster of n atoms on the surface is:

$$R_n^+ = N_1 l_n a\upsilon \exp\left(-\frac{Q_{\mathrm{D}}}{kT}\right) \tag{16.18}$$

where l_n is the capture length of the cluster, which is approximately the length of its periphery, and a is the atomic diameter, as illustrated in Figure 16.4. ν is a frequency, and Q_{D} is the activation energy for surface diffusion. An atom must be within a distance a from the cluster in order to join it in one jump. $N_1 l_n a$ is the number of adsorbed atoms within a distance a from the cluster.

The energy of a cluster of atoms on the surface can be written as:

$$E_n = -n\Delta E + b\sigma n^{1/2} \tag{16.19}$$

where ΔE is the energy change when an atom is removed from an infinite size cluster.

Atoms join and leave a cluster along a reaction path which has an energy profile as illustrated in Figure 16.5. The rate at which atoms leave a cluster of n atoms depends

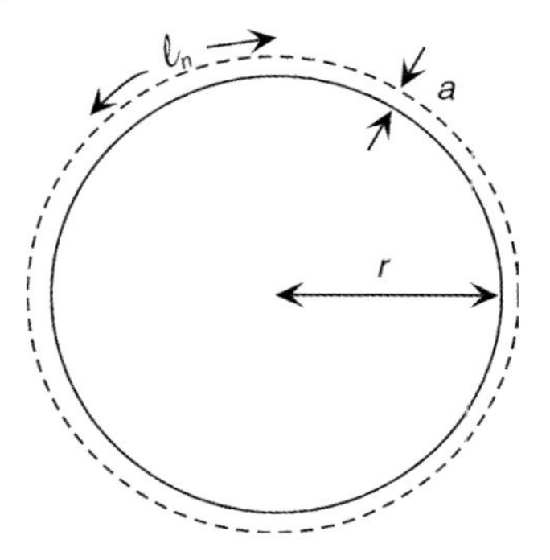

Figure 16.4 Cluster of atoms on a surface.

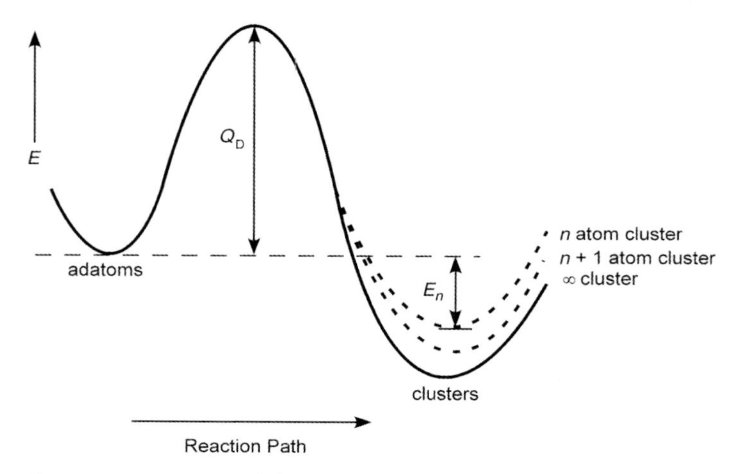

Figure 16.5 Reaction path for atoms joining a cluster.

on the height of the energy barrier along this path:

$$R_n^- = N\, l_n a \upsilon \exp\left(-\frac{Q_D + \Delta E_n}{kT}\right) \tag{16.20}$$

As before, the equilibrium distribution of clusters is given by:

$$N_n R_n^- = N_{n-1} R_{n-1}^+ \tag{16.21}$$

In this case,

$$\frac{R_n^+}{R_n^-} = \theta \exp\left(\frac{\Delta E_n}{kT}\right) \tag{16.22}$$

so that:

$$\frac{N_n}{N} \approx \theta^{n-1} \exp\left(\sum_{i=2}^{n} \frac{\Delta E_i}{kT}\right) = \theta^{n-1} \exp\left(-\frac{E_n}{kT}\right) \tag{16.23}$$

The nucleation rate is:

$$I = \frac{1}{2} R_{n^*}^{+} \frac{N_{n^*}}{N} = \frac{1}{2} R_{n^*}^{+} \left(\frac{J^+}{N\upsilon}\right)^{n^*} \exp\left[\frac{n^* E_{ad} + n^* \Delta E - b\sigma n^{*\frac{1}{2}}}{kT}\right] \tag{16.24}$$

The nucleation rate depends on the surface coverage, which in turn depends on the arrival of atoms at the surface, the adsorption energy, and the surface temperature. The derivation of this equation assumes that the atoms in the vapor phase are in equilibrium with the adatoms on the surface, and that there is an equilibrium distribution of clusters among the adatoms. The whole system is in equilibrium until nucleation occurs. After nucleation the growth of the clusters proceeds rapidly.

This derivation assumes that the density of adatoms on the surface is small. This same assumption is made to describe the nucleation of vapor droplets in a low density vapor, or the nucleation of a solid in a supercooled liquid, where it is usually valid. However, this assumption is often not valid during vapor deposition, where there can be a very large density of adatoms. In addition, if the substrate is cold enough that there is little evaporation of the impinging atoms, then there will not even be a steady state adpopulation. It is common to evaporate metals with melting points of about 1000 °C onto a substrate at room temperature, which is too cold for any significant evaporation to occur. The adatom density will increase with time. Indeed, the critical nucleus size can be as small as a few atoms under these conditions. But even without that, as the surface coverage increases, more and more of the atoms will have to occupy sites next to other atoms, so that the clusters will increase in size, and then they will begin to merge, so this analysis will not be relevant.

16.4.3 Rate Equations

The early stages of deposition can be described with time-dependent equations for the rate of build-up of the cluster distributions [4]. We will assume that most of the atoms are present as single atoms, so that the clusters grow and shrink only by the addition or subtraction of single atoms. The rate at which the number of single atoms on the surface increases is:

$$\frac{dN_1}{dt} = J^+ - N_1 \upsilon \exp\left(-\frac{E_{ad}}{kT}\right) - 2N_1 R_1^+ + 2N_2 R_2^- - \sum_{n=2}^{\infty} (N_n R_n^+ - N_{n+1} R_{n+1}^-) \tag{16.25}$$

and the rate of growth of a cluster of n atoms is:

$$\frac{dN_n}{dt} = N_{n-1}R_{n-1}^{+} + N_{n+1}R_{n+1}^{-} - N_nR_n^{+} - N_nR_n^{-} \tag{16.26}$$

The early stages of deposition can be modeled using these simultaneous differential equations which can be solved numerically assuming that there are initially no adsorbed atoms on the surface. The development of the cluster distribution can be followed with time as the adatom density inceases. The equations are only valid while the adatom density is small and the clusters are well separated so that the assumption that the clusters grow only by the addition of single atoms is valid. These rate equations do not include the coalescence of clusters.

The calculations require a knowledge of the adsorption energy, E_{ad}, the activation energy for surface diffusion, Q_D, and the free energy to form the clusters of various sizes, E_n.

These equations predict more or less correctly the build-up of the total adpopulation. They exhibit a change in the rate of increase of monomers when dimer formation becomes significant. But these equations are based on the assumption that there is a single configuration for a cluster containing n atoms, and on the assumption that clusters grow only by interactions with single atoms. They do not predict the correct cluster distributions.

For surface deposition, the cluster distributions can be measured with scanning tunneling microscopy (STM). Comparison of calculated results with these experimental data is a much more stringent test than is possible with nucleation in a liquid or a vapor, where the cluster distributions are unknown, and the only experimental datum is the nucleation temperature. Even more complex sets of rate equations than Equations 16.25 and 16.26 have been unable to produce cluster distributions similar to the experimental results.

Monte Carlo computer simulations, which we will discuss in more detail below, do give the experimental cluster distributions.

16.5 Thin Films

16.5.1 Epitaxy

Epitaxial deposition is the term used to describe the case where the lattice structure of the deposit is coherent with the lattice structure of the substrate. The substrate is usually a single crystal. The simplest version is when the deposit and the substrate are the same material. This is known as homoepitaxy. If the deposit is a different material than the substrate, it is called heteroepitaxy. Both of these are widely practised.

16.5.1.1 Homoepitaxy

The most widely practised form of homoepitaxy is silicon on silicon. The single crystals of silicon produced by Czochralski growth contain about 10^{18} oxygen atoms per cm^3, are boron doped, and have a resistivity of about 10 Ω cm. A much higher resistivity layer of silicon can be grown epitaxially on such a wafer. The doping level and type can also be changed.

Silicon is deposited by a chemical vapor deposition (CVD) process, using silane, SiH_4, or a chloro-silane, $SiH_{4-x}Cl_x$.

$$SiH_4 \rightleftharpoons Si + 2H_2\nearrow \qquad (16.27)$$

The substrate wafer is heated in a vacuum chamber, and when silane is introduced into the chamber, some of it deposits on the hot substrate, where the silane molecules break up, depositing the silicon, and the hydrogen and chlorine go off into the gas phase.

Silicon can also be deposited epitaxially on silicon by molecular beam epitaxy (MBE) in an ultrahigh vacuum (UHV) system.

In order for the deposit to be epitaxial, the surface of the wafer must be clean. The wafer is cleaned before it is put into the deposition chamber. In CVD, the hot wafer is cleaned with HCl vapor which etches the surface. In MBE, the surface is usually cleaned with an ion beam.

16.5.1.2 **Heteroepitaxy**

Heteroepitaxy refers to the growth of a single crystal layer of one material on a single crystal substrate of a different material. In some cases, the layers are components of a device structure.

Silicon on sapphire (SOS) is an example of heteroepitaxy. Silicon can be deposited on a sapphire wafer by the process described above. The epitaxial layer is far from perfect, but it is good enough to make CMOS devices. The sapphire (Al_2O_3) substrate is more rugged than a silicon wafer. Since it is not subject to electromagnetic impulse (EMP), the military likes to use devices made with this material.

The ubiquitous gallium arsenide lasers are made by depositing layers of various compositions of GaAlAs onto GaAs substrates. The lattice parameter of GaAs does not change when aluminum is substituted for gallium.

GaInAsP lasers are used for telecommunications because they emit at a wavelength of 1.55 μm, which is where silica fibers are most transparent. These lasers are grown on InP substrates, and the composition of the deposit is adjusted so that there is lattice match.

The new gallium nitride blue LEDs and lasers are another example. Gallium nitride is, as yet, impossible to grow as a single crystal. The GaN is usually deposited on a sapphire or other substrate.

A process for growing GaAs layers epitaxially on silicon was recently announced. Researchers have tried to do this for many years, but they have been unsuccessful because the lattice mismatch is large. Silicon wafers are much less expensive than GaAs wafers, and there would be a big cost advantage to using silicon substrates for GaAs devices. It was reported that strontium titanate could act as a transition layer between silicon and GaAs to produce epitaxial growth. Unfortunately, the GaAs grown by this process proved not to be of device quality.

Good epitaxy usually requires a small lattice mismatch, of less than a few percent, between the deposit and the substrate. This is a fairly easy criterion to apply, because tables of lattice parameters are available.

Good epitaxy also requires the right chemistry, although it is difficult to say exactly what that means.

If the chemistries are similar and the lattice constants match, then good epitaxy is likely. For example, for GaAlAs on GaAs, both the lattices and the chemistry match. The same can be said for the growth of lattice matched garnets on other garnets, as is done to make optical isolators.

Sometimes good epitaxy occurs without a good lattice match and sometimes it does not work with a good lattice match. An example of the latter is silicon on CeO_2.

A mysterious case of good epitaxy is gold on rock salt (NaCl). Gold will deposit epitaxially on a cleaved (100) face of rock salt. It is not perfect epitaxy, but the lattices align within a few degrees, but if the rock salt is cleaved in a vacuum, the deposit is not epitaxial. Exposure of the cleaved surface to air, so that a layer of water adsorbs on the surface, is necessary. The lattice match is not good in this case, and the chemistry is not at all obvious.

Increasingly, experimenters are introducing intermediate layers, or complex processing steps designed to modify surface layers, in order to promote good hetero-epitaxy. The intermediate layer, or the processing steps, which will work are usually not obvious, but must be found empirically.

16.5.2 Deposited Surface Layers

A deposited layer can be:

crystal, amorphous, or liquid
islands or layers
monolayer or multilayer
aligned with the substrate or not
small islands which are mobile or stationary

Just about anything that you can imagine will happen in some system or another.

16.5.2.1 Classes of Deposited Layers

Surface scientists have defined three classes of deposited layers [5, 6], each named for pioneers in the field.

16.5.2.1.1 Volmer–Weber Volmer–Weber growth is island growth, as illustrated in Figure 16.6. The islands are often three-dimensional rather than monolayers. The

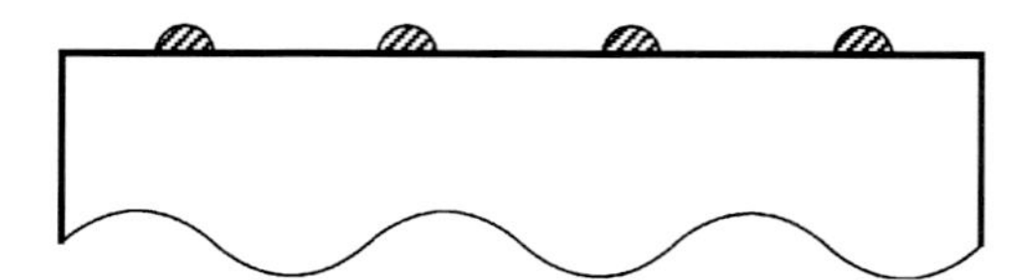

Figure 16.6 Volmer–Weber growth.

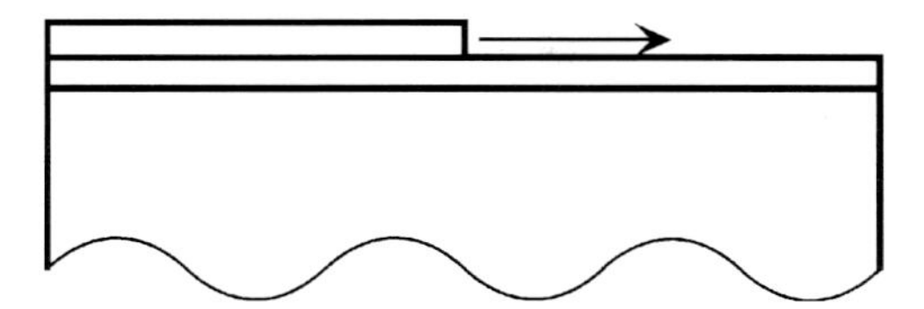

Figure 16.7 Frank–Van der Merve growth.

deposit nucleates as independent islands which may or may not be aligned with the substrate. Atoms can migrate across the surface to the islands. The islands grow, and then can merge or coalesce when they impinge on each other.

16.5.2.1.2 **Frank–Van der Merve** Frank–Van der Merve growth is layer by layer growth, illustrated in Figure 16.7. The atoms can migrate across the surface to the edge of the step. This is the growth mode which provides good epitaxy.

16.5.2.1.3 **Stranski–Krastanoff** Stranski–Krastanoff growth is island growth, but the islands form on top of a few monolayers of the same deposited material, as illustrated in Figure 16.8. An example is cadmium on tungsten. Tungsten has a much greater binding energy than cadmium. So the first few monolayers of cadmium which are deposited on tungsten bind more tightly to the tungsten, even though the lattice parameter is wrong, than the cadmium atoms that are bound together in cadmium. Then, after a few monolayers have been deposited, the effect of the tungsten on subsequent layers is reduced, so the cadmium deposits with its usual lattice parameter, which is different from that of the adsorbed layers. The cadmium forms islands on the adsorbed layers of cadmium.

16.6 Surface Reconstruction

The atoms at the surface of crystals are not bound to lattice sites in the same way that atoms are in the bulk, where regular packing of the atoms is required. The surface atoms can relax their positions in response to the different binding configuration at the surface, which is due to the absence of neighbors on one side.

Figure 16.9 shows the 7×7 reconstruction of the (111) surface of silicon. The periodicity of the surface rearrangement is seven times that of the underlying lattice. In low energy electron diffraction (LEED) and reflection high energy electron

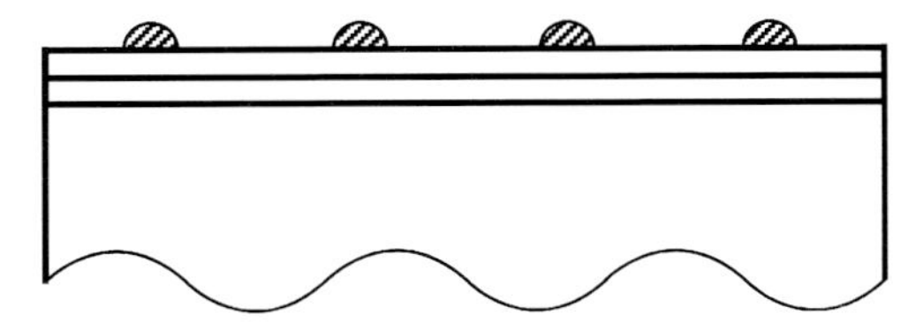

Figure 16.8 Stranski–Krastanoff growth.

Figure 16.9 STM image of 7×7 surface reconstruction on silicon (111).

diffraction (RHEED) patterns this shows up as six intermediate diffraction spots between the lattice peaks. Without reconstruction, the (111) surface of silicon would have many unsatisfied bonds sticking out into space, but these can pair up, and the atoms can move, in order to reduce the free energy of the surface. The presence of this reconstruction was known long before surface tunneling microscopy (STM), which was used to make the image in Figure 16.9. There were about 20 models explaining the surface reconstruction before STM. STM eliminated about half of these, but they have been replaced with new, improved models. The problem is that there are very accurate wavefunctions for bulk silicon based on planar wavefunctions. But the atom positions are fixed in these models, and there is no simple way to project the plane waves out into space where there are no atoms, and at the same time allow the surface atom positions to relax.

The surfaces of many crystals relax to form surface reconstructions. In most cases, the surface must be very clean in order for the reconstruction to be observed. The formation of the silicon 7×7 pattern on the (111) face at about 700 °C is used as a test for surface cleanliness.

If atoms are deposited onto a reconstructed surface, the surface atoms must rearrange into the bulk structure during deposition in order for the growth to proceed. The rearrangement process can be quite complex, as is the case for example on GaAs (211), which reconstructs in a 2×1 pattern.

16.7
Amorphous Deposits

Deposited layers can be amorphous. When silicon is deposited onto a substrate which is below about 400 °C, the deposit is amorphous. Germanium deposited at low temperatures is also amorphous. Materials which form metallic glasses, which are usually a metal alloyed with a non-metal, can be deposited as an amorphous layer. Electroless nickel deposits, which contain phosphorus, are often amorphous, and of course, deposited silica is amorphous. But, as is discussed in Chapter 20, pure metals do not form amorphous deposits.

Silicon and germanium can also be made amorphous by ion implantation. The amorphous phase of these two materials is distinct from both the liquid and the crystalline phases. The amorphous phase is a randomized version of the crystal, not a derivative of the liquid structure. The atoms all have four nearest neighbors in the amorphous phase. The liquid of these two materials is metallic, the atoms in the liquid have about nine nearest neighbors. The liquid is more dense than either the crystalline or amorphous phase. There is a first order phase transition between the liquid and the amorphous phase, as illustrated in Figure 16.10a.

The amorphous phase melts at a lower temperature than the crystal. It is not the lowest free energy phase at any temperature, but the transformation rate of the amorphous phase to the crystalline phase is negligible at low temperatures. This is also true for window glass, which is formed by quenching the liquid.

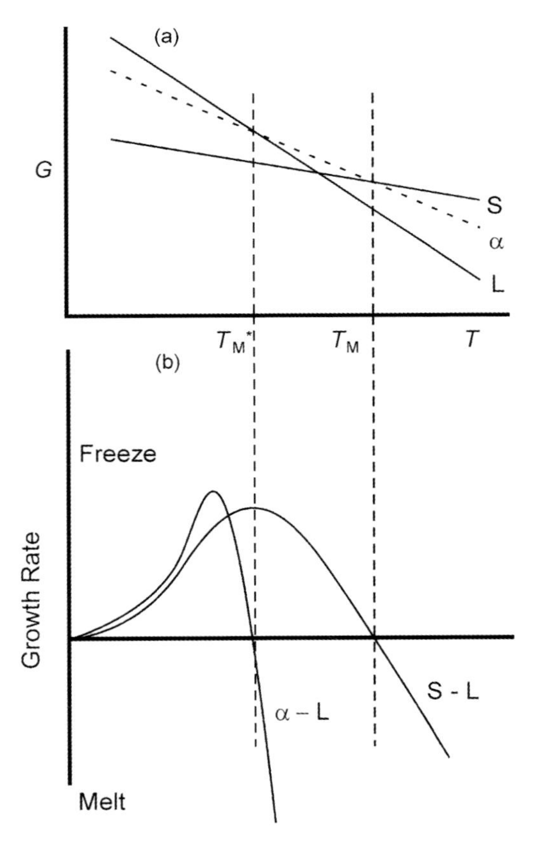

Figure 16.10 (a) Schematic free energies of the crystal, liquid and amorphous phases of silicon. (b) Schematic freezing and melting rates of the crystal and amorphous phases of silicon.

For very rapid cooling, for example, after melting with a short laser pulse, an amorphous layer of silicon can grow directly from the melt. This is illustrated schematically in Figure 16.10b. At a temperature some distance below the melting point of the amorphous phase, the amorphous phase will grow more rapidly from the melt than the crystalline phase. Formally, this occurs because the entropy difference between the liquid and the amorphous phase is smaller than that between the liquid and the crystal. This means that the term $\exp(-L/kT_M) = \exp(-\Delta S/k)$ in Equation 20.20 is larger for the liquid to amorphous transition, and so the growth rate is faster, as illustrated in Figure 16.10b. Another way to look at this is that for very rapid growth, the atoms do not have time to correct mistakes in their position, as is required in order to grow good crystal. For very rapid growth, the errors propagate and amplify, but in spite of this, each silicon atom still has four nearest neighbors.

Amorphous silicon can crystallize directly. The crystallization rate is a straight line on an Arrhenius plot over several orders of magnitude, as shown in Figure 16.11.

These rates are quite slow compared to the rates at which the interface between a crystal and a liquid can move, which can be meters per second.

A phenomenon called explosive crystallization has been observed in thin films of amorphous silicon and germanium. There is a temperature interval between the melting points of the amorphous phase and the crystal phase in Figure 16.10. In this

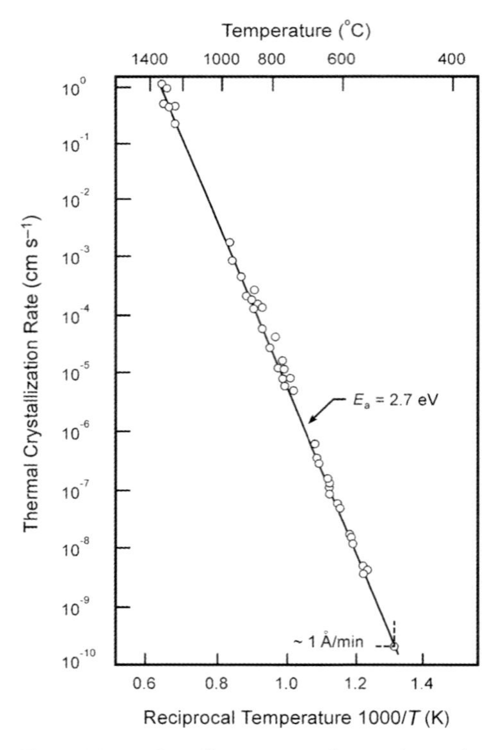

Figure 16.11 Crystallization rate of amorphous silicon [7].

interval, the amorphous phase can melt, and the liquid can crystallize. If a small volume of liquid is created in an amorphous layer, for example from a spark or a laser pulse, the liquid will crystallize rapidly. There is an evolution of heat when the liquid crystallizes, and the heat from the crystallization can melt more of the amorphous phase. The overall process is exothermic, since the crystal is a lower energy configuration than the amorphous phase. The melting front and the crystallization front are coupled together, and propagate throughout the sample at close to the speed of sound.

16.8 Surface Modification

The surfaces of materials are often modified, for example, to improve the surface hardness, to improve wear resistance, or to improve corrosion resistance. There are several techniques which are used.

A thin layer on the surface of a sample can be melted with a short laser pulse. For example, a 25 ns laser pulse with an energy density of about 1 j cm^{-2} will melt the surface of a silicon wafer to a depth of about 1–2 μm. The liquid layer will re-crystallize in a microsecond or so, regrowing from the substrate. This process is sometimes used to remove ion implantation damage in silicon, where the melted layer regrows as a single crystal. If the recrystallization rate is too fast, the melted layer will transform to the amorphous state. It is also used to transform a thin layer of deposited amorphous silicon into polysilicon for making thin film transistors on active matrix liquid crystal displays.

This method is more commonly used to modify the surface properties of alloys, for example, to harden the surface of gears. A thin layer of martensite can be created on the surface of a ductile steel matrix by rapidly heating the surface layer with a pulsed laser, and then letting it cool rapidly. In other cases, a thin layer of a different material can be deposited on the surface, and then melted with a laser pulse, so that the deposited layer becomes alloyed into the surface of the substrate. On refreezing the new surface layer will have quite different composition and properties than the bulk. The new surface layer will often have a very fine grain size, or even be amorphous, because of the fast quench.

The composition and properties of a surface can also be modified by ion implantation, perhaps followed by an anneal.

Electroless nickel containing phosphorus forms a very hard layer, either as a very fine grained polycrystal or as an amorphous phase. This is used to coat the interior surfaces of high precision steel molds to be used for precision casting of plastic parts.

Plasma spraying is used to coat surfaces. The plasma can be either an electric discharge or a flame. Particles are introduced into the plasma in a gas stream where they melt and then they impinge onto the surface to be coated. This is a very rapid and relatively inexpensive process. A major problem is with spalling, because the hot liquid droplets spread out when they hit the surface and then they freeze rapidly. This creates thermal stresses, resulting in adhesion problems between successive droplets.

16.9
Fractal Deposits

Atoms arriving at a surface in random locations do not form a compact deposit unless they have enough mobility to move around on the surface after they land [8]. The illustration in Figure 16.12 was generated by selecting an initial lateral position at random above the surface for each circle, and then moving the circle straight down onto the substrate until it runs into either the substrate or another atom, at which point it comes to rest.

It might be expected that this process would result in a relatively compact uniform layer, rather than a structure which has so much open space.

If the algorithm is changed, so that the circles do not stick where they first hit another circle, but rather roll over the surface of that circle until they run into a second circle, then the branches are thicker, but the structure is still very open. Adding an attractive force between the arriving atoms and atoms on the substrate, and letting the atoms move locally until they find a stable position makes the columns wider, but there is still a lot of open space.

If the circles are incident on the surface at an angle, the average direction of the stalks changes, but not as much as the incident angle was changed.

This process occurs during vapor deposition if the atoms cannot move around after they land on a surface.

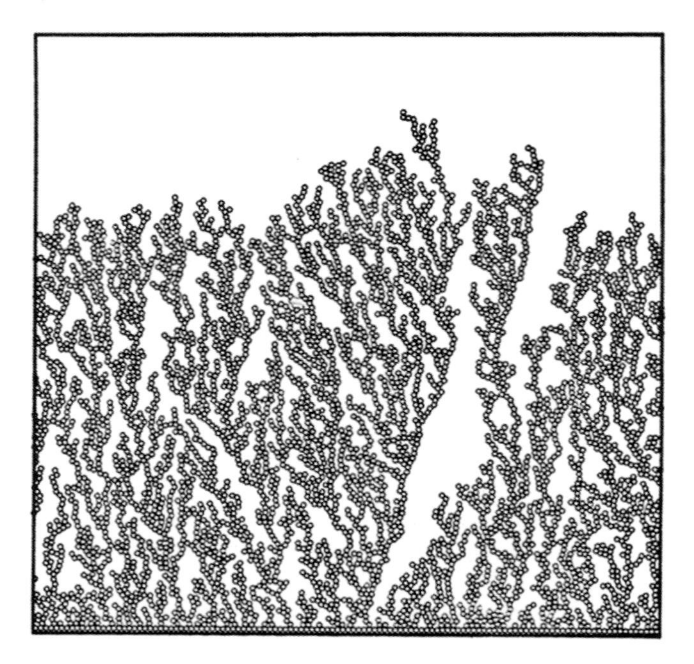

Figure 16.12 Fractal deposit.

When a material with a high melting point is deposited on a cold substrate, it can either form a polycrystalline deposit, or it can form a columnar structure, as described above. There is a transition between the two structures which depends on the temperature of the substrate. At low temperatures the deposit is columnar with open spaces between the columns, while at high temperatures the deposit is compact and polycrystalline. For plasma deposition, where the incident atoms have an energy which is much higher than usual thermal energies, the transition occurs at a lower substrate temperature.

Deposition in the polycrystalline region is discussed below. In the columnar region, the columns are typically 5 to 10 nm in diameter, and separated by gaps of about half that. A major fraction of the atoms in the deposit are at a surface of a column. There is evidence that amorphous silicon deposits have this columnar structure, but it is difficult to detect because of contrast problems in the electron microscope.

This process has been used to reduce spalling caused by thermal expansion mismatch between the thermal barrier coating (TBC), usually zirconium oxide, which is deposited on the nickel-based super alloy to make the blades in jet engines. The columnar structure is much more compliant in the lateral direction than a solid coating. The structure of the coating can be alternated between the columnar and the polycrystal modes during deposition, to provide a compliant coating which also prevents the hot gases from coming into contact with the nickel alloy.

16.10 Strain Energy and Misfit Dislocations

An adlayer can be coherent with the substrate, which is the desired condition for good epitaxial growth, as illustrated in Figure 16.13a. Alternatively, the deposit can have a structure which bears no relationship to that of the underlying substrate, in which

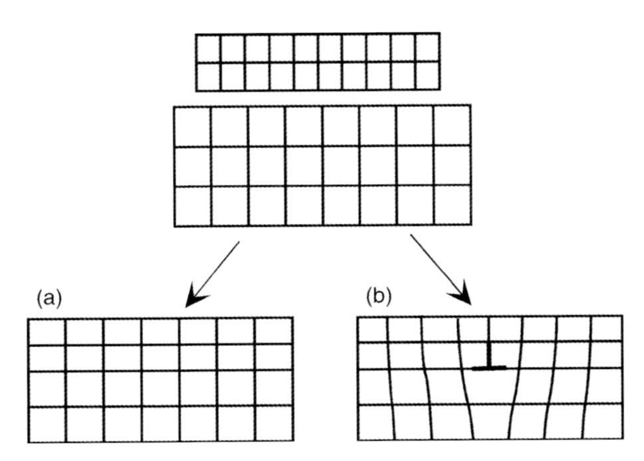

Figure 16.13 (a) Coherent strained layer and (b) a semi-coherent layer with misfit dislocations.

case the interface is termed incoherent. The interface can also be partly coherent if there are misfit dislocations in the interface, as illustrated in Figure 16.13b.

The elastic strain energy in an incommensurate adsorbed layer depends on the thickness of the layer. The elastic strain energy in the layer can be reduced by introducing misfit dislocations at the interface between the layer and the substrate. When the layer is thin, introducing the misfit dislocations increases the total energy of the layer. When the layer is thick enough, the total energy can be reduced by introducing misfit dislocations. Vandermerwe and Frank [9] calculated the critical thickness at which this should happen.

The misfit strain, f, is the difference between the lattice parameter of the substrate and the deposit:

$$f \equiv \frac{a_s - a_0}{a_0} \tag{16.28}$$

The total elastic energy per unit area in the epitaxial layer without misfit dislocations can be written as:

$$E_\varepsilon = h \int \sigma d\varepsilon \tag{16.29}$$

where h is the thickness of the layer, σ is the stress in the layer and ε is the strain in the layer. For a thin layer on a thick substrate, the strain will be uniform through the thickness of the layer, a condition known as plane strain. The stress in the layer is related to the strain by $\sigma = B\varepsilon$, where B is the appropriate elastic constant. So the total strain energy is:

$$E_\varepsilon = h \int B\varepsilon d\varepsilon = \frac{1}{2} h B \varepsilon^2 \tag{16.30}$$

Without misfit dislocations, the strain in the layer is the misfit strain:

$$\varepsilon = f \tag{16.31}$$

Introducing dislocations into the interface with a spacing S, as illustrated in Figure 16.14, reduces the strain in the layer by:

$$\delta = \frac{b}{S} \tag{16.32}$$

where b is the Burger's vector of the dislocations. The strain in the layer with dislocations is $\varepsilon = f - \delta$, so the total strain energy per unit area in the layer with dislocations is:

$$E_\varepsilon = \frac{1}{2}(f - \delta)^2 Bh \tag{16.33}$$

The energy of a dislocation line per unit length can be written as:

$$Db\left(\ln\frac{R}{b} + 1\right) \tag{16.34}$$

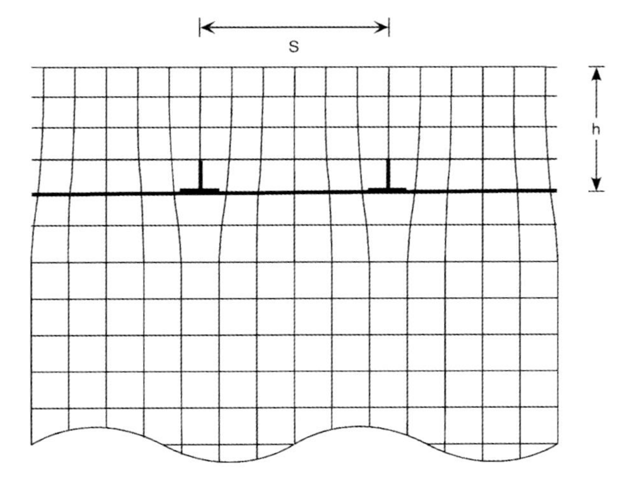

Figure 16.14 Misfit dislocations.

The first term in the parenthesis is the energy in the elastic strain field around the dislocation, and the second term represents the energy associated with the dislocation core. D is the appropriate elastic constant for the strain field around a dislocation. The elastic constants of crystals are dependent on the direction of the deformation, so in general the two elastic constants, B and D will be different. R is the upper limit on the size of the strain field around a dislocation, which we will take to be the thickness of the layer, h.

For an area of the interface $A = l^2$, the number of dislocation lines is l/S.

As illustrated in Figure 16.15, if the length of each dislocation line is l, the total length of dislocation in the area $A = l^2$ is:

$$\frac{l^2}{S} = \frac{A}{S} \tag{16.35}$$

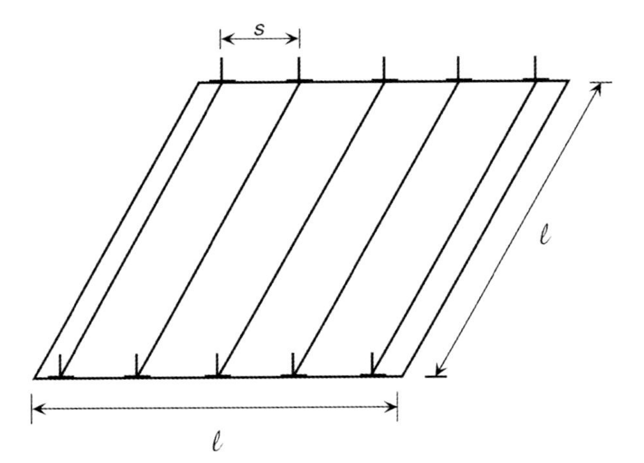

Figure 16.15 The total length of dislocation per unit area is $1/S$.

So the total length of dislocation per unit area is $1/S$.

The total energy per unit area due to the dislocations is:

$$E_d = D\frac{b}{S}\left(\ln\frac{h}{b}+1\right) = D\delta\left(\ln\frac{h}{b}+1\right) \tag{16.36}$$

The total energy per unit area due to the elastic stress in the film and the dislocations is:

$$E_{total} = E_\varepsilon + E_d = \frac{1}{2}(f-\delta)^2 Bh + D\delta\left(\ln\frac{h}{b}+1\right) \tag{16.37}$$

The spacing, S, which results in the smallest total energy can be found be minimizing the total energy with respect to δ:

$$\frac{dE_{total}}{d\delta} = 0 = -(f-\delta)Bh + D\left(\ln\frac{h}{b}+1\right) \tag{16.38}$$

The minimum total energy occurs for:

$$\delta = f - \frac{D}{Bh}\left(\ln\frac{h}{b}+1\right) \tag{16.39}$$

For small thickness, h, the second term will be large, and so the right-hand side is negative, which means there is no positive value of δ which minimizes the total energy. For

$$f > \frac{D}{Bh}\left(\ln\frac{h}{b}+1\right) \tag{16.40}$$

there is a finite value of δ which minimizes the total energy. The critical value of the thickness where this transition occurs is:

$$h_c = \frac{D}{Bf}\left(\ln\frac{h_c}{b}+1\right) = \frac{D}{B}\left(\frac{a_0}{a_s-a_0}\right)\left(\ln\frac{h_c}{b}+1\right) \tag{16.41}$$

There should be no misfit dislocations for layer thicknesses below this critical value. But this does not imply that misfit dislocations will always be present for thicknesses above this critical value. This is not a flaw in the analysis, there are refinements of this analysis which also do not predict when misfit dislocations are found to be present experimentally.

Misfit dislocations are usually found only at thicknesses considerably greater than h_C. They do not form spontaneously just because their presence would reduce the total energy. There is a nucleation barrier to their formation. Experimentally, there is a wide variation in the onset of misfit dislocations, even in the same material.

It has been shown that the misfit dislocations start at defects in the layer [10]. Small foreign particles in the stressed layer will act as stress concentrators: they will have larger stresses around them.

When the stress becomes large enough, these will blow out dislocation rings, which will expand laterally and vertically. At the top of the layer, a dislocation will run out of the surface, making a step. At the bottom of the layer, the dislocation will get trapped at the interface, where it will grow into a line as the dislocation loop expands laterally in the layer, as sketched in Figure 16.16. The dislocations in the interface are the misfit dislocations. The dislocations which expand towards the substrate due to the misfit stress have the correct Burgers vector to act as misfit dislocations to reduce the strain in the layer.

The formation of misfit dislocations has been discussed above using two-dimensional illustrations. In general, the stress in the substrate is not unidirectional, and so a grid of misfit dislocations in the interface is necessary to reduce all the misfit strain energy.

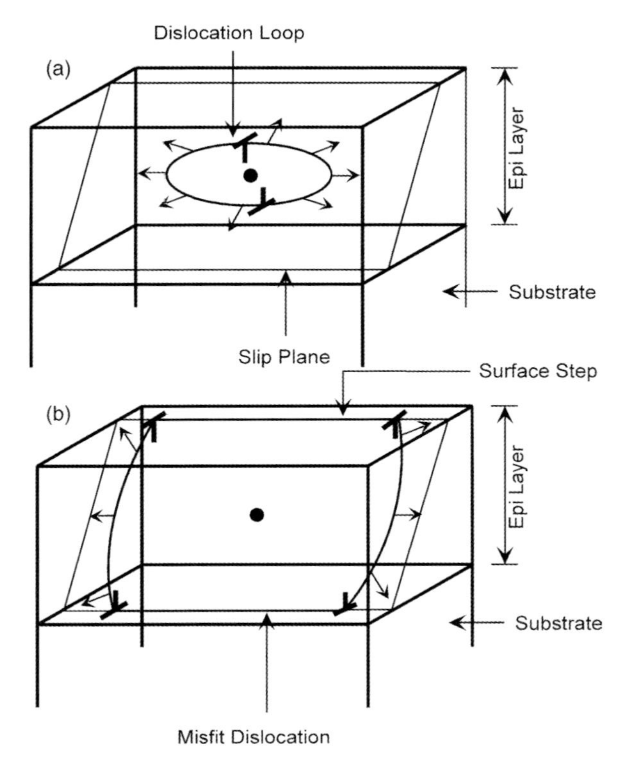

Figure 16.16 (a) A foreign particle initiating a dislocation loop. (b) The dislocation loop creates a misfit dislocation at the interface.

16.11
Strained Layer Growth

16.11.1
Surface Modulation

The total elastic energy in a strained layer can also be reduced by modulations of the surface, as shown in Figure 16.17. This is a purely elastic effect, and it occurs spontaneously.

16.11.2
Strained Layer Superlattice

The growth of alternating layers of the epitaxial material and the substrate material has been used to grow thick layers without misfit dislocation, as illustrated in Figure 16.18. This is known as a strained layer superlattice. Each epitaxial layer is kept below the critical thickness. The overall strain in the layers is an average of the two types of layers, and so the overall deposit does not exceed the critical thickness.

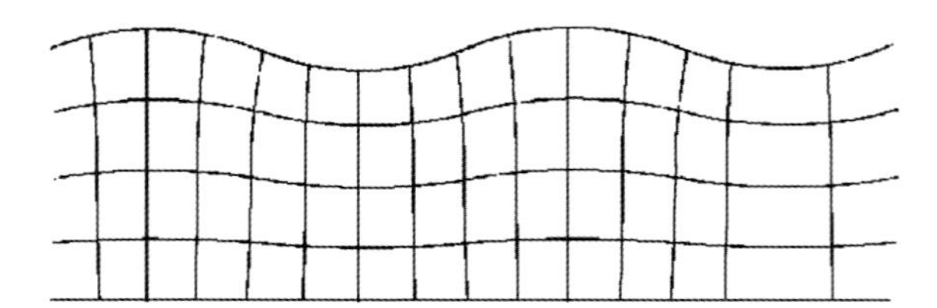

Figure 16.17 Surface modulation can reduce the total strain energy of a layer

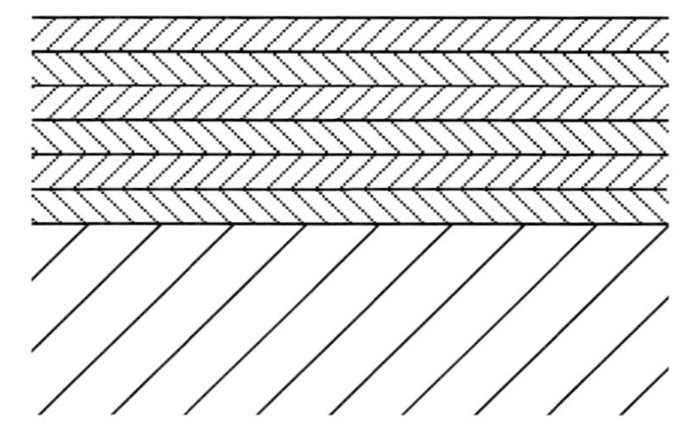

Figure 16.18 Strained layer superlattice.

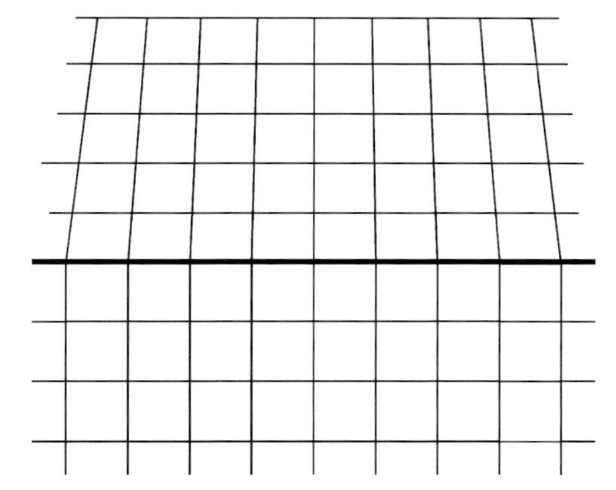

Figure 16.19 Graded strain layer produced by grading the composition.

16.11.3 Graded Strain Layers

Graded strain layers are made by grading the composition in order to change the lattice parameter gradually, in order to keep the stress below the critical value where misfit dislocations form, as illustrated in Figure 16.19.

This process is used to make silicon–germanium alloys on a silicon substrate. These alloys have a smaller bandgap than either silicon or germanium, and are used for IR detectors. Silicon–germanium alloys are also replacing GaAs for high frequency devices.

References

1 Langmuir, I. (1918) *J. Am. Chem. Soc.*, **40**, 1361.

2 Walton, D. (1962) *J. Chem. Phys.*, **37**, 2182.

3 Lewis, B. and Anderson, J.C. (1978) *Nucleation and Growth of Thin Films*, Academic Press, London.

4 Kuech, T.F. and Tischler, M.A. (2000) *Handbook of Semiconductor Technology*, Vol. 2, *Processing of Semiconductors* (ed. K.A. Jackson), Wiley-VCH, Weinheim, p. 111.

5 Ohring, M. (2002) *Materials Science of Thin Films*, 2nd edn, Academic Press, San Diego, CA, p. 197.

6 Venables, J.A. (2000) *Surface and Thin Film Processes*, Cambridge University Press, Cambridge, p. 146.

7 Olson, G.L., Kokorowski, S.A., Roth, J.A., and Hess, L.D. (1983) *Mater. Res. Soc. Symp. Proc.*, **13**, 141.

8 Leamy, H.J. and Dirks, A.G. (1979) *J. Appl. Phys.*, **50**, 2871.

9 Vandermerwe, J.H. and Frank F.C. (1949) *Proc. Phys. Soc. A*, **62**, 315.

10 Eaglesham, D.J., Kvam, E.P., Maher, D.M., Humphreys, C.J., and Bean, J.C. (1989) *Phil. Mag.*, **59**, 1059.

Problems

16.1. Computer simulation of Langmuir adsorption
Start with an array of 20×20 lattice sites.
Add atoms randomly at a rate $k^+ P$ to the empty sites, and remove atoms

randomly from the surface at a rate k^-, until the surface ad-population stops changing in time.
Choose k^+, k^- and P.
Select a site at random.
If the site is empty, add an atom if $k^+ P > \text{Rnd}$ (Rnd = a random number).
If the site is occupied, remove the atom if $k^- > \text{Rnd}$.
The surface coverage Θ will increase with time and will saturate at some value Θ_e.
Repeat this for various values of P.
Plot Θ_e vs. KP where $K = k^+/k^-$.
Compare your data with the Langmuir adsorption isotherm:

$$\Theta = KP/(1 + KP)$$

Note:
Most random number generators provide $0 \leq \text{Rnd} \leq 1$, so the simulation will not work if Rnd is compared with a number greater than 1.
For $KP = k^+ P/k^- \leq 1$, use $k^- = 1$, and various values of $k^+ P \leq 1$
For $KP = k^+ P/k^- > 1$, use $k^+ P = 1$, and various values of $k^- \leq 1$

16.2. Discuss the formation of new layers during epitaxial growth from the vapor phase.

17
Thin Film Deposition

17.1
Liquid Phase Epitaxy

Liquid phase epitaxy (LPE) is a solution growth process where the solution consists primarily of one of the components of the crystal [1]. For example, GaAs can be grown from a gallium rich solution of gallium and arsenic at a temperature far below the melting point of GaAs. The relevant phase diagram is illustrated in Figure 17.1.

As discussed in Chapter 7, the solution must be cooled below the equilibrium temperature of the liquid in order for growth to take place. In Figure 17.1 the melting point of GaAs is labeled T_1, and the stoichiometric composition is labeled A. A solution of composition B is in equilibrium with crystal of composition E at temperature T_2. LPE growth for this composition takes place at C, at the temperature

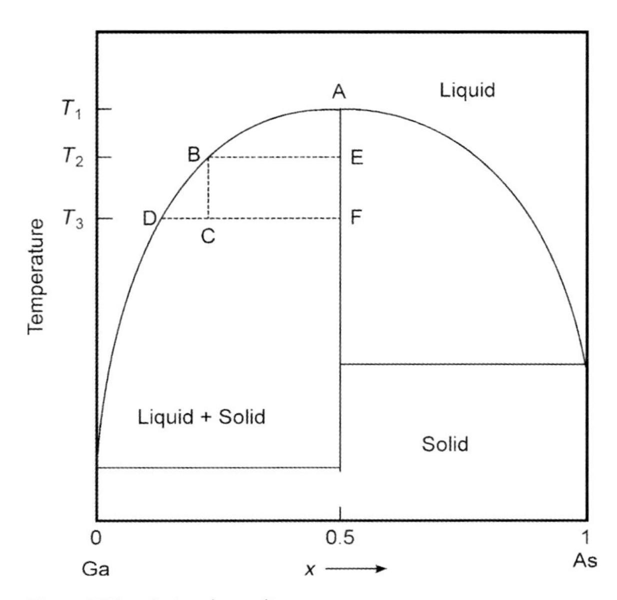

Figure 17.1 GaAs phase diagram.

Kinetic Processes: Crystal Growth, Diffusion, and Phase Transitions in Materials. Kenneth A. Jackson

ISBN: 978-3-527-32736-2

T_3, where the equilibrium liquid and solid equilibrium compositions are D and F, respectively. The growth temperature must be carefully controlled to limit fluctuations in the growth rate.

In the phase diagram, GaAs is represented as a line compound but the crystal which grows is likely to be slightly off the precise stoichoimetric composition, and this slight difference introduces a significant defect density on one of the sublattices in the crystal. For comparison the typical equilibrium vacancy concentration in a metal crystal at its melting point is typically about 0.1%. A binary crystal which is off-composition by only 0.1%, has a similar defect density. The defect density resulting from LPE growth depends sensitively on the position of the phase boundary.

Liquid phase epitaxy methods have the advantage that impurities with small *k*-values tend to stay in the liquid, so that the crystal can be purer than the starting materials.

17.2 Growth Configurations for LPE

There are three configurations which are used for liquid phase epi: tipping, dipping and a slider [2]. The first two are relatively simple methods, and the latter is a more complex set-up designed to grow multilayer structures.

17.2.1 Tipping

In tipping, the substrate is put into one end of a boat or container, and liquid is melted in the other end of the boat. The liquid is then slightly undercooled so that growth will occur when the liquid comes into contact with the crystal. The boat is then tipped so that the substrate is immersed in the liquid. After a time, during which a layer of the desired thickness grows, the boat is tipped back to remove the liquid from the substrate. The container can be sealed before starting the growth process, in order to control the atmosphere if one component is volatile.

17.2.2 Dipping

Dipping is also a very simple process. The substrate is held vertically in an open holder, and then immersed in a slightly undercooled liquid of the appropriate composition. A thin layer grows in a short time, and then the substrate is removed from the liquid. This method has been used very successfully for growing garnet layers of various compositions on gadolinium gallium garnet (GGG) substrates. GGG is relatively easy to grow as a single crystal, and the composition of garnets can be varied so as to obtain a garnet with desired properties which is also lattice matched to GGG. This method was used to grow epitaxial layers for magnetic bubble memories. The technology for making magnetic bubble memories was outstanding, but the product could not compete commercially with semiconductor memory.

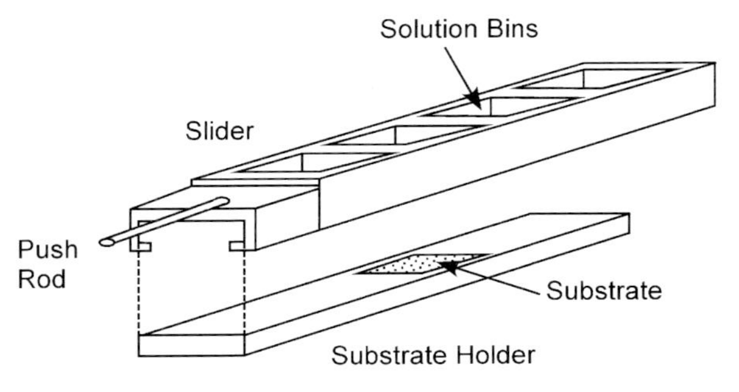

Figure 17.2 LPE slider.

Dipping is used today to make magneto-optical isolators, by growing layers of magnetic garnet of precise thickness on GGG substrates.

Multiple layers of varying compositions can be grown using tipping or dipping with liquids of different composaitions.

17.2.3 Slider

Multiple layers of different compositions are more readily grown using a slider, which is illustrated in Figure 17.2.

Each of the bins contains a solution of different composition. The substrate holder slides so that the substrate can be positioned under each bin for the desired growth period. The liquid is on top of the substrate, but since the slider is usually made of graphite, and the liquid does not wet graphite, the liquid does not leak out around the slider if the slider fits with a reasonable tolerance. The slider assembly is inserted into a furnace so that multiple layers can be grown without cooling the substrate. The temperature of the furnace can be ramped up or down, or the whole growth process can be carried out isothermally, by controlling the composition in each bin. The liquid in the first bin can be slightly superheated so as to dissolve or etch the substrate before growth, in order to clean it. A dummy substrate can be used to "condition" the liquid in each bin, ahead of the growth substrate. A solid source material can be floated in a bin to control the composition of the liquid in the bin. Several bins can have different compositions in order to grow the layers of a heterostructure.

The equipment required for LPE is relatively inexpensive. Experts can produce outstanding results with LPE, but it is a difficult process to control consistently. The first continuous working semiconductor lasers were made by LPE, but the method has not been used extensively in production.

17.3 Chemical Vapor Deposition

Chemical vapor deposition (CVD) [3] is commonly used to grow a layer of a different concentration or doping level or even different dopant type than the substrate. The process makes use of volatile molecules which carry the atoms to be deposited to the substrate. On the substrate, the molecules react or decompose, depositing the atoms to be added to the crystal and creating a volatile product which leaves the surface. Epitaxial layers can usually be grown more rapidly by CVD than by the MBE process which is discussed below. The growth is carried out in a vacuum chamber at moderate pressures which are determined to maximize the growth rate. In general, the hardware is much more expensive than LPE, and somewhat less expensive than MBE equipment. It is widely used as a production method.

For example, silicon epi layers are grown from silane or from a variety of chlorosilanes. Silicon tetrachloride ($SiCl_4$) and trichlorosilane ($SiHCl_3$) are liquid at room temperature. Dichlorosilane (SiH_2Cl_2), chlorosilane (SiH_3Cl), silane (SiH_4) and disilane (Si_2H_6) are gaseous at room temperature. All have been used to grow silicon epitaxial layers. The silanes transport silicon to the surface in the gas phase (the liquids are heated so that they evaporate), where they decompose on the hot substrate depositing silicon, and then the hydrogen and chlorine leave the surface. The chlorosilanes have the advantage that many metal chlorides have high vapor pressures, and so metallic impurities are not incorporated in the layers.

Figure 17.3 illustrates some usual configurations for CVD reactors.

In the horizontal reactor, the wafers are supported at an angle to obtain lamellar flow over the surface of the wafers. In a barrel reactor, the substrates lean against the barrel, which is rotated in order to obtain a uniform deposit. In a vertical reactor, also known as a pancake reactor, the wafers are placed on an rf heated susceptor, and the substrate holder is rotated to obtain a uniform deposit. In each of these reactors, the geometry and gas flow in the reactor is critical in order to obtain a uniform deposit. In a low pressure CVD (LPCVD) reactor, the pressure is reduced so that the mean free path of the gas atoms is less than the spacing between the wafers, so the uniformity of the deposit does not depend on controlling the gas flow. Figure 17.4 shows a planetary

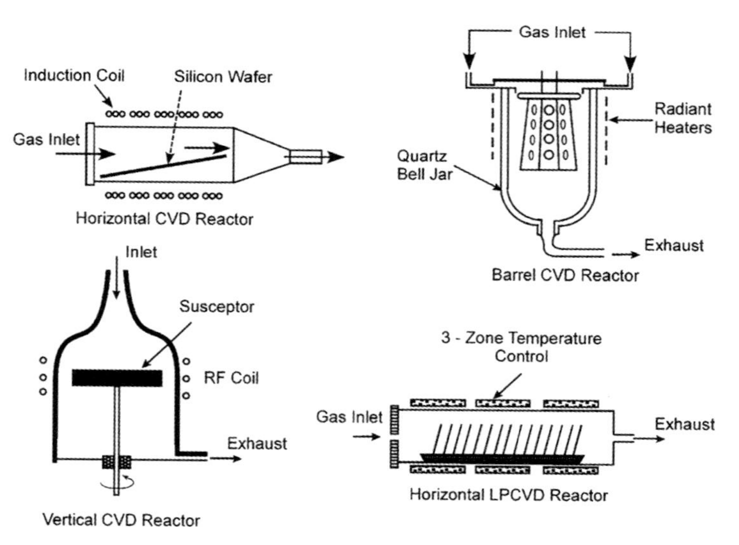

Figure 17.3 CVD reactors.

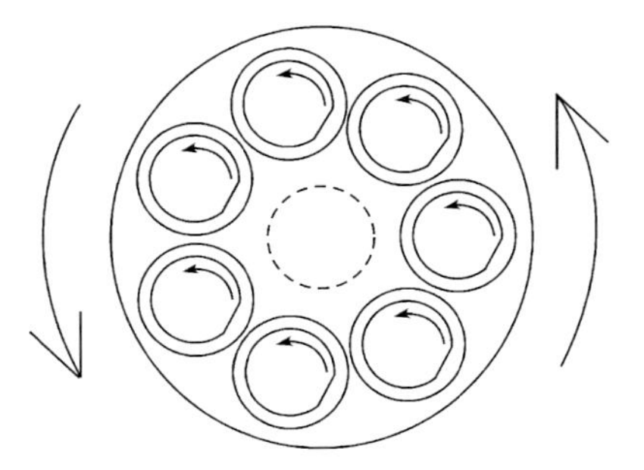

Figure 17.4 Planetary reactor.

reactor, where the substrate holder is rotated, and each wafer is rotated on the substrate in order to promote the uniformity of the deposit.

There are two processes which are involved in the growth. The first is the process by which the reactants are delivered to the surface by gas phase flow and diffusion. The second is the reaction on the surface which results in the addition of atoms to the crystal. The slower of the two processes is rate limiting, and controls the process. This is illustrated schematically in Figure 17.5.

$\Delta\mu^*$ is the total chemical potential difference between the inlet gas and the deposit. In Figure 17.5a, the growth is reaction rate limited: there is a large chemical potential

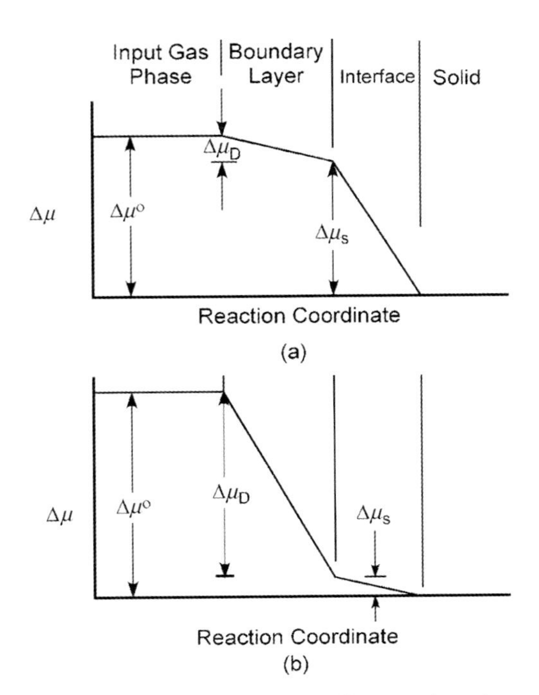

Figure 17.5 Chemical potential differences along the reaction path from the input gas to the crystal.

difference driving the interface growth process. In Figure 17.5b, the growth is boundary layer limited: most of the chemical potential difference is taken up by transport to the surface.

It is preferable to operate in a mode where the growth is transport limited since surface reactions are usually strongly temperature dependent. The growth is easier to control when operating in a mode where the growth rate is relatively independent of temperature. This means that the kinetic processes on the surface should be fast, so that they are not rate limiting. If the crystal growth process is near equilibrium, where the chemical potential driving the growth process is small, there are likely to be fewer growth defects since mistakes tend to be corrected by the reversibility of the process.

The control of the gas flow in the growth chamber is critical in reactor design. The chambers are designed to produce lamellar flow over the wafers, and then the reactant gases reach the surface by diffusion through a boundary layer above the wafer. The growth rate then depends on the flux to the surface:

$$J = D\left(\frac{dC}{dy}\right)_{\text{Surface}} \tag{17.1}$$

where D is the diffusion coefficient, and C/dy is the concentration gradient normal to the surface. Writing the concentration gradient as C/δ, where δ is the effective boundary thickness, and replacing $C = n/V$, the number of molecules per unit volume, with P/RT gives:

$$J \approx D\frac{P_{\text{Input}} - P_{\text{Interface}}}{RT\delta} \tag{17.2}$$

Epitaxial layers of III–V compound semiconductors are usually grown by CVD with an excess of the Group V element, often a ratio of 3 to 1. The growth rate then depends on the arrival rate of the Group III element at the substrate. So in this case, the pressure in Equation 17.2 would be the partial pressure of the Group III component. Figure 17.6 illustrates the growth rate of GaAs, AlAs, and InP, which are grown with an excess of arsenic or phosphorus, as a function of the flow rate of the alkyl component. The growth rate increases linearly with the flow rate of the alkyl.

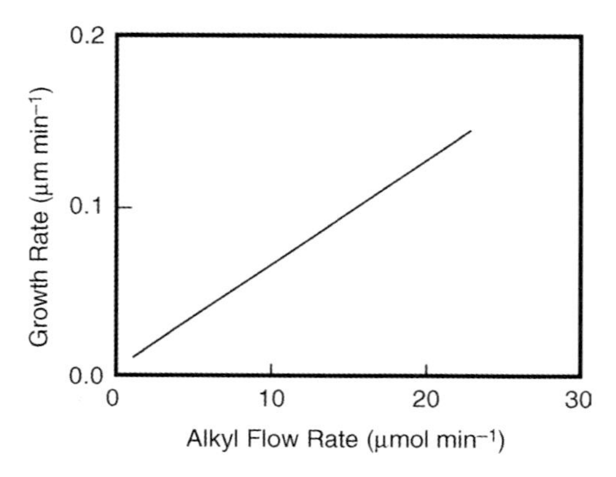

Figure 17.6 Typical growth rate of III–V compounds as a function of alkyl flow rate.

Epitaxial growth of II–VI compounds is usually carried out with an excess of the more volatile component.

17.4 Metal–Organic Chemical Vapor Deposition

Metal–organic CVD (MOCVD) or organo-metallic CVD (OMCVD) are different terms for the same process, where the carrier molecules used to deliver the reactant(s) to the surface are organic [4]. There is a wide variety of potential metal–organic compounds, and the choice of reactants has received a lot of attention. Perhaps the simplest and most widely used are trimethyl compounds, exemplified by the growth of GaAs from trimethyl gallium and arsene. Figure 17.7 presents the growth efficiency of GaAs as a function of temperature.

At high temperatures, the growth is mass transport limited, and the growth efficiency is limited by the desorption of the reactants. At low temperatures, there is low mobility on the surface, and the growth becomes limited by the reaction rate. Growth is carried out near the maximum on this plot.

Data such as illustrated in Figure 17.8 for the growth of GaAsSb from TMGa, TMSb, and TMAs must be accumulated for successful growth.

This figure indicates that droplets of either gallium or antimony can form on the surface, depending on the vapor pressures of the reactants. Good crystals cannot be grown in the regions where these droplets form.

17.5 Physical Vapor Deposition

For physical vapor deposition (PVD) the atoms to be deposited usually come from a material which has been heated to a temperature where it evaporates at a reasonable

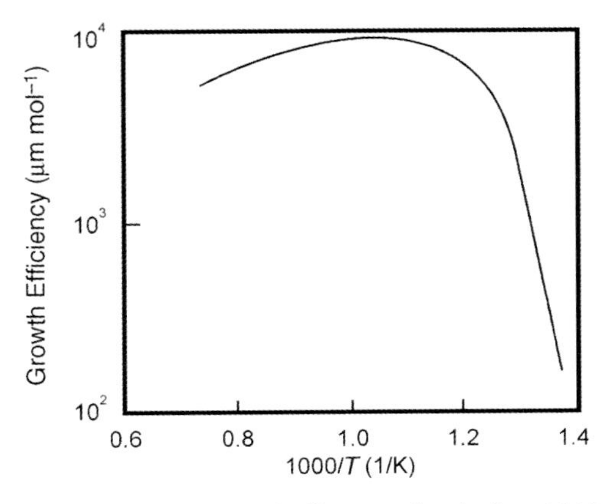

Figure 17.7 The growth efficiency of GaAs from TMGa and AsH_3.

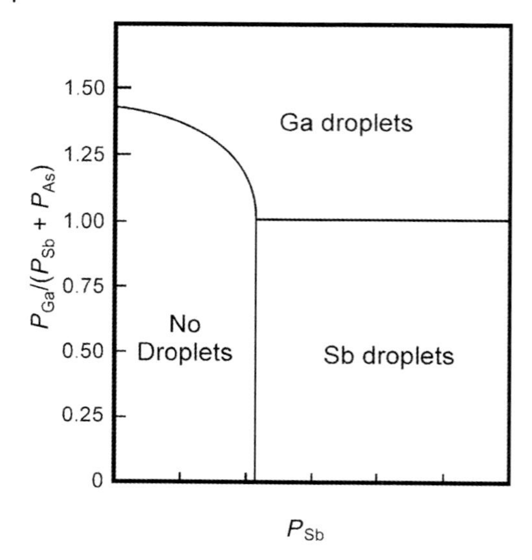

Figure 17.8 Gas-phase pressure diagram for the growth of GaAsSb.

rate. This usually means that the source material is melted. This method is not suitable for any compounds which decompose more readily than they evaporate. This is not a problem with metals. The material to be evaporated is often held in a small basket which is heated electrically. For better control of the evaporative flux, the source is often contained in a Knudsen cell. This is a cell as illustrated in Figure 17.9.

It is basically a can with a small aperture through which gas atoms can escape into a larger vacuum chamber where they will deposit on a substrate. Inside is the vapor source. The vapor pressure inside the chamber is determined by the temperature of the source material. The aperture is made small enough so that the vapor which is emitted from the cell does not significantly change the pressure inside the cell. The vapor pressure inside the cell is determined by the source temperature:

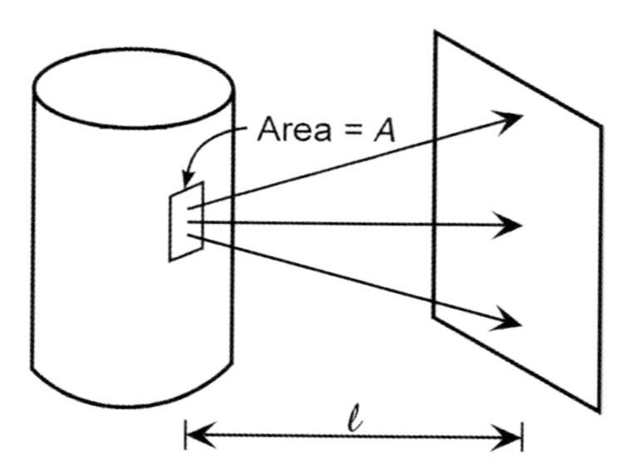

Figure 17.9 A Knudsen cell.

$$P = P_0 \exp\left(-\frac{Q_{evap}}{kT}\right) \quad (17.3)$$

The flux from the cell depends on the vapor pressure inside the cell, and on the size of the aperture. The flux of atoms arriving at a substrate from a Knudsen cell which is a distance l from the substrate is:

$$J^+ = \frac{PAN_0}{\pi l^2 \sqrt{2\pi mkT}} \quad (17.4)$$

where P is the pressure inside the cell, A is the area of the aperture of the cell, N_0 is Avagadro's number, m is the mass of the molecule, and T is the source temperature inside the cell.

Knudsen cells are frequently used in MBE machines to provide a source for metals, so that the metal atoms do not spray all over the interior of the chamber.

17.6 Sputter Deposition

Plasmas are discussed in Chapter 18 in terms of their use in plasma etching and chemical milling. Plasmas are also used for deposition, and this process is known as sputter deposition. The basic configuration is similar to a pancake reactor, as in Figures 18.11 and 18.12, or for magnetron sputtering as in Figure 18.13. The cathode is the target material. When it is bombarded with ions from the plasma, neutral atoms and ions are sputtered from it. The neutral atoms pass through the plasma and deposit on the anode, which is the substrate for deposition. The high energy ions in the plasma can remove atoms from just about any target. The atoms hitting the substrate are also relatively high energy, and so they produce damage at the substrate, typically a few atom layers deep. This process is not usually used for epitaxial growth because of the damage, but it is used commonly for the deposition of materials which have low vapor pressures and so are difficult to evaporate.

Both DC and rf plasmas are used for sputtering. DC plasmas can be used readily for metals because the target can be grounded. There is a problem with charge build-up with targets which are insulators, and so these are usually sputtered using rf.

Oxides of metals can be sputter deposited using DC and metal targets, by introducing oxygen into the plasma or near the target surface.

17.7 Metallization

The conductor stripes on semiconductor wafers are usually deposited either by CVD, PVD or sputtering. PVD has the advantage of being relatively simple. However, in both PVD and sputtering, the atoms tend to be directed towards the surface in a limited range of angles. As a result, the coverage over steps tends to be poor. The flux onto vertical features on the surface is less than on flat surfaces. Step coverage with CVD is much better, because the decomposition of the carrier molecules occurs on the surface. Aluminum has traditionally been deposited using PVD, and it has been difficult to find an appropriate carrier molecule for it. Copper can also be deposited by PVD, but CVD is preferred for metallization.

The metallization is polycrystalline, and during the early stages of metallization, many nuclei of diverse orientation form on the surface. These merge and coalesce into a polycrystalline deposit. The deposited metals are pure, since pure metals have lower resistivities. The grain boundaries in pure metals are quite mobile, especially when the grain size is very small so grain growth occurs during the deposition process. The final grain size is usually much larger than the initial distance between nuclei on the surface.

During growth there are also two other effects, in addition to grain growth, which can influence the final grain size.

1) Orientations with low surface energy will create a groove at the interface which favors the expansion of the low energy surface over higher energy surfaces.
2) The growth rate is orientation dependent. Rapidly growing orientations will tend to overgrow slower growing orientations, and so crowd out the slower growing grains.

These two orientation dependent effects not only increase the grain size as the deposited layer grows in thickness, but they also produce a preferred orientation texture in the deposit.

The relative importance of these three mechanisms varies with material, with growth conditions and with substrate temperature.

It is often the case that some combination of these grain coarsening effects produces a final grain size in the deposit which is comparable to the thickness of the deposit.

17.8 Laser Ablation

Intense laser pulses can be used to volatilize atoms on the surface of a target. The incident radiation heats the surface locally to a temperature where atoms have enough thermal energy to escape from the surface. Typically, the boiling point of a material is less than a factor of two above its melting point, so the laser pulse needs to be only about twice as energetic to evaporate atoms from the surface as to melt the surface.

The laser light in an opaque material such as a metal is absorbed very near the surface, and so it is possible to heat a thin layer near the surface very rapidly to the boiling point. The atoms coming from the surface have relatively high thermal energies, but these are typically only a small fraction of an electron volt, unlike a sputtered atom, which will typically have about one kilovolt of energy from the plasma. These relatively low energy atoms can impinge on a nearby substrate without creating damage. This process is used for expitaxial growth of materials which are difficult to evaporate.

The laser must be pulsed in order to get a lot of energy into the surface in a short time so that the surface heats rapidly. Also, the evaporated atoms tend to form a vapor cloud over the surface which absorbs the laser light and prevents the light from reaching the surface. Pulsing the laser gives this cloud time to dissipate, so that the next pulse can reach the surface.

17.9 Molecular Beam Epitaxy

Molecular beam epitaxy is very simple in principle [5–7]. The growth takes place in a high vacuum environment. The vapor pressure in the chamber is low enough so that the atoms or molecules go from their source to the substrate without a collision: they form a molecular beam. A simple schematic of a growth chamber is shown in Figure 17.10.

Because the main chamber is under an ultrahigh vacuum (UHV), it is seldom opened to the atmosphere. Instead, samples are introduced through load locks and buffer chambers, which are pumped down before the sample is transferred from them. The load lock is typically pumped down to 10^{-6} to 10^{-7} torr, while growth takes place in the growth chamber at 10^{-10} to 10^{-12} torr. At 10^{-10} torr, the number of gas atoms hitting the substrate can form a monolayer in half an hour. Layers are typically

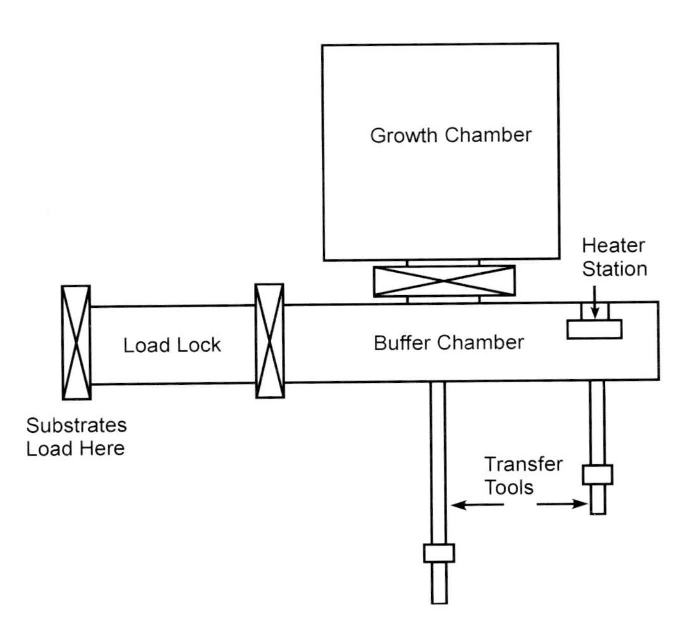

Figure 17.10 Schematic of an MBE machine.

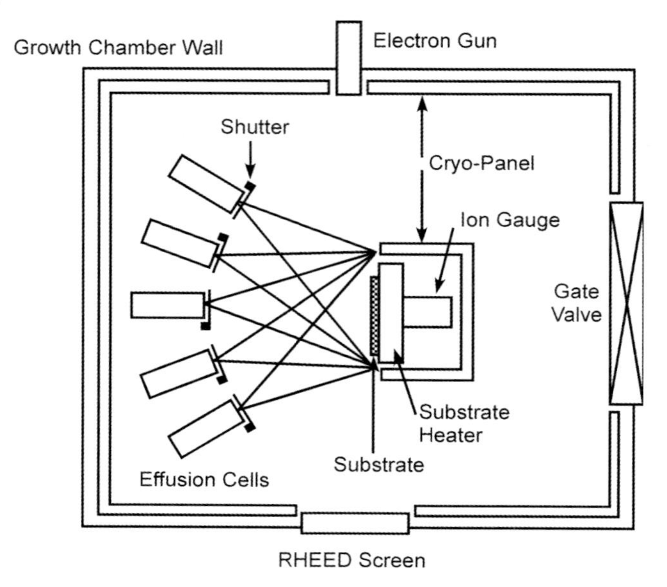

Figure 17.11 Growth chamber of an MBE machine.

grown at rates of about 1 μm h^{-1}. The pressure rises to about 10^{-6} torr during the deposition process.

No rubber gaskets or vacuum grease can be used in a UHV system. The gaskets are usually made from soft copper. The load lock is typically baked at 200 °C, and the buffer chamber at 600 °C to drive off water and other adsorbed gases, and then run at room temperature. Cryopanels are used to condense gases.

Typically, effusion cells, which contain a heater to evaporate source material are used, as illustrated in Figure 17.11. The molecular beams incident on the sample are controlled with shutters to grow various layers of different compositions.

The growth can be monitored *in situ* with reflection high energy electron diffraction (RHEED), where the electrons are incident on the sample at a grazing angle.

The growth rate in MBE is quite slow. On the other hand, it permits exquisite control over the thickness of deposited layers, and is used to make multilayer quantum structures.

A variant on MBE has been termed chemical beam epitaxy (CBE), which uses molecular species to carry the atoms to be deposited to the surface. The environment is UHV, so that the incident molecules form a molecular beam, but there is a chemical reaction or decomposition on the surface to deposit the desired species.

17.10
Atomic Layer Epitaxy

Atomic layer epitaxy (ALE), also termed atomic layer deposition (ALD) is a process where atomic layers of alternating composition are deposited. This process was first used for zinc oxide, for which it works very well. Zinc vapor is introduced into the chamber, and a monolayer of zinc will deposit, but only a monolayer, and then the deposition stops. Next oxygen is introduced into the chamber, and a monolayer of oxygen deposits, and then the deposition stops. So the deposition can be controlled, a monolayer at a time, by alternating zinc and oxygen in the deposition chamber. The process is relatively easy for ZnO, because the deposition of each layer is self-limiting, although the growth of a thick layer will take a long time.

The process has also been used for GaAs, for example, using a compound such as GaCl which is introduced into the chamber to make monolayer coverage of the surface. Enough arsene is then introduced into the chamber to cover the GaCl with a monolayer of an arsenic-containing compound. The gases are then cycled in controlled amounts. The chemistry is complex, and the process is tricky because the deposition of alternating layers is not self-limiting, but it can be made to work.

References

1 Logan, R.A. (1987) *J. Cryst. Growth*, **83**, 233.

2 Kuech, T.F.and Tischler, M.A. (2000) *Handbook of Semiconductor Technology*, Vol. 2, *Processing of Semiconductors* (ed. K.A. Jackson), Wiley-VCH, Weinheim, p. 111.

3 Panish, M.B.and Temkin, H. (1989) *Annu. Rev. Mater. Sci.*, **19**, 209.

4 Stringfellow, G.B. (1999) *Organometallic Vapor-Phase Epitaxy*, 2nd edn, Academic Press, New York.

5 Arthur, J.R. (1968) *J. Appl. Phys.*, **39**, 4032; (2002) *Surf. Sci.*, **500**, 189.

6 Cho, A.Y. (1975) *J. Electrochem. Soc.*, **122**; Parker, E.H. (ed.) (1985) *The Technology and Physics of Molecular Beam Epitaxy*, Plenum, New York.

7 Joyce, B. (1988) *J. Phys. Chem. Solids*, **49**, 237.

Problems

17.1. Physical vapor deposition is a very simple process. Why are all thin films not deposited using this method?

17.2. When is chemical vapor deposition preferred over MBE?

18 Plasmas

Plasmas are used in many phases of semiconductor processing. They are used for sputtering to remove material from a target, and to deposit it on a substrate. This is a common method for depositing a metal. Silicon nitride, Si_3N_4, which is used to encapsulate microelectronic chips is deposited from a plasma. Other elements and compounds are deposited using plasma-enhanced chemical vapor deposition, PECVD. And plasmas are used for reactive ion etching, RIE.

The plasmas typically used in semiconductor processing are low pressure plasmas. Most of the atoms or molecules in these, typically weak, plasmas are neutrals, but some of them are ionized. The electrons which have left the ionized atoms or molecules gain energy from the electric fields in the plasma. The plasma is an electrical conductor, and the electrons in the plasma can move at high speeds through the plasma. These energetic electrons then collide with the atoms and molecules in the plasma, resulting in:

$$\text{Ionization, creating more ions and electrons:}\quad \begin{array}{lcl} \mathrm{A} & \rightleftharpoons & \mathrm{A}^{+} + \mathrm{e} \\ \mathrm{AB} & \rightleftharpoons & \mathrm{AB}^{+} + \mathrm{e} \end{array}$$

$$\text{Dissociation, creating radicals}:\quad \mathrm{AB} \rightleftharpoons \mathrm{A} + \mathrm{B}$$

$$\text{Excitation:}\quad \begin{array}{lcl} \mathrm{A} & \rightleftharpoons & \mathrm{A}^{*} \\ \mathrm{AB} & \rightleftharpoons & \mathrm{AB}^{*} \end{array}$$

The excited atoms and molecules emit light when they return to the ground state. The radicals are likely to be ionized, and are very reactive chemically.

18.1 Direct Current (DC) Plasmas

A plasma set-up is illustrated schematically in Figure 18.1.

The vacuum pump creates the low pressure inside the chamber which is necessary to maintain the plasmas useful in semiconductor processing. The gas inlet is used to

Kinetic Processes: Crystal Growth, Diffusion, and Phase Transitions in Materials. Kenneth A. Jackson

ISBN: 978-3-527-32736-2

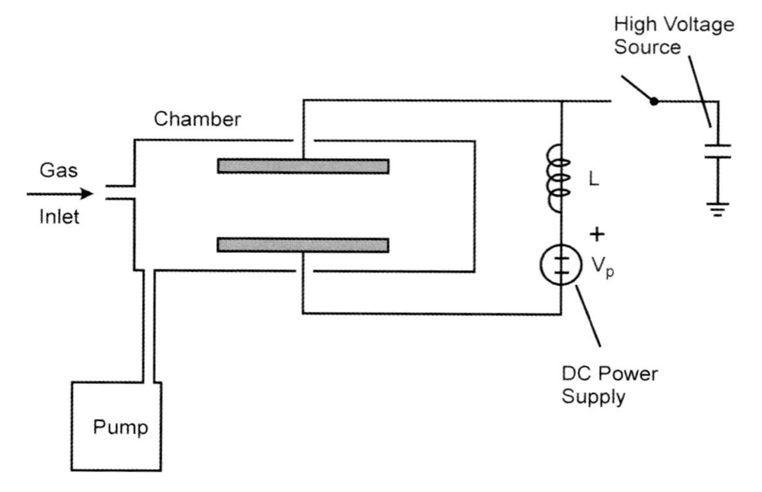

Figure 18.1 Plasma set-up.

introduce atoms or molecules into the chamber. Without the plasma, the gas in the chamber is an insulator. A plasma is initiated with the high voltage source which is used to create an electrical discharge in the chamber. This discharge creates ions and electrons in the chamber. The electrons are then accelerated by the DC power supply voltage to maintain the plasma in the region between the two electrodes.

The amount of energy which an electron can gain from an electric field depends on the average distance between collisions, which is called the mean free path. A high energy electron loses most of its kinetic energy when it collides with an atom. The low pressure in the chamber increases the mean free path of the electrons, so that they can accelerate to higher energies. In order to maintain the plasma, some of the electrons must be sufficiently energetic so that they will ionize the atoms or molecules during a collision, which typically requires at least 15 eV. Argon is often used in a plasma or to start a plasma because it ionizes readily, is inert, has a high mass, and is relatively inexpensive. Other elements or molecules are then introduced after the plasma has been ignited.

In the plasma, ions and some of the electrons have the same average kinetic energy, but the electrons move much more rapidly because of their smaller mass. There are also some high energy electrons, typically having an energy of several eV, which were originally accelerated by the cathode sheath potential, and have not been "thermalized". The electrons go to the anode at high velocities. The ions move toward the cathode, much more slowly. When the ions hit the cathode, they eject electrons from the cathode, and these electrons accelerate across the cathode sheath, where most of the potential is located. This results in a distribution of charges in the plasma, as illustrated in Figure 18.2. These charges modify the local electrical field, as illustrated.

In the plasma glow region, the density of ions and electrons is the same, so there is little net charge in this region. There are more ions near the cathode, because the

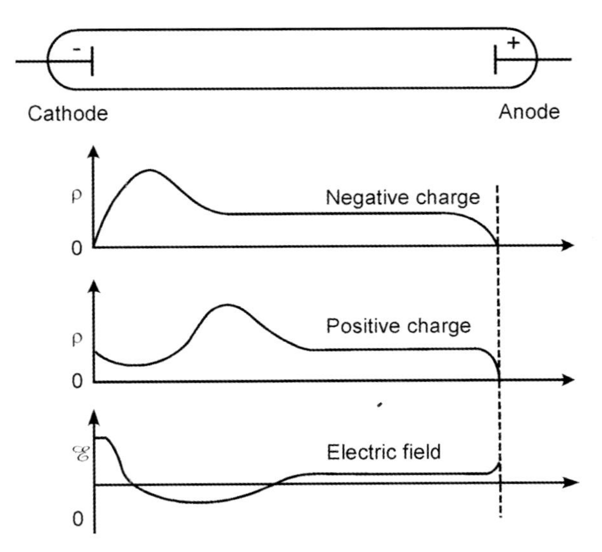

Figure 18.2 Distribution of charges and the resultant electrical field in a plasma.

electrons accelerate rapidly away from there. This creates a positively charged region near the cathode, and a negatively charged region further from the cathode. Some distance from the cathode, the electrons have enough energy to ionize atoms, and so there is a region of increased ion density, which contains a net positive charge. These ions shield the field from the cathode, so the ion density drops off farther from the cathode.

Moderate energy electrons, less than about 15 eV, can excite atoms or molecules, so that they emit light. Electron energies greater than about 15 eV are sufficient to create ions. This results in the emission of light from the plasma, as illustrated in Figure 18.3.

The width of the Crook's dark space depends on the mean free path required for electron–atom collisions. The potential across this region gets bigger and dark space gets wider for lower pressures. In the anode dark space, there are no ions. In the Faraday dark space, where there is little field, electrons have enough energy to ionize the atoms, rather than just exciting them, so there are no excited atoms to emit light.

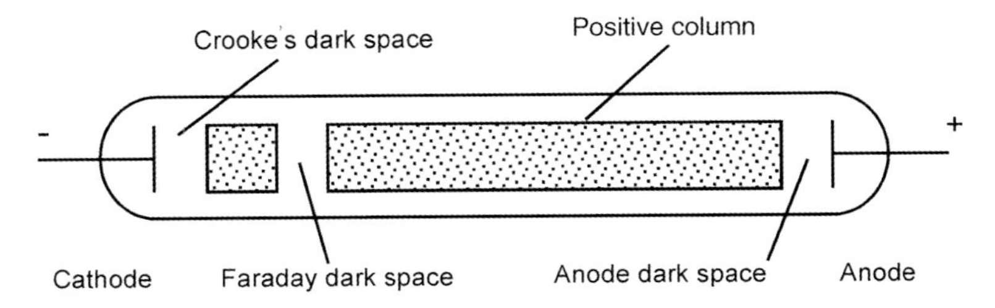

Figure 18.3 Pattern of light emission from a plasma.

There is a large field in the Crook's dark space, and so ions accelerate towards the cathode in this region. The width of the Crook's dark space depends on the pressure in the chamber, and the energy of the ions hitting the cathode depends on the potential across this dark space, and so the energy of the ions hitting the cathode can be controlled with pressure. This is typically about 0.1–1 torr for DC plasmas, so that the ions can sputter atoms off the cathode.

18.2 Radio Frequency Plasmas

A DC voltage creates a build-up of charge on the cathode if it is an insulator, and this makes it difficult for a plasma to light, and so radio frequency (rf) plasmas are used to sputter atoms from insulators.

A typical configuration is shown in Figure 18.4. A frequency of 13.56 MHz is typically used in order to stay away from broadcast frequencies. Above about 40–50 khz, the ions cannot follow the field, but the electrons can. This creates a dark space at both electrodes, as illustrated in Figure 18.5. There is no net charge in most of the region between the electrodes, and the electrons slosh back and forth in this region in the rf field. Ions are accelerated towards the surfaces of the electrodes by the net negative surface charges at the electrodes, which are created by the rapidly moving electrons.

The voltage V_1 can be increased by more than an order of magnitude over the voltage V_2 by grounding the chamber and electrode 2, and by increasing the relative

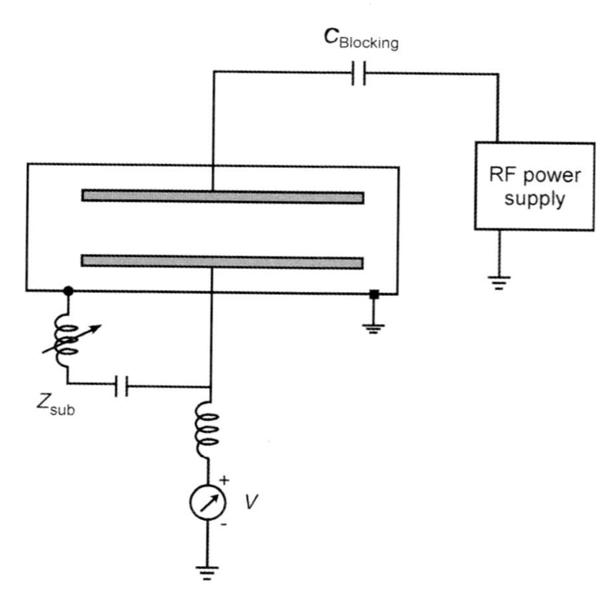

Figure 18.4 Set-up for an rf plasma.

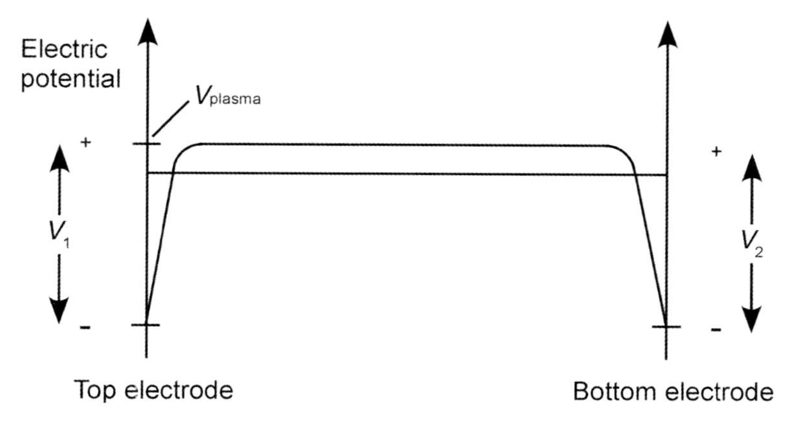

Figure 18.5 Electric potential at the electrodes in an rf plasma.

size of electrode 2. This results in much more energetic positive ions bombarding electrode 1.

18.3 Plasma Etching

There are different regimes for removing material from a surface using ions which are generated in plasmas. At one extreme is sputtering, where atoms are literally knocked from the surface by high energy incident ions. At the other extreme is plasma etching, where bombarding reactive ions which are generated in the plasma react chemically with surface atoms to create volatile species which evaporate from the surface. This is also referred to as chemical sputtering, since there is still an important kinetic element to the material removal. In between these two extremes is reactive ion etching (RIE) which is a balanced process, combining both kinetic and chemical removal of material. These three regimes are outlined in Figure 18.6.

The energy of the ions in the plasma depends on the gas pressure due to both the resultant electrical fields of the discharge and the resultant mean free path of the ions. The lower the gas pressure, the longer the mean free path of the electrons in the plasma, and so the larger the electric fields in the vicinity of the electrodes, as illustrated in Figures 18.2 and 18.5. This field accelerates the ions towards the surface. The lower operating pressures also reduce the amount of kinetic energy which is dissipated in collisions. As indicated in Figure 18.6, at low pressure, the energy of the ions striking the electrode is high, and material is removed primarily by physical impact. At high pressures, the ions are incident at the surface of the electrode with low energies, and so material is removed primarily by chemical reactions.

This difference has consequences in selectivity and in the anisotropy of the etching process, as illustrated in Figure 18.7.

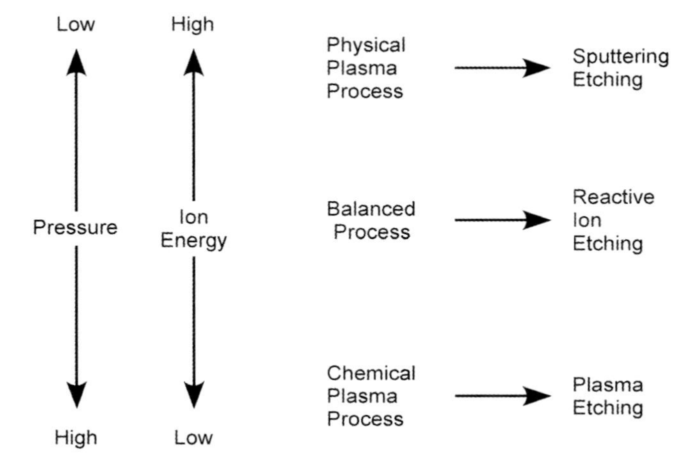

Figure 18.6 Regimes of material removal in plasmas.

Selectivity is the ability of the etching process to remove different materials from the surface at different rates. In some cases, it is desirable to remove one material selectively from the surface, while not removing other materials, for example to remove a metal but not silica, or vice versa. The choice of appropriate chemical species to create desired ions in the plasma can accomplish this. However, this process will usually remove the target material uniformly, because the ion energy is low and the mean free path is short. The active ions arrive at the surface from random orientations and adsorb on the surface to react with the atoms there. Since the incident ion energy is low, the ions are unlikely to damage the substrate.

In sputter etching, high energy ions are incident on the surface. These ions physically remove atoms from the surface by transferring some of their large kinetic energy to the surface atoms, which then have enough kinetic energy so that they fly off the surface. This process is relatively independent of the type of atom in the surface. Any type of surface atom can be knocked from the surface, and so this process has

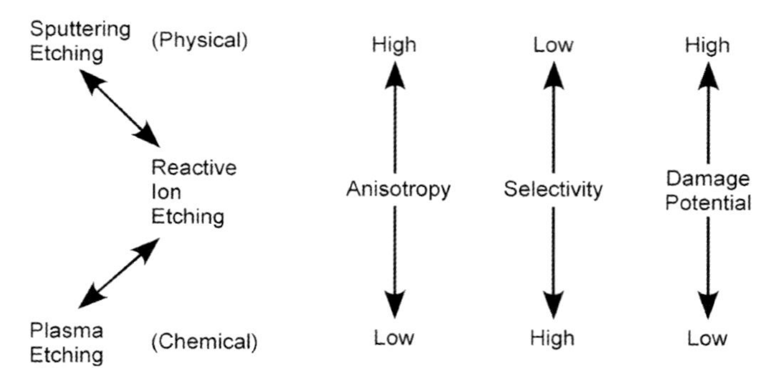

Figure 18.7 Selectivity and anisotropy of plasma etching.

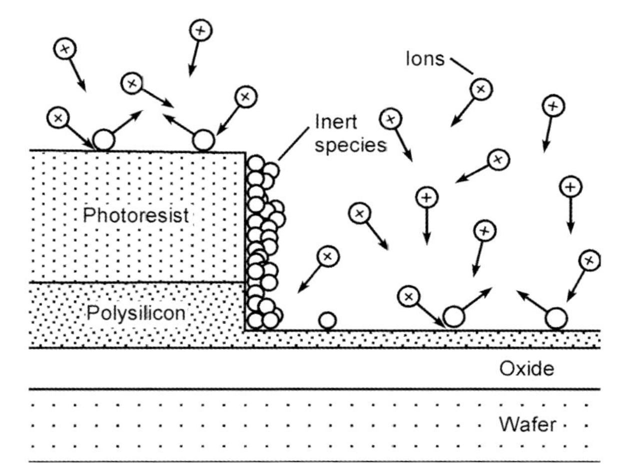

Figure 18.8 Anisotropic etching.

low selectivity. On the other hand, the ions are accelerated towards the surface by the large field there, and so their velocities have a strong component which is towards the surface. So, if there is a step on a surface, or a hole in the surface, material will be removed from the riser of the step or from the side walls of the hole at a much slower rate than from the flat parts of the surface. By masking part of the surface, a hole or a trench of a desired shape can be etched into the surface. This is anisotropic etching. Because of the high energy of the incident ions, sputtering is quite likely to displace atoms in the substrate, creating damage and defects.

The anisotropy of the etching process is increased by the accumulation of inert species, sometimes called "peanut butter", on the sidewalls of the etched features, as illustrated in Figure 18.8.

Chemical etching will attack the sides of the hole at the same rate as it attacks the bottom of the hole, so it cannot be used to create a deep narrow trench.

The selectivity and anisotropy of the plasma etching process can be controlled by changing the pressure and the species in the plasma, in the intermediate regime called reactive ion etching.

The typical pressure regimes are presented in Table 18.1. The barrel and Plasma configurations are usually used for selective etching. The magnetron and ECR

Table 18.1 Typical parameters for plasma etching.

	Pressure range (torr)	Min/Max eV at Surface	Typical ion density (cm^{-3})
Barrel	0.1–10	3/20	10^{12}–10^{13}
Plasma	1–5	100/1000	5×10^{12}
RIE	0.05–0.5	100/1000	10^{9}–10^{10}
Magnetron	0.01–0.1	50/1000	10^{9}–10^{11}
ECR	0.001–0.2	5/500	10^{11}–10^{13}

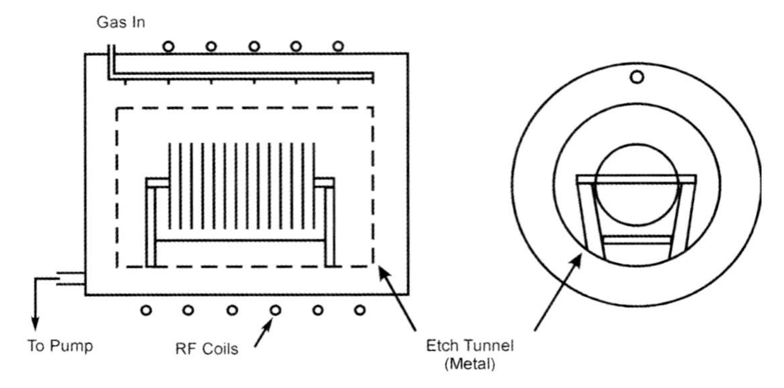

Figure 18.9 Barrel reactor.

configurations are used for sputtering. Several of the different configurations which are used for plasma etching are listed in Table 18.1. Typical pressure ranges, the range of incident ion energies and the typical ion densities are indicated in the table.

18.4 Plasma Reactors

A typical barrel reactor is illustrated in Figure 18.9. Many wafers are stacked in a rack in this type of reactor, which is typically used to strip off a photoresist after it has been used to pattern the features in the substrate. The reactor is run at relatively high pressures, but even so, there is a problem with the uniformity of etching.

A typical plasma etcher is illustrated in Figure 18.10. This type of reactor is also usually run at relatively high pressure for a vacuum chamber, up to 10 torr. The two electrodes are closely spaced, and the reactor is run at high power so that the etching rate is rapid. It is used to etch dielectrics.

Two configurations for reactive ion etching are shown in Figure 18.11. The one in Figure 18.11a is known as a pancake reactor. It is similar to the plasma etcher in Figure 18.10 except that the substrate is rotated. The barrel configuration in Figure 18.11b allows the simultaneous etching of several wafers.

RIE reactors are usually run at pressures in the 20–400 mtorr range. They are used for high resolution etching. In the physical etch regime, they are used to etch oxides, nitrides, and to make etch trenches in silicon for isolation, and in the chemical regime, they are used to etch metals, such as aluminum. In the mixed regime, they are used to etch polysilicon and silicides, as well as aluminum/copper metallizations. RIE provides a lot of control over the etching process.

Some of the chemical species which are introduced in the plasma for etching various materials using RIE are listed in Table 18.2

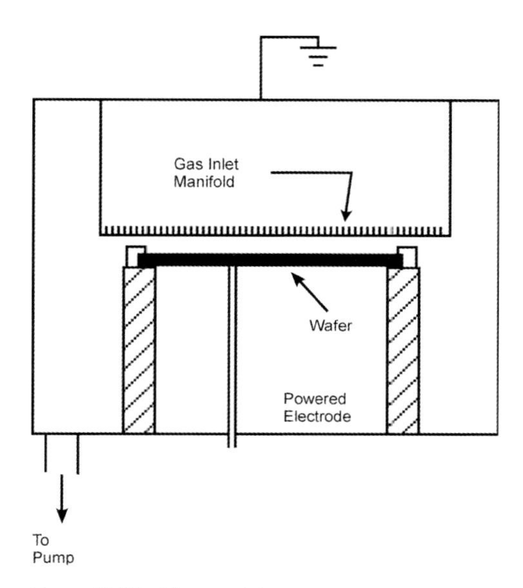

Figure 18.10 Plasma etcher.

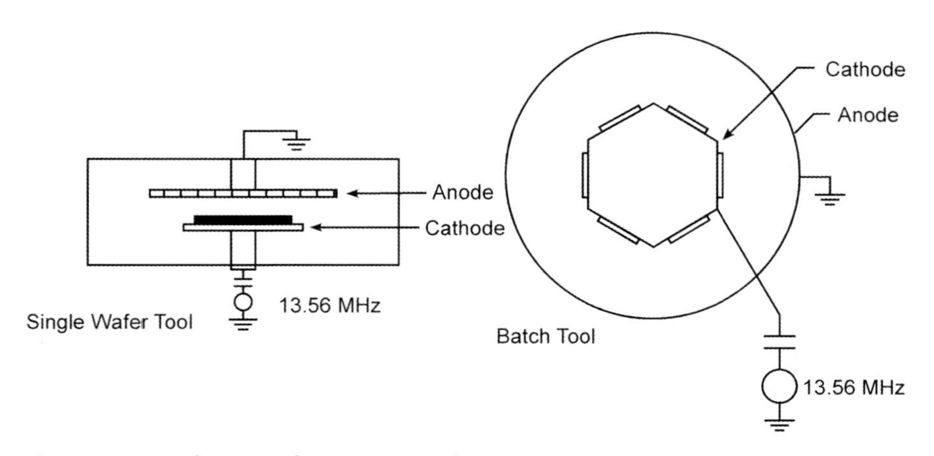

Figure 18.11 Configurations for reactive ion etching.

The primary species used to etch both silicon and silica is the fluorine ion. Fluorine ions are obtained from a variety of fluorine-containing compounds for RIE. The etching rates depend on the species present, and on the concentrations and chemistry in the plasma.

At low concentrations, oxygen helps to break down the CF_4 molecules, so the etch rate initially increases with oxygen concentration, but at higher concentrations, it makes stable COF_2, which ties up the fluorine, and reduces the etch rate.

Table 18.2 Chemical species used for RIE.

Silicon	CF_4/O_2, CF_2Cl_2, CF_3Cl, $SF_6/O_2/Cl_2$, $Cl_2/H_2/C_2F_6/CCl_4$, C_2ClF_5/O_2, Br_2, SiF_4/O_2, NF_3, ClF_3, CCl_4, CCL_3F_5, C_2ClF_5/SF_6, C_2F_6/CF_3Cl, CF_3Cl/Br_2
SiO_2	CF_4/H_2, C_2F_6, C_3F_8, CHF_3/O_2
SiN_4	$CF_4/O_2/H_2$, C_2F_6, C_3F_8, CHF_3
Organics	O_2, CF_4/O_2, SF_6/O_2
Aluminum	BCl_3, BCl_3/Cl_2, $CCl_4/Cl_2/BCl_3$, $SiCl_4/Cl_2$
Silicides	CF_4/O_2, NF_3, SF_6/Cl_2, CF_4/Cl_2
Refractories	CF_4/O_2, NF_3/H_2, SF_6/O_2
GaAs	BCl_3/Ar, $Cl_2/O_2/H_2$, $CCl_2F_2/O_2/Ar/He$, H_2, CH_4/H_2, $CClH_3/H_2$
InP	CH_4/H_2, C_2H_6/H_2, Cl_2/Ar
Au	$C_2Cl_2F_4$, Cl_2, $CClF_3$

18.5 Magnetron Sputtering

A magnetic field produces a force on a moving electron which is perpendicular to both the path of the electron and to the magnetic field.

$$\vec{F} = q\vec{v} \times \vec{B} \tag{18.1}$$

$\vec{F}$ is the force on the electron, q is the electronic charge, $\vec{v}$ is the velocity of the electron and $\vec{B}$ is the magnetic field. The magnetic field makes the electrons follow a helical path, with a radius r, given by:

$$r = \frac{mv}{qB} \tag{18.2}$$

This increases the path length for electrons to go from the cathode to anode, which increases the density of ions and radicals in the plasma. It has been suggested that the magnetic field can also be used to return electrons to the cathode, as illustrated in Figure 18.12, but it is speculation how many of the secondary emitted electrons make it back to the surface, and whether their role is significant.

Magnetically enhanced RIE (MERIE) is illustrated in Figure 18.13. MERIE is used to increase the path length of electrons going from the cathode to the anode, which increases the density of ions and reactive species.

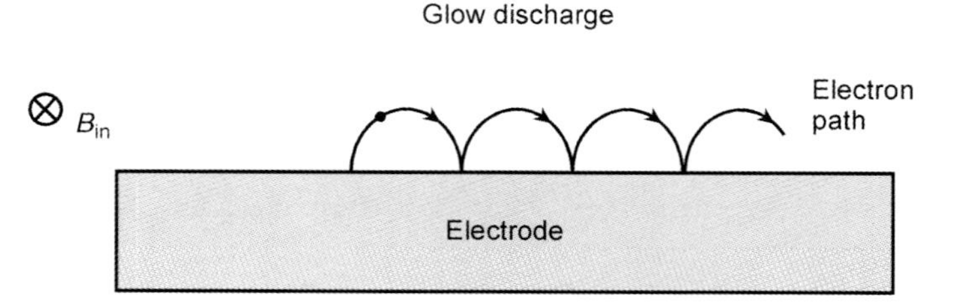

Figure 18.12 A magnetic field is used to control the path and to increase the path length of electrons in the plasma.

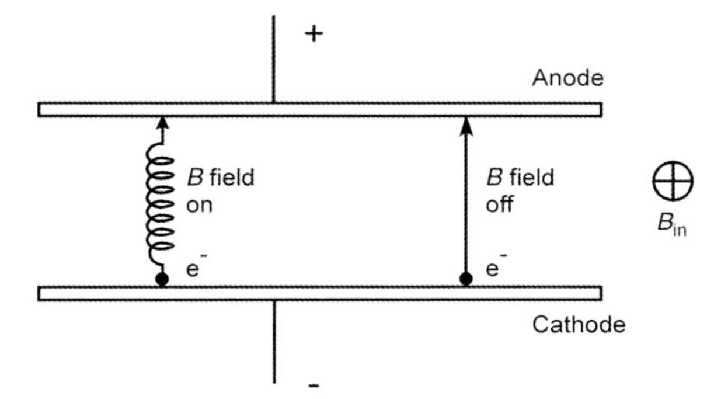

Figure 18.13 Magnetically enhanced reactive ion etching. The electrons spin around the magnetic field vector.

18.6
Electron Cyclotron Resonance

Electron cyclotron resonance (ECR) requires a high power microwave source. A typical set-up is illustrated in Figure 18.14

A plasma is created remotely from the substrate by a microwave field typically at a frequency of 2.45 GHz. Ions extracted from the plasma are often used for photoresist

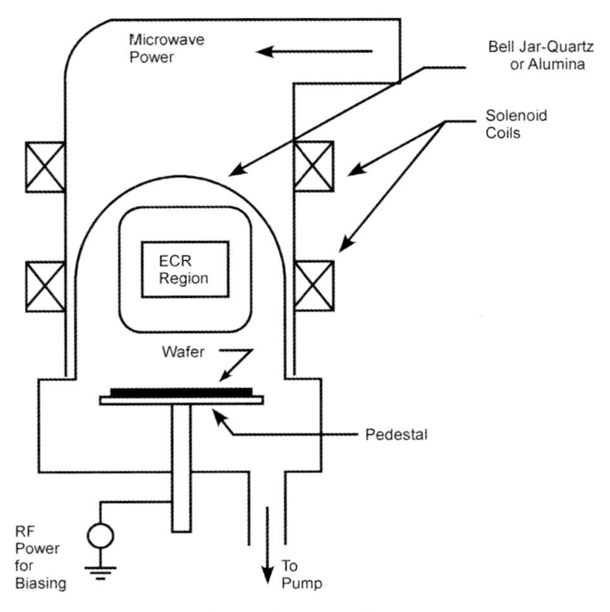

Figure 18.14 Set-up for an electron cyclotron resonance plasma.

stripping. The electron resonant frequency is given by:

$$\omega = \frac{eB}{m} \tag{18.3}$$

where m is the electron mass. For a magnetic field of 875 G, the resonant frequency is about 1.6×10^{10} rad s^{-1}, which corresponds to 2.45 GHz, the industry standard. The particles move in circles, and gain energy from the microwave field throughout the circle if the circumference of the circle is greater than the electron mean free path. The accelerating electrons generate Bremsstrahlung and can emit in the deep UV or soft X-ray region, so shielding is required. The efficient coupling of power to plasma electrons and absence of high fields required in the earlier electrode-coupled discharges allows high densities of ions and radicals in the plasma.

Electron cyclotron resonance is also used to create a remote plasma from which a high density of low energy ions can be extracted in order to reduce wafer damage.

18.7 Ion Milling

Ion milling is similar in principle to plasma sputtering, but it is used to remove material especially uniformly from a surface. Instead of a plasma in contact with the surface, an ion beam is generated so that ions are incident on the surface at a small angle. The glazing angle of incidence tends to remove any bumps from the surface. But atoms are removed from the surface primarily by physical sputtering processes. Ion milling is used to prepare specimens for transmission electron microscopy, and also to clean substrates in vacuum chambers prior to deposition.

A Kaufman ion source is often used to generate the ion beam. Electrons from a heated filament are accelerated towards an anode through a field of about 40 V so that they can ionize argon atoms. The argon ions are then extracted and accelerated by 500–1000 V to make an ion beam. The system is run at relatively low pressure, and the erosion rate is relatively independent of the type of material in the target.

Further Reading

Donohoe, K.G., Turner, T., and Jackson, K.A. (2000) *Handbook of Semiconductor Technology*, Vol. 2, *Processing of Semiconductors* (ed. K.A. Jackson), Wiley-VCH, Weinheim, p. 298.

Murarka, S.P. and Peckerar, M.C. (1989) *Electronic Materials*, Academic Press, San Diego, CA, p. 510.

Plummer, J.D., Deal, M.D., and Griffin, P.B. (2000) *Silicon VLSI Technology*, Prentice-Hall, Upper Saddle River, NJ, 527, 619.

Problems

18.1. Why is sputtering not used to deposit epitaxial layers?

18.2. Discuss the differences between RIE, sputtering and ion milling.

19
Rapid Thermal Processing

Rapid thermal processing (RTP) [1, 2] is being used increasingly during semiconductor processing, in situations where it has an advantage over furnace annealing.

19.1
Introduction

In general, a short time anneal at a high temperature will produce an effect which is equivalent to a longer anneal at a lower temperature, but different processes have different activation energies, so that they depend differently on temperature. If a desired process proceeds at a relatively faster rate than an unwanted process at higher temperature, then it will be advantageous to do a higher temperature anneal.

For example, after ion implantation, many of the dopant atoms are not on substitutional sites in the lattice where they will be electrically active. And so the p/n junction is not at the metallurgical junction. It is usually desired to make these dopant atoms electrically active with minimal diffusive motion, and 15 s at 900 °C is often sufficient to do this. The atoms can move the short distance necessary to find a lattice site, and there is no long range motion. The short anneal time must be accompanied by rapid heating and cooling in order to limit the total time at a high temperature.

The rapid cooling in an RTP apparatus is also used to increase the doping level. At higher temperatures the solubility of the dopants in silicon is greater, and, using rapid cooling, these high concentrations of electrically active dopants can be quenched in. A high concentration of dopant is needed in the channel under the gate in CMOS, and concentrations of active dopants which are significantly higher than the equilibrium concentration are obtained this way.

Silicide conductors stripes and contacts are also made with RTP. For example, titanium can be deposited onto an oxide which has been patterned to expose some of the underlying silicon. The titanium can be diffused into the exposed silicon to make titanium silicide, using RTP. The titanium on top of the oxide does not form a silicide, and can subsequently be removed by etching.

Kinetic Processes: Crystal Growth, Diffusion, and Phase Transitions in Materials. Kenneth A. Jackson

ISBN: 978-3-527-32736-2

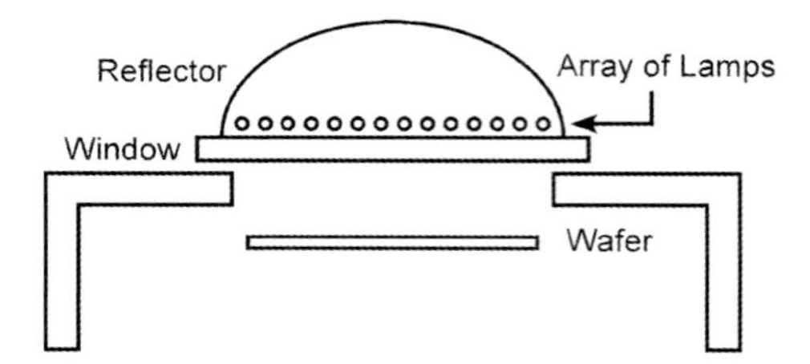

Figure 19.1 Rapid thermal heating systems.

Also, as device features shrink, the distances dopants need to diffuse are shorter, and so annealing times decrease. For furnace annealing, there is a limitation to how rapidly the wafers can be heated without cracking, and so the furnace temperatures must be ramped on heating and cooling. This precludes very short annealing times. It is possible to heat wafers much more rapidly with RTP than in a furnace, as will be discussed below.

19.2
Rapid Thermal Processing Equipment

RTP makes use of high powered tungsten halogen lamps to heat the wafer radiatively, as shown schematically in Figure 19.1. These lamps are about the size of a pencil. They have a tungsten filament in a silica envelope. The lamps are tubular in form, and run very hot. Tungsten evaporates from the filament, and deposits on the inside wall of the tube. A halide, often $PNBr_2$, picks up the tungsten from the wall and re-deposits it back on the filament. These lamps can put out about 100 W per centimeter length, and can be assembled into an array which can output tens of kilowatts of optical power.

A wafer is supported under the array of heat lamps on silica pins. Heating rates up to hundreds of degrees per second can be achieved. Wafers are heated to temperatures in the range 850–1050 °C, held there for times of the order of 10–60 s, and then cooled equally rapidly when the lamps are turned off.

Plasma discharge lamps using krypton or xenon are also used. These are more efficient and can output up to about 700 W cm^{-1}, but they need to be ignited, and they require a regulated DC power source.

19.3
Radiative Heating

For radiative heating, the spectral radiance for a black body is given by:

$$M = \varepsilon\sigma T^4 \text{ W cm}^{-2} \tag{19.1}$$

Here ε is the emissivity and $\sigma = 5.6697 \times 10^{-8}\text{ W cm}^2\,°\text{C}^{-4}$ is Stefan's constant. A black body is defined as a body with an emissivity of one, independent of

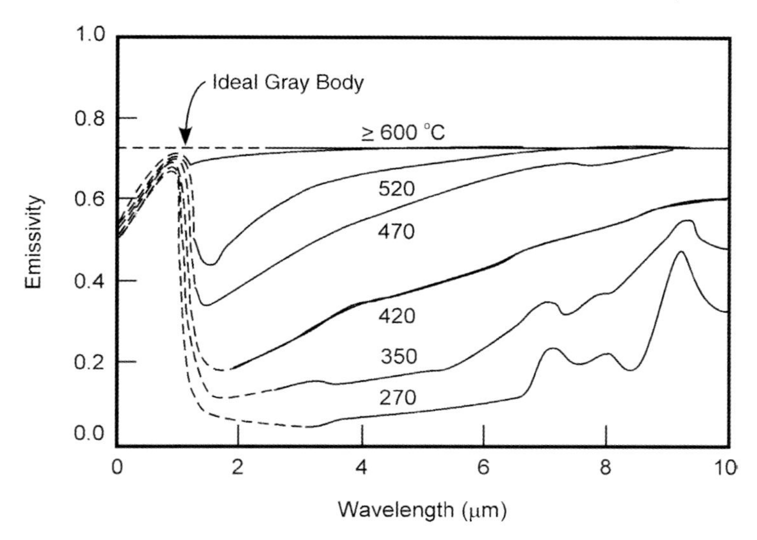

Figure 19.2 The emissivity of silicon depends on both wavelength and temperature. (After Sato *et al.* [3]).

wavelength. An ideal gray body has an emissivity which is independent of wavelength, but less than one. As illustrated in Figure 19.2, the emissivity of a material, in general, depends on both the wavelength and the temperature:

$$\varepsilon(\lambda, T) = 1 - \varrho(\lambda, T) - \tau(\lambda, T) \tag{19.2}$$

where ϱ is the reflectivity and τ is the transmission. τ is zero for an opaque material.

The power transferred by radiation from a body 1 at temperature T_1 to body 2 at temperature T_2 is:

$$\sigma\left(\varepsilon_1 T_1^4 - \varepsilon_2 T_2^4\right) A_1 F \tag{19.3}$$

where A_1 is the surface area of body 1, and F is the view factor. Different parts of a wafer experience different view factors, as illustrated in Figure 19.3.

19.4 Temperature Measurement

Measuring the temperature of a sample that is being radiatively heated is tricky. The heating rate and final temperature of a sample depend on a balance between the incident radiation energy absorbed by the sample and the energy radiated from the sample. The radiative energy absorbed depends on the emissivity of the sample, and the energy radiated from the sample depends on both its temperature and emissivity. The apparent temperature as viewed by a pyrometer depends on the temperature of

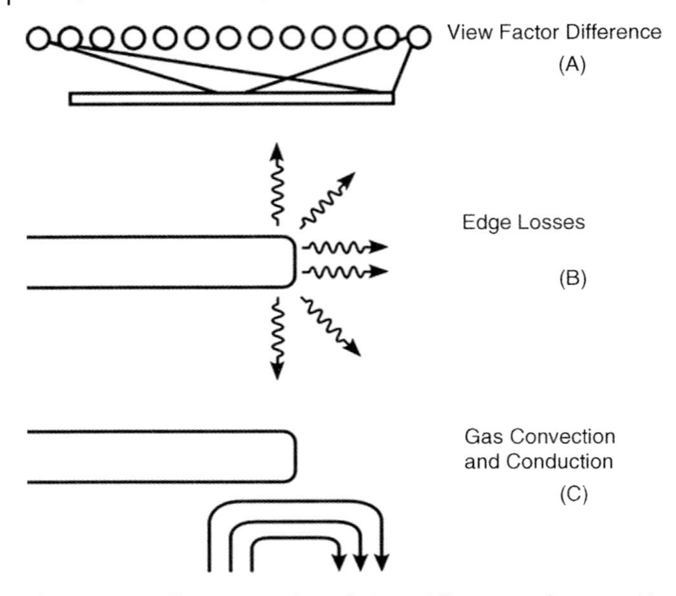

Figure 19.3 Different parts of a wafer have different view factors and heating rates.

the sample and on its emissivity. Increasing the emissivity of the surface will lower the temperature of the sample by reducing the net power into the sample, but will simultaneously tend to increase the apparent temperature by increasing the radiation from the sample.

A thermocouple near the wafer will not correctly indicate the temperature of the wafer. If the thermocouple touches the wafer, it will indicate the correct temperature only if there is good thermal contact, in which case the thermocouple is likely to act as a local heat sink, changing the temperature locally. The thermocouple can also deposit unwanted elements on the wafer. A non-contact thermopile can be used if its emissivity can be matched to that of the wafer. A pyrometer can be used, but the direct and indirect radiation from the heat lamps must be filtered out. Using two-color pyrometers, where the emissivity at both wavelengths is known, the temperature can be determined from the ratio of the measured intensities at the two wavelengths. This can be used to control the temperature of the wafer to about $\pm 10^{\circ}$. Transmission through layers can result in interference effects, as illustrated in Figure 19.4, so that pyrometers cannot be used with layered structures.

Diffraction gratings have been put on the surface of wafers to measure the thermal expansion, from which the temperature can be derived. However, the grating material can result in contamination, and it must be subsequently removed. It is also possible to estimate the temperature from the change in size of the wafer.

The velocity of sound in a wafer depends on its temperature. Acoustic waves have been launched into wafers through the support pins. A temperature profile can be obtained by using several pins. This method has been used for the calibration of multizone heaters.

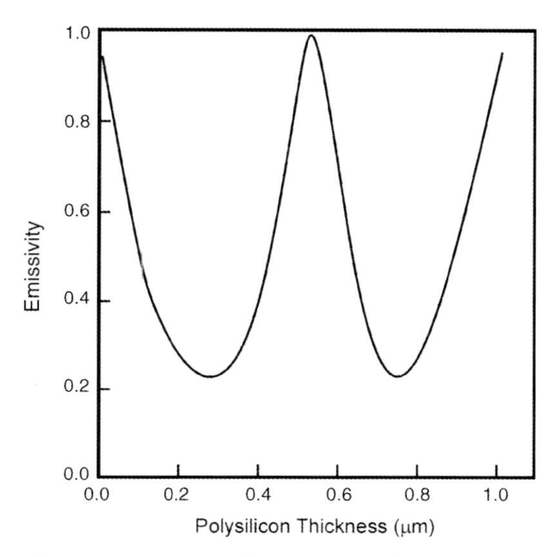

Figure 19.4 Emissivity of a $Si/SiO_2/PolySi$ layered structure at a wavelength of 3.5 μm, as a function of the thickness of the polysilicon layer.

19.5
Thermal Stress

Non-uniform temperature distributions create stresses. A sample changes size with temperature, but, if it is unconstrained, this change in size does not create a stress. When a linear temperature gradient is imposed on a free-standing sample, the expansion at one end of the sample does not create much stress on the other end of the sample. Only a small stress arises because there must be some bending of the lattice planes to accommodate the different expansions.

However, if a sample is constrained in length and then heated, large stresses can be generated. If a sample of length l is heated through a temperature change ΔT, its length will change by $\Delta l = \alpha \Delta T l$, where α is the thermal expansion coefficient. In order to constrain such a sample from changing length, a stress σ must be applied, where σ is given by the strain, $\Delta l/l = \alpha \Delta T$, times the appropriate elastic constant, E:

$$\sigma = E\alpha\Delta T \tag{19.4}$$

If there is a radial temperature gradient on a wafer, then the outside part of the wafer tries to constrain the inside part from changing size. This introduces a stress of magnitude given by Equation 19.4. Dislocations multiply rapidly when the shear stresses on the slip systems in the crystal reach a critical value, so it is important to minimize radial temperature gradients on the wafer. The yield stress varies widely for different crystals, and is temperature dependent. Silicon has a fairly high yield stress, and can withstand much higher stresses that GaAs, for example.

In RTP, radial temperature differences are minimized by carefully controlling the heat input so that it is very uniform across the wafer. A linear temperature gradient through the thickness of the wafer, on the other hand, does not generate much stress. When these conditions are met, the wafers can be heated and cooled very rapidly with RTP.

During furnace annealing, many wafers which are held more or less vertically on a rack are inserted into the furnace at the same time. The edges of a wafer are closer to the wall of the furnace, and so heat and cool more slowly than the center. So the furnace temperature must be ramped up and down in order to heat and cool the wafers slowly, in order to minimize the radial temperature gradients. The larger the wafer diameter, the slower must be the ramping of the temperature.

Many wafers can be annealed in a furnace at one time, so it is inherently a cheaper process than RTP, where wafers are processed one at a time. There must be a significant advantage in resultant properties in order to justify the extra cost of RTP.

19.6 Laser Heating

Pulsed lasers can also be used to heat a thin layer on the surface of a sample very rapidly, and even to melt a layer on the surface of a silicon wafer. A large pulse of energy can be deposited at the surface in a time which is short compared to the time it takes for the deposited heat to diffuse into the sample. However, the energy in a pulse is somewhat difficult to control precisely, so this is used only where the energy density deposited is not critical. Alternatively, a cw laser beam can be raster scanned or spirally scanned over a surface.

Laser heating is now used in the fabrication of active matrix liquid crystal displays (AMLCDs). Active matrix displays have a small thin film transistor (tft) controlling each pixel. The display is built up on a glass substrate and viewed with light transmitted though the structure. The first tfts were made in amorphous silicon which had been deposited at a low temperature. Tfts with improved properties are now made in polysilicon but the substrate cannot withstand the temperature needed to deposit polysilicon, so amorphous silicon is deposited and then converted to polysilicon by rapidly heating only a thin layer at the surface with a laser.

References

1 Campbell, S.A. (1996) *The Science and Engineering of Microelectronic Fabrication*, Oxford, New York, p. 126.

2 Hill, C.S. and Boys, D. (1989) *Reduced Thermal Processing for ULSI* (ed. R.A. Levy), Plenum, New York.

3 Sato, T. (1967) *Jap. J. Appl. Phys.*, **6**, 339.

Problems

19.1. Why is the measurement of the temperature of a wafer difficult to measure in RTP? What factors does the apparent temperature depend on?

19.2. Why is RTP coming into increasing use in semiconductor processing?

20
Kinetics of First Order Phase Transformations

20.1
General Considerations

The crystal growth process is usually reversible: the net growth rate is the difference between the arrival rate and departure rate of atoms at the crystal surface, as is implied by Equation 13.33, which is based on chemical reaction rate theory. At equilibrium these two rates are equal. When both phases are present, the crystal will grow when the interface is below the equilibrium temperature, and it will melt or dissolve when the temperature is above the equilibrium temperature.

In general, crystals grow by the addition of atoms or molecules from the mother phase, one at a time. The path of the atoms or molecules between the two phases is reversible, which is known as "microscopic reversibility". The rate at which an atom or molecule joins or leaves the crystal depends on the local environment of the atom or molecule in the growth phase, as well as on how many of the nearest neighbor lattice sites are occupied with atoms of the crystal. This determines how tightly the solid atom or molecule is bound to the crystal, and how readily an atom can join or leave the crystal. This is discussed in more detail in Section 20.8 below.

In growth from the vapor phase or from a solution, it is fairly obvious whether an atom or molecule is part of the crystal or not. Surface atoms or molecules can arrive from the vapor phase or from the solution, move around on the surface by diffusion, and then perhaps leave the surface again. Indeed, if it were not possible to distinguish the atoms of the crystal from those of the neighboring phase, then it is difficult to imagine how there could be a nucleation barrier to the formation of new layers, as there usually is in vapor phase or solution growth.

In melt growth the situation is not so obvious. For metals, there is only a small percentage difference in density between the melt and the crystal, so it is possible that there could be a gradual transition from the crystal to the melt. But molecular dynamics simulations indicate that atoms which belong to the crystal can be distinguished from those which are part of the melt. The distinction can be made in two ways.

Kinetic Processes: Crystal Growth, Diffusion, and Phase Transitions in Materials. Kenneth A. Jackson

ISBN: 978-3-527-32736-2

One is using the time trajectory of the atoms. Atoms which are part of the crystal vibrate about a lattice site. Atoms which are part of the liquid migrate randomly. An individual atom at the interface will wander randomly for a while, then perhaps join the crystal where it vibrates about a fixed position, and then perhaps leave the crystal again to wander randomly. There is a clear distinction between these two modes. When an atom joins the crystal, it typically stays there for many vibrational periods, and when it is in the liquid, it wanders for many vibrational periods. The vibrational mode can be clearly distinguished from the wandering mode.

The other method of distinguishing between solid and liquid atoms is using the radial distribution function of the atom, which describes the locations of neighboring atoms. The radial distribution function of solid atoms has a distinct minimum which is not present for liquid atoms. That the state of individual atoms at a crystal/melt interface can be distinguished is perhaps a surprising result, since a metal crystal and its melt have such similar properties.

From molecular dynamics simulations, it is evident that the atoms of the crystal at the interface create a potential well because of their regular positions. Atoms in the liquid can fall into this well, stay there for a while and then hop out again when they have enough energy. The latent heat of the transformation is associated with the potential energy of the atom falling into or hopping out of this energy well.

There has been some discussion in the literature about the importance of the observation in molecular dynamics simulations that there is density variation in a liquid next to a plane wall. The period of the density variations is the atomic diameter, and they extend several atomic diameters into the liquid away from the wall. However, there is no evident dependence of the radial distribution function of the atoms on their distance from the wall. These density variations are apparently due to minor variations in the atom positions near the flat wall. Similar density variations are found in the liquid next to crystal surfaces, and it has been suggested that these density waves influence the crystallization of the liquid atoms. However, it is difficult to understand how these minor variations in position of the atoms could influence the energy or momentum distributions of the atoms in such a way as to influence the rate at which the atoms join the crystal.

Although atoms are either solid or liquid, there can be a gradual transition from the liquid to the solid in one sense. For the simple metals and some other materials, the interface is rough at the atomic level, as discussed in Chapter 21. There are a few atomic layers at the interface in which there are both solid and liquid atoms. The atoms are either solid or liquid, and these layers contain both.

So, in general, we can assume that there are atoms which belong to each phase, and that the transition between the two phases occurs atom by atom or molecule by molecule. But this statement is not meant to imply that there is no interaction between adjacent atoms during the crystallization process.

Some crystal interfaces are smooth at the atomic level, and some are rough at the atomic level. This distinction is the subject of Chapter 21. The theory of the surface roughening transition is based on the assumption that the atoms of the crystal can be distinguished from the atoms in the phase it is in contact with. It concludes that some interfaces are smooth at the atomic level, and that some are rough at the atomic level, depending in large measure on the difference between the properties of the two phases. If the density of the atoms or molecules is very different in the two phases, as it is for vapor phase growth or solution growth, then the interface is likely to be smooth. If the difference is small, as it is for a metal in contact with its melt, then the interface is likely to be rough.

After a brief comment on the shape of crystalline materials, there will be a discussion of growth on rough interfaces, followed by models for growth on smooth interfaces.

20.2 The Macroscopic Shape of Crystals

Some crystals grow with a crystalline morphology. They have a polyhedral shape. This is the basis for the ancient Greek ideas about matter consisting of atoms.

As was pointed out by Gibbs [1] many years ago, the surface free energy of a macroscopic crystal is a very small contribution to its total free energy. The size of this contribution can be seen from the relationship between the critical nucleus size and the corresponding undercooling, Equation 15.11. For a crystal containing a few hundred atoms, the equilibrium temperature is reduced significantly. For a crystal of micron dimensions, the equilibrium temperature will be reduced by about one degree. This undercooling represents the amount of free energy which is available to drive a diffusion process which could change the shape of the crystal. If the crystal is of millimeter dimensions, then the undercooling due to the surface free energy will be millidegrees. The corresponding driving force will produce little diffusion over distances of a millimeter. Temperature fluctuations of millidegrees are present in most experimental systems. So macroscopic crystals which have a "crystalline" shape do not owe their shape to equilibrium surface tension, but rather to anisotropy in the growth rate.

The crystalline shape develops because the rapidly growing faces disappear leaving the crystal bounded by the slow growing crystal faces, as illustrated in Figure 20.1.

Gibbs in a footnote in one of his papers pointed out that it was possible that there could be a problem in the formation of new layers during crystal growth, and left it at that, but we will explore this in detail. The growth rate is much more rapid on smooth interfaces than on rough interfaces. Crystals which have all rough interfaces, such as the metals growing from their melts, have rapid growth rates which are fairly isotropic. The growth rate is usually limited by diffusion, and so the crystals do not look like crystals.

20.3 General Equation for the Growth Rate of Crystals

The rate at which a first order phase transformation proceeds was discussed in Chapter 13, where it was shown that the rate of crystallization depended on the

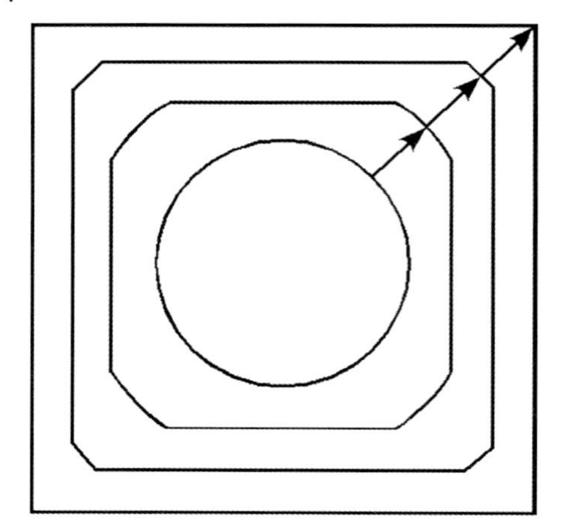

Figure 20.1 The rapidly growing faces disappear, leaving only the slowest growing faces.

difference in free energy between the two phases. But that discussion included only a formal representation of the crystallization process. The rate of crystallization can be expressed as the product of four terms: a length, a frequency, a term which depends on the structure of the interface, and the free energy difference between the two phases.

$$v = a\,v^{+} f\, u_{\mathrm{k}} \tag{20.1}$$

The first term, a, is a distance which is related to the atomic or molecular diameter of the growth unit. The second term, v^{+}, is the rate at which atoms join the crystal at active growth sites on the surface. The third term, f, depends on the roughness of the interface, and is the fraction of interface sites which are active growth sites. The local free energy or chemical potential difference between the two phases which is driving the transformation is contained in the fourth term u_{k}:

$$u_{\mathrm{k}} = 1 - \exp\left(\frac{\Delta G}{k_{\mathrm{B}} T}\right) \tag{20.2}$$

as in Equation 13.31. ΔG can be expressed as a pressure difference using Equation 20.11, as an undercooling or Equation 20.7 as a supersaturation, as discussed in Chapter 9.

The atoms or molecules at the interface join and leave the crystal at rates which depend only on their local environment and on the local departure from equilibrium. Their motion depends on their individual kinetic energies and the local potentials to which they are subject. There is no action at a distance.

There is an ambiguity in defining the area occupied by a growth site on the interface for different orientations of the crystal. We can get around this ambiguity by defining the distance a to be the cube root of the atomic volume, $a^3 = \Omega$, and the area of a growth site on the interface to be a^2. Then if an atom is added to each growth site, the interface will advance a distance a. This is numerically equivalent to adding a volume Ω to the crystal with each atom, and then defining f to be the density of growth sites on the interface.

Not all of the surface sites on a crystal are active growth sites. The factor f is the probability of finding an active growth site in an area a^2 of the interface. Atoms leave and join the crystal surface at many sites, but net growth occurs only at active growth sites. The density of active growth sites can vary widely, and is strongly temperature dependent if the growth depends on the nucleation of new layers. This will be discussed in more detail below. f is typically about $^1/_4$ for a rough surface, but can be very small on a smooth surface

The rate at which atoms or molecules join the crystal at active growth sites, ν^+, depends on the phase into which the crystal is growing, and on the mobility of growth units in that phase.

The classical theory for vapor growth was developed by Herst [2] and by Knudsen [3] and that for melt growth by Wilson [4] and Frankel [5]. In these models, the density of growth sites, f, was not discussed. It was assumed implicitly that atoms or molecules could join a crystal at any surface site. But these models contain the essence of the physical description of the rate at which atoms or molecules can join the crystal at active growth sites, ν^+, and so they are discussed below in Section 20.5.2 and Section 20.6.

This chapter also contains a brief review of two models for growth on smooth surfaces: surface nucleation, Section 20.9, and growth aided by screw dislocations, Section 20.10. These are the two classical models for f, the fraction of interface sites which are active growth sites. These models apply only to growth on smooth surfaces, and their applicability depends on how smooth the surface is.

Reliable information about the distribution of active growth sites on a surface can be obtained from Monte Carlo computer simulations, and detailed information about growth rates for specific materials and orientations requires molecular dynamics modeling.

20.4
Kinetic Driving Force

The kinetic driving force, ΔT_K or ΔC_K in Chapter 9, depends on the difference between the free energy of the two phases at the interface. This is a local free energy difference, and it usually varies from point to point along the interface. This is the chemical potential difference which the atoms at the interface see, and it determines their net rate of motion of atoms across the interface.

The free energies of the two phases are illustrated in Figure 20.2.

The difference between the free energies of the two phases can be written as:

$$\Delta G = (E_1 - E_2) + P(V_1 - V_2) - T(S_1 - S_2) \tag{20.3}$$

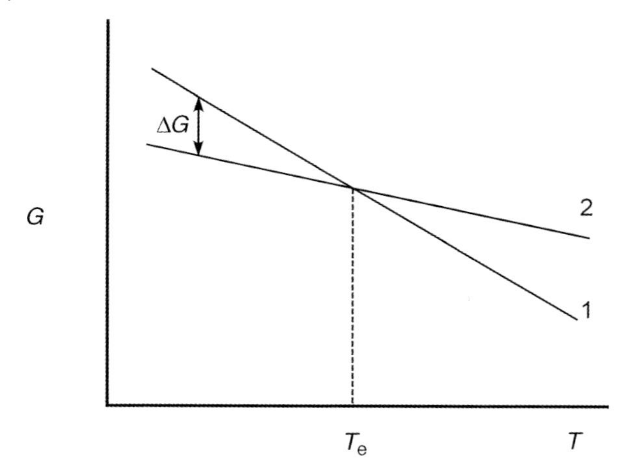

Figure 20.2 Free energies of the two phases.

Here T is the temperature and P is the pressure. The internal energy, E, the volume, V, and the entropy, S, of each phase indicated by the subscript. At equilibrium, $\Delta G = 0$, so that:

$$\Delta E + P_e \Delta V - T_e \Delta S = 0 \tag{20.4}$$

where $\Delta E = E_2 - E_1$, $\Delta V = V_2 - V_1$, and $\Delta S = S_2 - S_1$. The enthalpy difference, $\Delta H = \Delta E + P\Delta V$, which is the also known as the latent heat of the transformation, L, is:

$$\Delta H = L = T_e \Delta S \tag{20.5}$$

For variations in temperature and pressure, the free energy difference between two phases can be written as:

$$\Delta G = -\Delta P \Delta V + \Delta T \Delta S \tag{20.6}$$

where $\Delta P = P_e - P$, and $\Delta T = T_e - T$. This is the difference between the free energies of two phases which are not in equilibrium.

At constant pressure, the free energy difference is linearly proportional to the undercooling:

$$\Delta G = \Delta T \Delta S = \frac{L\Delta T}{T_e} \tag{20.7}$$

Inserting Equation 20.7 into Equation 13.33 suggests that the transformation rate should be linearly proportional to the undercooling at constant pressure for small departures from equilibrium.

Nominal changes in pressure have little effect on the temperature of equilibrium between condensed phases, since $\Delta V \approx 0$. For example, Equation 20.7 is usually used to evaluate the free energy difference between a crystal and its melt.

The free energy curves in Figure 20.2 are not precisely linear with temperature or pressure, and so these relationships are not precisely linear, but they are a very good approximation for small departures from equilibrium.

20.5 Vapor Phase Growth

20.5.1 Equilibrium

For equilibrium between the two phases, $\Delta G = 0$, Equation 20.6 can be combined with Equation 20.5 to give the change in equilibrium pressure with temperature:

$$\frac{dP_e}{dT_e} = \frac{L}{T_e \Delta V} \tag{20.8}$$

which is known as the Clausius–Clapeyron equation. If only one of the phases is a vapor, the volume of the vapor phase is much greater than that of a solid or liquid, then $\Delta V = V_{vapor} - V_{solid} \approx V_{vapor} \approx RT/P$ for an ideal gas, so Equation 20.8 can be written:

$$\frac{dP_e}{P_e} = \frac{L}{R}\frac{dT_e}{T_e^2} \tag{20.9}$$

which can be integrated to give:

$$P_e = P_e^0 \exp\left(-\frac{L}{kT_e}\right) \tag{20.10}$$

Which describes the temperature dependence of the equilibrium vapor pressure of many materials. This temperature dependence, given by a Boltzmann factor, is just the fraction of atoms on the surface which have enough kinetic energy to overcome the binding energy of the solid in order to escape into the vapor phase.

20.5.2 Kinetics of Vapor Phase Growth

The free energy difference between two phases is linearly proportional to the pressure difference at constant temperature:

$$\Delta G = -\Delta P\, \Delta V = -kT\frac{\Delta P}{P} \tag{20.11}$$

where the latter approximation assumes that one of the phases is an ideal gas. Combining Equtation 13.33 with Equtation 20.11 suggests that the transformation rate should be linearly proportional to the pressure difference at constant temperature:

$$\Delta R \approx R_0 \frac{\Delta P}{P} \tag{20.12}$$

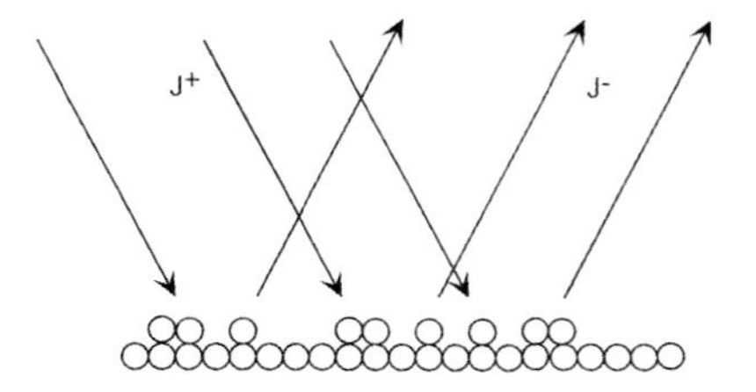

Figure 20.3 Flux of atoms to and from a surface.

A simple model for the kinetics of vapor phase growth was developed by Herst [2] and Knudsen [3] about 100 years ago. Knudsen's name is still associated with an evaporative source used for vapor deposition, the Knudsen cell.

The vapor pressure of a material which is in equilibrium with a solid or liquid is called the equilibrium vapor pressure.

When the rate at which atoms or molecules join the solid or liquid is the same as the rate at which they are evaporating, as illustrated in Figure 20.3, then there is equilibrium. The rate at which atoms or molecules strike the surface of the solid or liquid depends on the density of the atoms or molecules in the vapor phase.

In Chapter 1, we developed an expression for the pressure generated by atoms hitting a piston. It is the product of the number of atoms striking the piston per second times the momentum change when the atoms bounced off the piston:

$$P = (\text{change in momentum per molecule}) \times (\text{number of molecules per second striking the piston per unit area})$$

so that:

$$P = (2mv_x)J \tag{20.13}$$

In Equation 1.4, $^1/_2\, nv$ was used instead of J for the flux of atoms hitting the piston. We assumed that the atoms did not interact with the surface of the piston, they just bounced off it, as would be the case, say, for argon atoms hitting a stainless steel surface. But here we are discussing growth from the vapor phase, so the surface is the same material as the atoms or molecules. The atoms or molecules striking the surface may bounce, or they may stick on the surface, and then evaporate at a later time. This difference in behavior is described by the sticking coefficient, which is the fraction of atoms which adsorb onto the surface. For various combinations of vapor phase species and substrates, the sticking coefficient can be anywhere between 0 and 1. The flux of atoms to the surface from the vapor is independent of the sticking coefficient, and here we will assume that the sticking coefficient is 1.

A detailed analysis gives a more precise value than Equtation 20.13 for the relationship between the vapor pressure and the flux of atoms hitting a surface:

$$P = \sqrt{\pi}\, mv\, J \tag{20.14}$$

Using

$$v = \sqrt{\frac{2\, kT}{m}} \tag{20.15}$$

gives:

$$J^{+} = \frac{P}{\sqrt{2\pi\, mkT}} \tag{20.16}$$

As in Equation 1.28, at equilibrium, the flux of atoms to the surface is equal to the evaporating flux of atoms:

$$J^{-} = \frac{P_{\mathrm{e}}}{\sqrt{2\pi mkT}} \tag{20.17}$$

For non-equilibrium conditions, the net flux of atoms to the surface can be written:

$$\bar{J} = J^{+} - J^{-} = \frac{P - P_{\mathrm{e}}}{\sqrt{2\pi mkT}} \tag{20.18}$$

So the net flux to the surface is proportional to the difference between the pressure and the equilibrium pressure. The equilibrium vapor pressure depends on temperature, as indicated by Equation 20.12.

This expression assumes that all the atoms on the surface are in equivalent positions, and that there is no nucleation barrier to the formation of new layers of the crystal. These are good assumptions for the interaction between a liquid and a vapor, but they are not valid in general for a crystal, as will be discussed in more detail below.

20.6 Melt Growth

A crystal invariably has a lower enthalpy than its melt. The difference is the latent heat of fusion. The atoms of a crystal must have sufficient energy to leave the crystal and join the melt, just as they must have sufficient energy to leave the crystal to join the vapor phase, as in Equation 20.10.

The rate at which atoms leave the solid to join the liquid at active growth sites, ν^{-} in Equation 20.1, must contain this energy difference as a Boltzmann factor:

$$\nu^{-} = \nu_0^{-} \exp\left(-\frac{L}{kT}\right) \tag{20.19}$$

At the equilibrium melting point, T_{M}, the rate at which atoms join the crystal must contain a similar factor:

$$\nu^{+} = \nu_0^{-} \exp\left(-\frac{L}{kT_{\mathrm{M}}}\right) \tag{20.20}$$

$\Delta S = L/TM$ is the entropy difference between the two phases. This entropy factor for atoms joining the crystal compensates for the increase in energy when atoms leave the crystal. It derives from the difference in order between the crystal and its melt. An atom in the crystal occupies a volume which is very similar to the volume it will occupy in the melt. For example, the difference in specific volume between a crystal and its melt is typically about 3% for metals. So an atom in the crystal must increase its energy to leave, and an atom in the liquid increases its state of order to join the crystal. It must be going towards a lattice site in order to join the crystal. These two effects, one energetic, and the other geometrical, compensate to make the net rate of transition equal to zero at equilibrium. Of course there can also be a geometrical factor involved in the rate at which atoms leave the crystal, but then the entropy factor is the ratio of these two geometrical factors.

For melt growth of metals, $\Delta S/k$ is typically about one, so this factor is about 1/3. For small organic molecules, $\Delta S/k$ is typically about 6. The difference in order between the crystal and the melt should account for one unit of this, as for a metal. The other five units come from the rotational disorder of the melt. The molecule must have the correct orientation in order to join the crystal. This suggests that less than 1% of the molecules are in the right orientation to join the crystal.

20.6.1
Early Models for Melt Growth

The first analysis of the growth rate of a crystal from a liquid was published by Wilson [4] in 1900. He related the rate at which atoms join the crystal to the diffusion coefficient of the liquid, which he assumed was of the form:

$$D = \frac{1}{6} a^2 \upsilon_D \exp(-Q/kT) \tag{20.21}$$

He assumed that that $(G^* - G_1)$in Equation 13.33 was the Q in Equation 20.21. So the Wilson expression for the growth rate is:

$$v = \frac{6D}{a} \frac{L\Delta T}{kT_M T} \tag{20.22}$$

About 30 years later, Frenkel [5] expressed the growth rate in terms of the viscosity of the liquid. The two treatments are essentially equivalent since the viscosity of a liquid is related to the diffusion coefficient by the Stokes–Einstein relationship:

$$D = \frac{kT}{3\pi\eta a} \tag{20.23}$$

The Frenkel expression for the growth rate can be written:

$$v \approx \frac{2kT}{\pi a^2 \eta} \frac{L\Delta T}{kT_M T} \tag{20.24}$$

It is found experimentally that the diffusion coefficient and the reciprocal of the viscosity have the same temperature dependence. Experiments on glass-forming systems, where it is relatively easy to measure the growth rate as a function of temperature, indicate that the same activation energy also applies for crystal growth. That is, all three have the same activation energy, Q.

But the prefactor is not correctly predicted by these equations.

The entropy factor of Equation 20.20 should be included in both Equations 20.22 and 20.24.

In addition, the diffusion jump distance in the liquid is not the atomic diameter, a, as suggested by Equation 20.21. An atom in the liquid diffuses by a series of small motion,s as discussed in Chapter 2. In Chapter 2, Section 2.1, a distinction was made between the mean free path, λ, of an atom in a liquid, and Λ, which is the average distance which an atom moves during a diffusive motion.

In some cases, a rearrangement of the configuration and positions of other atoms in the liquid is required in order for the atom or molecule to join the crystal. In these cases, it is the average time required for an atom to move a diffusion jump distance, Λ, which determines the rate at which an atom or molecule can join the crystal. Replacing a^2 with Λ^2 in Equation 20.21, and adding the factor f as well as the entropy term from Equation 20.20 to Equation 20.22 results in a modified version of the Wilson expression for the growth rate:

$$v = \frac{6\,a\,D}{\Lambda^2} f \exp\left(-\frac{\Delta S}{k}\right)\left[1-\exp\left(-\frac{L\Delta T}{kT_{\mathrm{M}}T}\right)\right] \tag{20.25}$$

Similar modifications can be made to the Frenkel expression for the growth rate, Equation 20.24.

This is the appropriate expression for the crystallization rate for many glass forming materials, where the mobility of the atoms limits the growth rate. Crystallization rate measurements can be made readily for glass forming materials, and at large undercooling, the temperature dependence of the growth rate is dominated by the temperature dependence of the mobility and so the growth rate follows Arrhenius behavior. This equation also agrees with the results of molecular dynamics simulations of the crystallization of silicon, as discussed in the next section.

20.6.2 Melt Growth Rates

Metal crystals grow very rapidly at very small undercoolings and so growth usually occurs at small undercoolings. In this case, the growth rate is usually limited by diffusive processes. For pure metals, the growth rate is limited by heat flow, that is, by how fast the latent heat can diffuse away from the interface. For alloys, chemical diffusion is also important.

Over the limited temperature range where growth occurs, a simple expression for the temperature dependence of the growth rate can be obtained by approximating the square bracket in Equation 20.25. The relationship between the growth rate and the undercooling can then be written in the simple form:

$$v = \mu\,\Delta T \tag{20.26}$$

where μ is called the kinetic coefficient.

For a typical metal growing from the melt,

$$v \sim 50\,\Delta T\ \mathrm{cm\,s^{-1}} = 0.5\Delta T\ \mathrm{m\,s^{-1}} \tag{20.27}$$

It is important to remember that growth processes take place only at the interface. The atoms at the interface are subject to the conditions at the interface: the local temperature, the local composition, the local curvature of the interface. The thermal fields and compositional fields in the sample as a whole influence the local conditions at the interface, but the kinetic processes at the interface on the atomic scale depend only on the local conditions there. The atoms at the interface respond only to their immediate environment.

20.7 Molecular Dynamics Studies of Melt Crystallization Kinetics

The diffusion coefficients as determined from molecular dynamics (MD) simulations for liquid argon, using a Lennard-Jones (LJ) potential, and for silicon, using the Stillinger–Weber (SW) potential are shown in Figure 2.2. The diffusion processes in both liquids are very similar. The activation energies for diffusion scale with the melting points of the two materials. Both liquids are very similar, but the crystals which form are quite different.

For the LJ potential, the force which any two atoms exert on each other depends only on the distance between them, and not on the location of any other atoms. This is called a pair potential. With this potential, the atoms crystallize with a face centered cubic structure, which is a close-packed structure.

The Stillinger–Weber [6] potential is a three-body potential developed especially for silicon. Each atom in crystalline silicon has four nearest neighbors, and the angle between the bonds to the nearest neighbors is important. So the potential depends not only on the distance from each neighboring atom, but also on the angle between each pair of neighboring atoms. The potential provides a good approximation for the structure and properties of liquid silicon, and, below the melting point, a crystal forms with the diamond cubic structure, as the potential was designed to do.

The crystallization behavior of the two is quite different.

MD simulations using the SW potential [7] are shown in Figure 20.4. The solid line is the Wilson–Frenkel expression, Equation 20.25, fitted to the data using the activation energy derived from the diffusion data for liquid silicon. This correctly describes the computer simulation data. This curve is similar to the experimental data for the crystallization of silica, Figure 9.1.

MD simulations using the LJ potential [8] are shown in Figure 20.5.

The open circles are the simulations results. The solid line is the Wilson–Frenkel growth rate, Equation 20.25, using the activation energy for liquid diffusion from simulation. It is obvious that the data are not described by the Wilson– Frenkel equation. The growth is not thermally activated.

This result suggests that an atom can move a small distance to join the crystal without interference from other atoms. In this case, it is the mean free time, which is the time required for the atom to traverse its mean free path, λ, which determines the growth rate.

An atom can traverse its mean free path to join the crystal at the average thermal velocity, $(3\,kT/m)^{1/2}$. υ^+ in Equation 20.1 is then the reciprocal of the mean free time between collisions in the liquid, multiplied by the entropy factor, as in Equation 20.20,

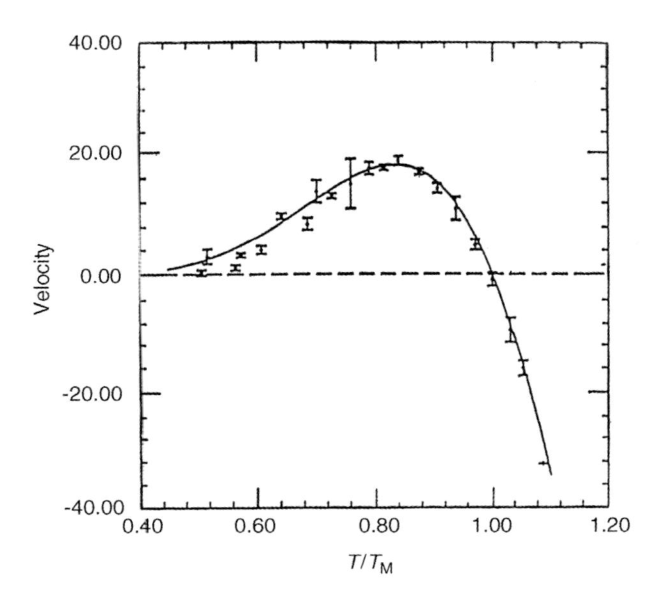

Figure 20.4 Molecular dynamics growth rates for (100) silicon using the Stillinger–Weber potential.

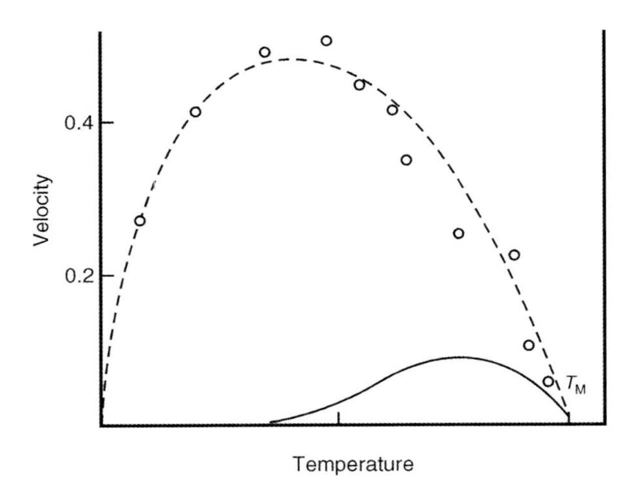

Figure 20.5 Molecular dynamics growth rates for (100) using a Lennard-Jones potential.

which gives the fraction of atoms moving in the right direction:

$$v_0^- = 1/\tau = \bar{v}/\lambda = \sqrt{3\,kT}/m/\lambda \tag{20.28}$$

The growth rate can be written:

$$v = \frac{a}{\lambda}\sqrt{\frac{3\,kT}{m}}\exp\left(-\frac{\Delta S}{R}\right)\left[1-\exp\left(-\frac{L\Delta T}{kT_{\mathrm{M}}T}\right)\right]f \tag{20.29}$$

The dashed line through the data was calculated using Equation 20.29, with $\lambda = 0.1a$, and $f = 0.25$.

This implies that the atoms run unobstructed from their positions in the liquid at the interface into the neighboring solid site at the interface, with the average thermal velocity of the atoms in the liquid.

There are potential wells formed at the interface by the regular positions of the solid atoms at the interface. In the simulations, a liquid atom falls into one of these wells, stays there for a while, vibrating about an average position, and then hops to move randomly in the liquid for a while until it falls back into the same, or a different, well.

This growth rate does not depend on the rate at which atoms can move around in the liquid by diffusion. Unlike the diffusion jump distance, Λ, the mean free path, λ, is not strongly temperature dependent. This is the appropriate expression for the solidification rate of metal crystals, and also for the crystallization of the inert gases. These materials can crystallize at very low temperatures, and they do not form glasses.

For silicon, and for molecular materials in general, there is an activation energy associated with rearrangement of the liquid structure. This activation energy determines the temperature dependence of the diffusion coefficient in the liquid, the viscosity of the liquid, and the crystallization rate. This is unlike the results from simulations with the LJ potential, where the atoms are essentially spheres. In this case, the activation energy for liquid diffusion and viscosity are the same, but the crystallization process is not thermally activated.

These results suggest that materials can be divided into two groups based on their crystallization behavior:

> One group contains materials for which no rearrangement of the liquid structure is required for any individual atom to join the crystal. The crystallization is not thermally activated. These materials have not been made into glasses.
> The other group contains materials where there must be some structural rearrangement of the liquid around an atom or molecule before it can join the crystal. The structural rearrangement process is thermally activated with the same activation energy as liquid diffusion and viscosity. These materials will not crystallize at very low temperatures, so that they can, in principle, be made into glasses.

In the discussion above, the number of active surface sites, f, has been assumed to be constant. This is usually the case for metals and the inert gases. However, for many materials, surface nucleation is required for growth, in which case f is strongly temperature dependent. This will be discussed in detail in Chapter 21.

20.8
The Kossel–Stranski Model

Stranski [9] suggested that the reason crystals grow with a crystalline shape is because there is a barrier to the formation of new layers on the closest packed faces. He suggested that the barrier exists because the rate at which individual atoms

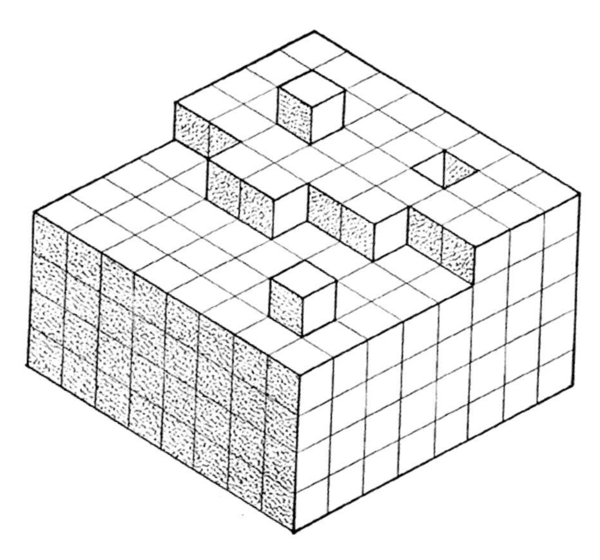

Figure 20.6 The rate at which an atom leaves the surface of a crystal depends on how many nearest neighbors it has.

leave the surface of a crystal depends on how tightly they are bound to the surface. The surface of a simple cubic crystal where the surface atoms have various numbers of nearest neighbors in the crystal is illustrated in Figure 20.6.

Stranski assumed that the strength of binding of an atom to the surface depends on its number of nearest neighbors. He applied this to a simple cubic crystal, which is also known as a Kossel crystal [10], and so this is now known as the Kossel–Stranski model. This is, of course, a simple approximation to how the binding depends on the number of neighbors, but this model does contain the essence of an explanation of why there is a strong anisotropy in the growth rate of some crystals. This model is still the basis for modern statistical mechanical models of crystal growth.

In this model, it is assumed that the rate at which atoms join the crystal is independent of the local configuration, and that the rate at which an atom leaves the crystal depends on how many of its nearest neighbor sites are occupied by atoms of the crystal. The normalized net rate at which an atom joins and leaves the crystal is:

$$1-\exp\left(\frac{L}{kT_{\mathrm{M}}}-\frac{n\phi}{kT}\right) \tag{20.30}$$

Here ϕ is the bond energy, given by $\phi = 2L/Z$, where Z is the number of nearest neighbor sites in the crystal. n is the actual number of nearest neighbors of the atom at the surface of the crystal. Because each bond is shared between two atoms in the bulk of the crystal, the binding energy per atom, L is equal to $Z\phi/2$.

An important concept introduced by Stranski is the repeatable step site, which is also called a "kink site" because it is the site at a kink in a surface step. It is a site, as can be seen in Figure 20.6, where an atom has half of its nearest neighbors, $n = Z/2$. An atom in the kink site breaks these $Z/2$ bonds when it leaves the surface, so the energy for it to leave is exactly the average latent heat per atom. When an atom joins or leaves a kink site, the repeatable step site moves along the step, so the free energy associated with the surface configuration does not change. Except for edge effects, the entire crystal can be built up by the motion of these kink sites.

At equilibrium, the rate of arrival of atoms at a kink site is the same as the rate of departure, as can be seen from the Equation 20.30.

A general expression for the growth rate was written in Equation 20.1 as:

$$\upsilon = a\upsilon^{+} f\, u_{\mathrm{K}} \tag{20.31}$$

f is the fraction of surface sites which are active growth sites. All active growth sites are kink sites. But on very rough surfaces, not all of the kink sites are active growth sites. For growth on a rough crystal surface, the factor f is more or less a constant, approximately $^1/_4$, a value which depends on the configuration of a rough surface. A surface cannot be constructed which has only repeatable step sites, and computer modeling suggests that the value of $^1/_4$ for f is about the best that can be done. The surface can be made rougher, but the effective number of growth sites does not increase much.

If we knew the site distribution function for the surface, that is, the probability of an atom on the surface having n nearest neighbors, we could replace the terms fu_{K} in Equation 20.31 with a sum of the probability for each type of site, times the net rate of addition of atoms at that type, as in Equation 20.30. The site distribution function can be estimated at equilibrium, but it is best determined by computer simulation.

20.9 Nucleation of Layers

The Kossel–Stranski model implies that when the surface of a crystal is essentially atomically flat, with only a few adatoms in the next atomic layer, there is a kinetic barrier to the formation of new layers. After his success with developing nucleation theory, Becker applied it to evaluate the rate at which a circular disc of atoms will form on an atomically flat surface. Following the model for the nucleation of a three-dimensional spherical cluster of atoms, the free energy of a monolayer disc on a surface, as illustrated in Figure 20.7, can be written as:

$$\Delta G_{\mathrm{r}} = -a\pi r^2 \Delta G_{\mathrm{V}} + a2\pi r\sigma \tag{20.32}$$

where a is the height of the disc, r is the radius of the disk, ΔG_{v} is the change in free energy per unit volume to add atoms to the crystal, and σ is the surface free energy per unit area.

Becker used the surface free energy of a flat surface for the edge free energy of the disc. We will see later that this is wrong. This equation for surface nucleation is similar to the change in free energy to form a sphere, except it is for a two-dimensional disc on a surface. The critical size for such a disc is given by:

$$\frac{1}{r^*} = \frac{\Delta G_{\mathrm{V}}}{\sigma} = \frac{L\Delta T}{\sigma T_{\mathrm{M}}} \tag{20.33}$$

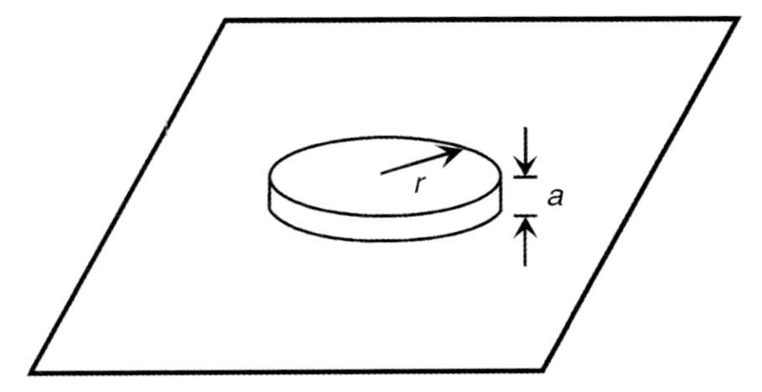

Figure 20.7 Disc-shaped nucleus of a new layer on a crystal surface.

and the rate of nucleation of new layers on the surface is given by:

$$I = I_0 \exp\left(-\frac{\Delta G_{r^*}}{kT}\right) \qquad (20.34)$$

On a smooth surface, the factor f, the density of active growth sites in Equation 20.31, depends on the density of growth sites provided by the nucleation process.

Equation 20.34 is very similar to the expression for three-dimensional nucleation, Equation 15.33, and so the undercooling required for nucleation is similar in both cases. The undercooling should be of the order of 20% of the melting temperature in order to get a reasonable nucleation rate. This implies that there should be a major nucleation barrier to the formation of new layers on a crystal. But crystals can grow at very small undercoolings.

Figure 20.8 shows the crystallization rate of Salol as a function of temperature. Salol can be quenched into the glassy state readily because its crystallization rate is so

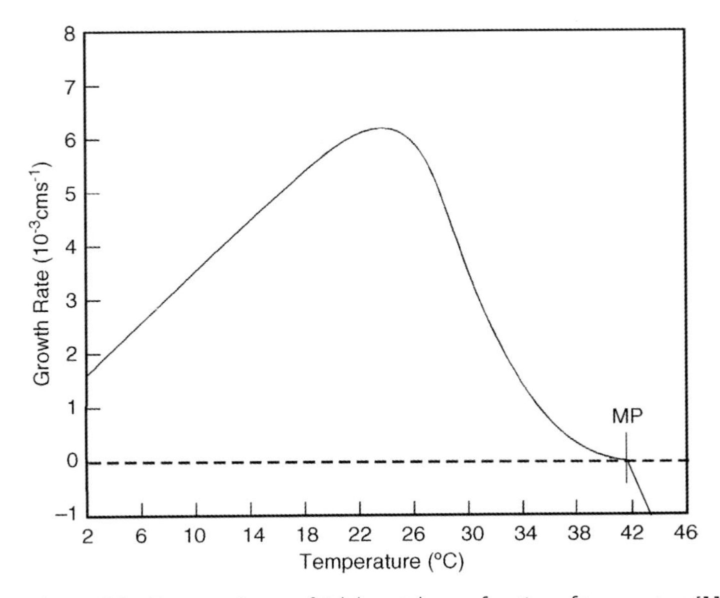

Figure 20.8 The growth rate of Salol crystals as a function of temperature [11].

slow. The growth rate at small undercooling exhibits nucleation limited behavior: the growth rate comes into the melting point horizontally.

The growth rate decreases at lower temperatures because the mobility of the atom, as indicated by the diffusion coefficient as in Equation 20.22, or by the viscosity, as in Equation 20.24, decreases as the temperature drops, but the growth rate is not linear with undercooling at small growth rates.

Figure 20.8 indicates an asymmetry in the crystallization rate and the melting rate at the melting point. On melting, the crystals become rounded as steps move in from the edge of the crystal. The melting rate is not limited by the nucleation of new layers. Experiments on the growth of voids internally in a crystal have shown the opposite effect on the morphology when the crystal is growing or shrinking. The growing void is faceted because the nucleation of steps is necessary, whereas when the void is shrinking, it becomes rounded.

20.10
Growth on Screw Dislocations

Charles Frank (see [12]) suggested that defects in the crystal structure could help to form new layers, so that a nucleation process was not necessary for growth. He pointed out that a screw dislocation which ended at a surface could provide a continuous step on a surface for growth. He suggested that the surface step would wind up into a spiral, as illustrated in Figure 20.9.

The minimum radius of curvature on the spiral should be the critical radius, as in Equation 20.33, which would result in the spacing between the arms of the spiral also being equal to the critical radius. Thus the step density on the surface will be proportional to the undercooling. This means that the factor f is proportional to ΔT, and since u_K is also proportional to ΔT at small undercoolings, Frank predicted that the growth rate of a crystal should be proportional to ΔT^2.

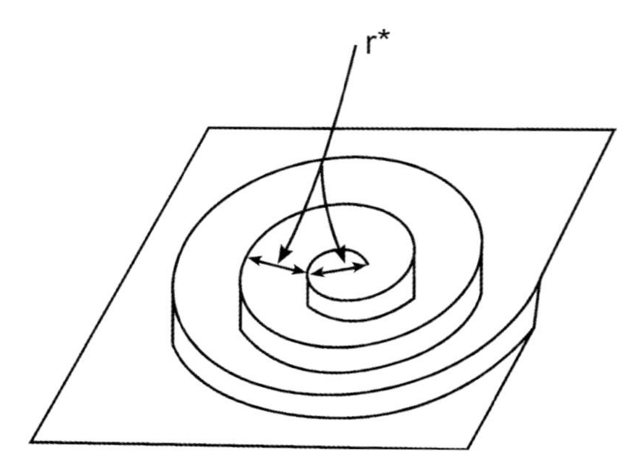

Figure 20.9 Spiral step on a crystal surface generated by a dislocation.

Dislocations are not necessary for crystal growth: large dislocation-free crystals of silicon are grown every day.

However, growth on defects can be important when there is a barrier to the formation of new layers. When the rate at which new layers nucleate is very slow, then growth can occur at the growth sites provided by defects. A growth rate which is proportional to ΔT^2, as Frank predicted, is often observed at small undercoolings on surfaces where there is a nucleation barrier to growth. At larger undercoolings, the nucleation rate increases so that the nucleation of new layers takes over the growth process. An example is shown in Figure 20.10 for the crystallization of lithium disilicate [13]. The reduced growth rate is the actual growth rate, times the viscosity, divided by u_k. Multiplying by the viscosity removes the temperature dependence of the mobility, and dividing by u_k removes the temperature dependence due to the difference in free energies of the two phases. So this is a plot of the density of active growth sites, the factor f in Equation 20.1, as a function of undercooling. At small undercooling, as in the inset, the active growth site density increases linearly with undercooling, up to an undercooling of about 100 °C, in accordance with the screw dislocation model. The density of active growth sites is then relatively constant from about 100 °C undercooling to 400 °C undercooling. At still larger undercooling, surface nucleation takes over, and the growth rate increases dramatically.

The defect density varies from crystal to crystal, and the nucleation rate depends on the crystal face as well as the crystal structure, as will be discussed below, so the

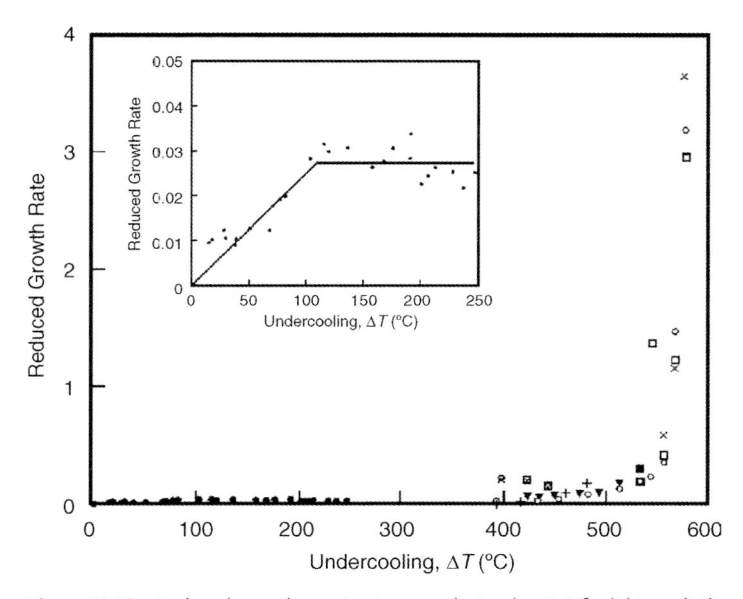

Figure 20.10 Reduced growth rate (active growth site density) for lithium disilicate [13]. At small undercoolings the defect density determines the density of active growth sites, and at large undercooling surface nucleation takes over.

magnitude of the rates, and the cross-over point from defect dominated growth to surface nucleation growth vary widely.

The barrier to the formation of new layers is not as large as that calculated by Becker, and there is no barrier to the formation of new layers on a rough surface. Becker was wrong to assume that the edge of a step has the same specific free energy as a flat surface of the crystal. The edge of a step is much rougher than a crystal surface, and the entropy associated with this roughness decreases the free energy of the step.

In order to treat this in more detail we will examine the equilibrium configuration of a surface in the next chapter, and we will do this using the Kossel–Stranski model, which is illustrated in Figure 20.6.

20.11 The Fluctuation Dissipation Theorem

This section is included to demonstrate the applicability of the fluctuation dissipation theorem to crystal growth. The kinetic coefficient can be determined by measuring the fluctuations of the interface.

20.11.1 Determination of the Kinetic Coefficient

The measurement of fluctuations in the interface position at equilibrium can be used to determine both the kinetic coefficient and the surface tension [14]. This correlation makes use of the Onsager fluctuation dissipation theorem [15], which can be stated as follows.

> The rate at which a system, which has been displaced from equilibrium, returns to equilibrium is the same as the rate at which fluctuations in the system decay at equilibrium.

How the kinetic coefficient can be obtained by studying fluctuations at equilibrium [16] is outlined below.

We will start by examining the rate of decay of fluctuations in a system at equilibrium that is part solid and part liquid. An example is shown in Figure 20.11, where the data are taken from a Monte Carlo simulation [17].

The number of solid atoms fluctuates with time, and we will denote the number of solid atoms at any given time, t, as $N_S(t)$. Similarly, the number of liquid atoms at time t is $N_L(t)$. The total number of atoms, N, is fixed: $N = N_S(t) + N_L(t)$. Over time, there is an average number of solid atoms, which we will denote $<N_S>$. The number of solid atoms fluctuates about this average value, and we will define the instantaneous departure from the average value as $\Delta N_S(t) = N_S(t) - <N_S>$. The decay rate of fluctuations can be determined using a time correlation function:

$$C(t) = \frac{\langle \Delta N_S(t) \cdot \Delta N_S(0) \rangle}{\langle \Delta N_S^2(0) \rangle} \tag{20.35}$$

The carets indicate average values.

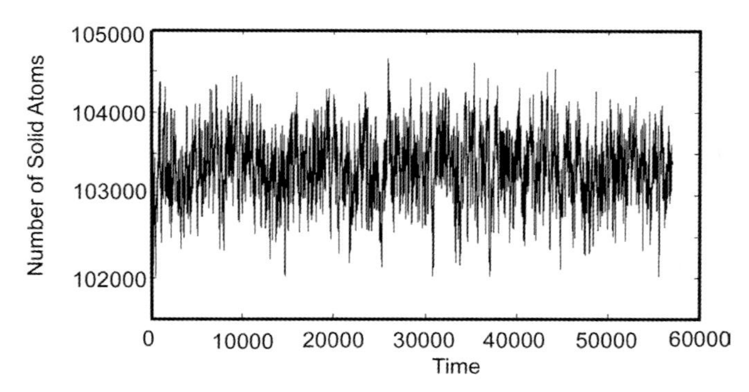

Figure 20.11 Number of solid atoms in a Monte Carlo simulation of a crystal at equilibrium with its melt.

The time correlation function is evaluated by designating some arbitrary time to be $t = 0$. The number of solid atoms at that time is recorded, and then the number of solid atoms is recorded as a function of time after this initial time. In order to do this, there must be a method to determine the number of solid atoms as a function of time, as there is, for example, in Monte Carlo simulation data in Figure 20.11. For molecular dynamics simulations, a solid atom can be identified by its immediate surroundings, such as the positions of its nearest neighbors.

The product $\Delta N_S(t).\Delta N_S(0)$ can then be calculated for each time, t. Then another time is picked as $t = 0$, and the process is repeated. Then the values of $\Delta N_S(t).\Delta N_S(0)$ for each starting time are averaged. If $\Delta N(0)$ is positive, then at short times, $\Delta N_S(t)$ is also likely to be positive, and if $\Delta N_S(0)$ is negative, then at short times, $\Delta N_S(t)$ is also likely to be negative, but the product $\Delta N_S(t).\Delta N_S(0)$ will be positive in either case.

For $t = 0$, the time correlation function, $C(0) = 1$. At some long time later, the number of atoms in the solid will be randomly greater than or less than the average, and so the product $\Delta N_S(t)\cdot\Delta N_S(0)$ will be randomly positive and negative, with an average value of 0. So $C(t) = 0$ for large t. The time correlation function should have the form:

$$C(t) = \exp(-t/\tau) \tag{20.36}$$

If $C(t)$ decays exponentially as indicated, then the time constant, τ, can be determined from the data. If $C(t)$ does not decay exponentially with time, then this analysis will not work. The time correlation function shown in Figure 20.12 is for the data in Figure 20.11. The slope of the solid line is $1/\tau$.

The fluctuations in the number of solid atoms in Figure 20.11 seems random, but there is a time constant buried in the fluctuations which can be extracted using time correlations, as shown in Figure 20.12.

In order for this scheme to work, there must be a fixed average number of solid atoms, and the number of solid atoms fluctuates about this average. If the interface is unconstrained, the number of solid atoms can wander randomly. So there must be a restoring force which drives the number of solid atoms back to the average value. We will impose a temperature gradient, $G = \mathrm{d}T/\mathrm{d}z$, on the system, to provide such a

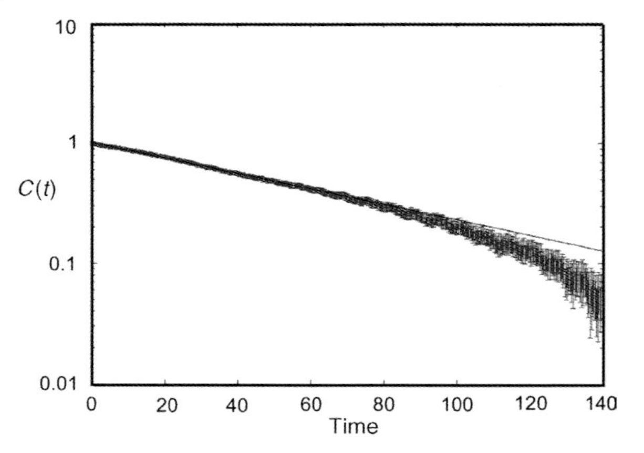

Figure 20.12 Time correlation function, $C(t)$, for the data shown in Figure 20.11. The line is an exponential fit to the data for short times.

restoring force. The location of the melting point isotherm then defines the equilibrium position of the interface, which in turn defines the average number of solid atoms.

$$T = T_{\mathrm{M}} + Gz \tag{20.37}$$

We could equally well have a constant volume system, and use the pressure generated by the volume change to define the equilibrium number of solid atoms. Or we could have an adiabatic system, and use the temperature change caused by the latent heat to establish the equilibrium number of solid atoms.

The time constant, τ, is the rate at which fluctuations in the system decay, which is half of the Onsager relationship. We must next determine the rate at which our system will return to equilibrium when it is displaced from equilibrium.

If the number of solid atoms changes by ΔN_{S}, then the solid–liquid interface is displaced by an average distance Δz:

$$\Delta z = \frac{\Omega}{A} \Delta N_{\mathrm{S}} \tag{20.38}$$

where Ω is the atomic volume, and A is the area of the interface.

Due to the change in position, the average temperature of the interface changes by $G\Delta z$:

$$T_{\mathrm{M}} - T = \Delta T = -\frac{\Omega}{A} G \Delta N_{\mathrm{S}} \tag{20.39}$$

The rate of motion of the interface, v, depends on the rate at which atoms join or leave the solid:

$$v = \frac{\mathrm{d}\Delta z}{\mathrm{d}t} = \frac{\Omega}{A} \frac{\mathrm{d}\Delta N_{\mathrm{S}}}{\mathrm{d}t} \tag{20.40}$$

Assuming that the growth rate of the crystal depends linearly on the undercooling, we can write:

$$v = \mu \Delta T \tag{20.41}$$

where μ is the kinetic coefficient. If the crystal growth rate does not depend linearly on the growth rate, then this analysis will not work. Combining Equations 20.40 and 20.41:

$$\frac{d\Delta N_S}{dt} = \frac{A}{\Omega} \mu \Delta T \tag{20.42}$$

Replacing ΔT with the value in Equation 20.39:

$$\frac{d\Delta N_S}{dt} = -\mu G \Delta N_S \tag{20.43}$$

The time dependence of the interface displacement is thus:

$$\Delta N_S(t) = \Delta N_S(0)\exp(-\mu G t) \tag{20.44}$$

This gives the rate at which the system will return to equilibrium when it is displaced from equilibrium, which is the other half of the Onsager relationship.

Onsager's fluctuation dissipation theorem states that the value of the time constant, τ, derived from the time correlation function should be the same as the time constant, $1/\mu G$, in the expression for the rate at which the interface returns to equilibrium when it is displaced from its equilibrium position. And so, the kinetic coefficient, μ, can be calculated from value of τ, which was obtained from the time correlation function.

$$\mu = \frac{1}{\tau G} \tag{20.45}$$

In the above analysis, it was assumed that the interface velocity is linear with undercooling. A different analysis would be needed if this was not the case.

The fluctuations in pressure in a constant volume system could be monitored to give the number of solid atoms, or the temperature in an adiabatic system could be monitored to give the number of solid atoms. In any case, the total number of atoms must be large enough so that statistically significant fluctuations are observed, and small enough so that the fluctuations are not averaged out.

Figure 21.13 shows the kinetic coefficient determined this way from Monte Carlo simulations of the (100) face of a simple cubic crystal.

20.11.2
Experimental Determination of Surface Tension

The surface tension can be determined by measuring the time correlation function for the intensity of laser light scattered from a surface or an interface. The light from a laser will scatter from surface fluctuations. The light which is scattered at some particular angle has been diffracted by fluctuations in the surface having a specific wavelength which depends on the angle of incidence, on the angle of the detector, and on the wavelength of the light. The rate of decay of a sinusoidal fluctuation of the surface shape depends on the wavelength of the fluctuation, the surface tension and the kinetic coefficient. If the rate at which a displacement of the interface decays is linear with its amplitude, and the kinetic coefficient is known, then the surface tension can be calculated from the time correlation of the diffracted intensity. This will work for a liquid–vapor or a liquid–liquid interface. It will also work for a liquid–solid interface, if the interface is rough, so that it behaves like a fluid–fluid interface. This is a neat experiment: the surface tension can be determined by light scattering.

References

1 Willard Gibbs, J. (1957) *Collected Works*, vol. 1, Yale University Press, New York.
2 Hertz, H. (1882) *Ann. Phys.*, **17**, 177.
3 Knudsen, M. (1909) *Ann. Phys.*, **34**, 593.
4 Wilson, H.A. (1900) *Phil. Mag.*, **50**, 238.
5 Frenkel, J. (1932) *Phys. Z. Sowjet Union*, **1**, 498.
6 Stillinger, F.H.and Weber, T. (1985) *Phys. Rev. B*, **31**, 5262.
7 Grabow, M.H., Gilmer, G.H.and Bakker, A.F. (1989) *Mater. Res. Soc. Symp. Proc.*, **141**, 349.
8 Broughton, J.Q., Gilmer, G.H.and Jackson, K.A. (1982) *Phys. Rev. Lett.*, **49**, 1496.
9 Stranski, I.N. (1928) *Z. Phys. Chem. (Leipzig)*, **136**, 259.
10 Kossel, W. (1927) *Nachr. Ges. Wiss. Göttingen*, 135.
11 Neumann, K. and Micus, G. (1954) *Z. Phys. Chem.*, **2**, 25.
12 Burton, W.K., Cabrerra, N. and Frank, F.C. (1951) *Phil. Trans. Roy. Soc.*, **A243**, 299.
13 Burgner, L.L.and Weinberg, M.C. (2001) *J. Non-Cryst. Solids*, **279**, 28.
14 Hoyt, J.J., Sadigh, B., Asta, M. and Foiles, S.M. (1999) *Acta Mater.*, **47**, 3181.
15 Reichl, L.E. (1991) *A Modern Course in Statistical Mechanics*, University of Texas Press, pp. 545–560.
16 Briels, W.J.and Tepper, H.L. (1997) *Phys. Rev. Lett.*, **79**, 5074.
17 Bentz, D.N.and Jackson, K.A. (2003) *Mater. Res. Soc. Symp. Proc.*, **778**, 255.

Problems

20.1. For $L/kT_M = 5$, what is the normalized net departure rate as given by Equation 20.30 for $n = 1, 2, 3, 4, 5$, for a crystal structure with $Z = 6$.

20.2. If the unit of time in Figure 20.12 corresponds to 3 ps, and the temperature gradient in the simulation was $100\,^{\circ}\mathrm{C}\,\mu\mathrm{m}^{-1}$, what is the kinetic coefficient?

21
The Surface Roughening Transition

21.1
Surface Roughness

Some surfaces are smooth on the atomic scale, and others are rough. This has a dramatic influence on the crystal growth process. The surfaces which are rough on an atomic scale usually have isotropic properties, and so are rounded on a microscopic scale. The surfaces which are smooth on an atomic scale have anisotropic properties, and form microscopic or macroscopic facets. The transition between these two modes is known as the surface roughening transition, and it is related to the order–disorder transition in two dimensions. The nature of this transition has been studied using the Ising model. Interfaces which are smooth on the atomic scale and rounded on the microscopic scale are shown in Figures 11.2, 12.4, 26.1. Figure 21.1 is a photograph of the growth front of a Salol crystal, which has a smooth interface on the atomic scale and faceted on a microscopic scale, growing under similar conditions.

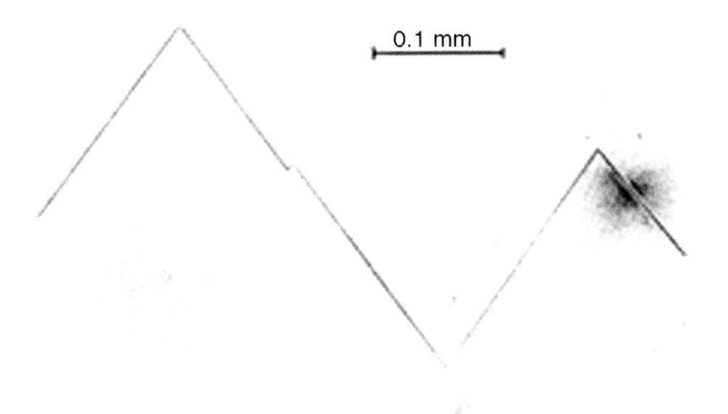

Figure 21.1 Faceted growth of a Salol crystal.

Kinetic Processes: Crystal Growth, Diffusion, and Phase Transitions in Materials. Kenneth A. Jackson

ISBN: 978-3-527-32736-2

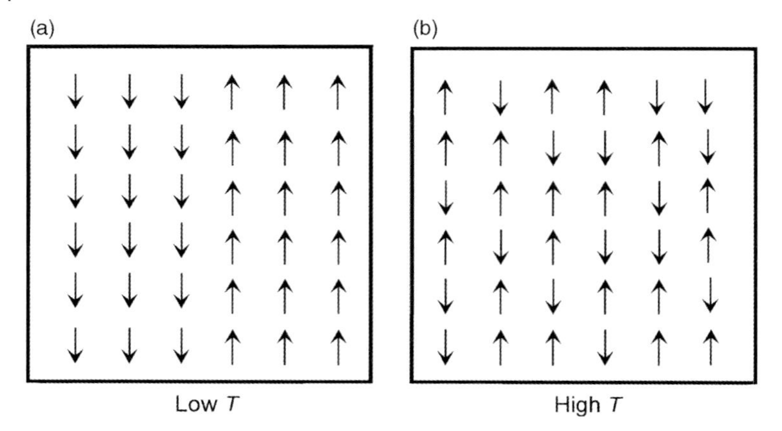

Figure 21.2 (a) Ordered and (b) disordered states of a ferromagnet.

21.2 The Ising Model

It turns out that the mathematics of the Kossel–Stranki model [1, 2] (see Figure 20.6) is identical to the mathematics of the model which was originally developed by Ising for the ferromagnetic Curie point. This model is simple, in principle, but becomes very complex in detail. It is a very important model which has been extensively studied in statistical mechanics.

The Ising model for a ferromagnet assumes that the spin on each atom points either up or down.

It is assumed that the spins interact only with neighboring spins. There is one interaction energy if the two spins are parallel, and a different interaction if the two spins are anti-parallel. This model predicts that all the spins in each domain will align at a low temperature, as shown in Figure 21.2a. The nearest neighbor interactions, which are very short range, produce long range order. At high temperatures, when the thermal energy is large compared to the interaction energy, the spins are random, as illustrated in Figure 21.2b. There is a second order phase transition between these two regimes, at a temperature called the critical point, which for the ferromagnet is the Curie temperature.

However, the mathematics for this model is identical to that for ordering or phase separation in an alloy, where there is one interaction between atoms of the same kind, and a different interaction between dissimilar atoms. At low temperatures, there are regions where similar atoms occupy adjacent sites and there is a critical temperature above which the occupancy of sites is random, as illustrated in Figure 21.3. The mathematics of the Ising model applies equally well to ordering in an alloy.

The same mathematical model applies to the Kossel–Stranski model. An atom has one interaction energy with an occupied neighboring site, and a different interaction

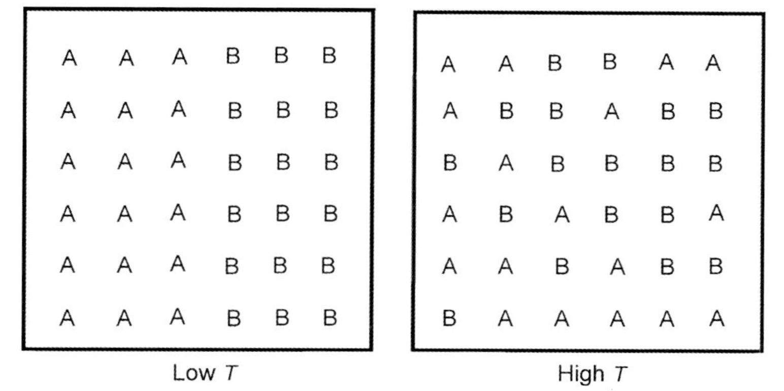

Figure 21.3 Phase separated and disordered states in an alloy.

energy with an empty neighboring site. So the extra atoms on a surface tend to cluster at low temperatures, and the atoms distribute randomly at high temperatures.

The same mathematics applies to each of these cases: the ferromagnet, the phase separating alloy, and the surface of a Kossel–Stranski crystal, and so they are said to belong to the Ising Universality Class.

21.3 Cooperative Processes

The phenomena we are discussing, where there are long range ordering effects which are due to local interactions, are known as cooperative processes. The fact that many different systems can be described by the same mathematics has led to the defining of universality classes. Systems in the same universality class have similar behavior in the vicinity of their critical points [3]. The behaviors of systems near their critical points are known as critical phenomena. The behavior of many properties of the system, such as the order parameter, can be described by an expression of the form:

$$\left(\frac{T_c - T}{T_c}\right)^{\beta} \tag{21.1}$$

Here T_c is the critical temperature, and β is known as a critical exponent. The critical exponents can be calculated or estimated for a variety of system parameters, such as the order parameter, the sound velocity, the density, the viscosity, the diffusion coefficient, the coherence length, and so on. Relationships between the critical exponents of various of these system parameters have been found.

Members of the same universality class all have the same critical exponents. It is surprising what the members of a particular universality class do and do not have in common.

The critical temperature depends on the specific material system and on the details of the interactions between the atoms, but the critical exponent,
β is independent of:

details of the local interactions
lattice type
anisotropy of bond strengths
second nearest neighbor interactions
(these can be folded into the first neighbor interactions)
whether there is a lattice.

β depends on:

number of components in the order parameter
dimensionality of the system.

For example, the critical exponent, β, for the order parameter in the Ising system is 0.324. The experimental critical exponent for the order parameter in β-brass, a copper–zinc alloy is 0.324. The experimental critical exponent for the difference in particle density between the liquid and vapor phases along the coexistence curve approaching the critical point in argon is 0.324. These are both in the Ising universality class and so have the same critical exponents.
Some of the universality classes are:

two-dimensional Ising
three-dimensional Ising
x–y model, where the interaction energy depends on the angle between the spins: the superfluid transition in liquid ^{3}He is in this universality class
spin-1 Ising model which has four spin states: binary alloy solidification is in this universality class
spin glass model, which has nearest neighbor interactions of random magnitudes.

At first it seems strange that the critical exponents should be so universal. But look at the list which β does not depend on and compare it to the way we apply nucleation theory. The standard nucleation theory assumes that there are clusters of the new phase in a parent phase. The equations are applied to the nucleation of a crystal in a liquid, a precipitate in a crystal, or a liquid in a vapor, and so on, independent of all the things that critical exponents do not depend on. Similarly, the cluster distributions derived by analysis on a lattice can be applied universally to all the members of the Ising class.
Nucleation theory does not predict melting points, and the Ising model does not predict critical temperatures.
Some people have objected to using the Ising model to analyze crystallization because the liquid atoms are assumed to be on lattice sites. But universality says that this is not an issue. Similarly, the structure of the liquid never comes up in writing the equations which describe the nucleation of a crystal in a liquid. In applying the Ising model to crystal growth, it is assumed that there is a lattice in order to enable the analysis, but that does not affect the universality of the result.

It is worth emphasizing that the critical temperature of a member of a universality class cannot be predicted from an analysis of the cooperative behavior, because in the past it has been assumed that the critical point for the surface roughening temperature in the Ising model can be used to predict the surface roughening transition for real crystals. The Ising model describes the behavior of the system in the vicinity of the critical point, which is similar for all members of the same universality class. However, for example, the Curie temperature of a ferromagnetic material cannot be predicted from the Ising model. Nor can the roughening transition of the surface of any particular crystal be predicted from the Kossel–Stranski model.

The Ising model is a simple model which provides insight into crystallization processes, but there is a problem. Exact analytical solutions for the three-dimensional Ising model do not exist, even for equilibrium, and we would like to model crystallization kinetics, which requires more than knowledge of the equilibrium properties. Onsager, who developed the Fluctuation Dissipation Theorem discussed in Chapter 20, also obtained an exact solution for the Ising model in two dimensions. He received the Nobel prize for his work. This model was applied to the roughening of a crystal surface by Burton, Cabrerra, and Frank [4]. There are many approximate solutions for the Ising model, which provide significant insights. For example, Kenneth Wilson, at Cornell, received the Nobel prize for developing Renormalization Group Theory. This is a method for analyzing groups of atoms in the Ising system, enumerating the properties of the groups and their interactions with other groups of atoms, and doing this self-consistently so as to provide an expansion scheme in terms of cluster size for the properties of the system. But exact analytical solutions for the three-dimensional Ising model do not exist. In fact, it was recently reported in *Nature* that the three-dimensional Ising model belongs to a class of problems which are insoluble. The author concluded that there are some things we will never know. Perhaps he doesn't believe in computer modeling, because a great deal is known about the Ising model from computer simulations.

Monte Carlo computer simulations can be readily performed on the Ising system [5], and then the cluster distributions and the other behavior obtained from the computer simulations can be scaled and used to describe the cluster distributions and the behavior of real systems. This seems to be difficult to accept for people who believe that true understanding comes only from equations. They are happy to use computers to solve their equations, but unwilling to accept computer simulations as an alternative to equations.

The Ising model contains nucleation behavior as one limit. For example, in a magnetized ferromagnet at a low temperature, all the spins are aligned. In a demagnetized sample, all the spins are locally aligned, but there are magnetic domains of the opposite alignment, so there is no net magnetization of the sample. If a magnetic field is applied to a magnetized sample in a direction which requires the spins to flip, then there will be a nucleation barrier to the formation of the new magnetic domains. At a low temperature, there will be a critical size for new domains, which depends on the strength of the applied magnetic field. The formation of domains of reverse magnetization can be described by the standard equations for nucleation. But at temperatures approaching the Curie temperature, many flipped spins occur spontaneously, and small sub-critical regions of flipped spins will increase in size and number. These will interact and merge. Close to the Curie temperature, standard nucleation theory does not apply to the formation of a domain of reverse magnetization, because the density of flipped spins is much too high. However, the Ising model does incorporate this behavior, even though it is so difficult to describe analytically.

21.4 Monte Carlo Simulations of Crystallization

In these simulations, which are based on the Kossel–Stranski model, atoms arrive randomly at sites on the surface at some specified rate, and atoms leave the surface at a rate which is determined by a probability factor which depends on the number of adjacent crystalline atoms. If the departure probability is bigger than a random number, then the atom leaves, and if not, it does not. The arrival probability, P^+, and departure probability, P^-, can be written as:

$$\begin{aligned} P^+ &= \upsilon^+ \\ P^- &= \upsilon^+ \exp\left(\frac{L}{kT_M} - \frac{n\phi}{kT}\right) \end{aligned} \tag{21.2}$$

Here L is the latent heat of the transformation, and ϕ is the bond energy. These two are related by $L = Z\phi/2$, where Z is the number of nearest neighbor sites in the lattice. n is the actual number of nearest neighbors of the surface atom. At a repeatable step site where $n = Z/2$, the arrival rate is equal to the departure rate at equilibrium. In the computer, atoms join and leave the crystal according to these simple rules and so the computer simulation provides a statistical analysis of this model. The statistics are never exact, since the simulated systems are finite in size and simulation time, but the atoms form clusters and nucleate new layers, and so on, just as we suppose happens in the real world. When the bond energy is comparable to kT then there are many adsorbed atoms on the surface, according to Equation 21.2, there is no nucleation barrier to the formation of new layers. When the bond energy is large compared to kT, a single atom on the surface has a high probability of leaving the surface, and so there are few on the surface. The formation of nuclei on the surface is difficult. Many atoms have to get together on the surface to form a cluster of critical size and the growth of new layers involves the lateral spreading of these clusters. In this regime, growth occurs by nucleation and the motion of steps on the surface.

21.5 Equilibrium Surface Structure

21.5.1 Thermodynamic Model for Surface Roughness

The density of growth sites on the surface of a crystal depends above all on the roughness of the crystal surface. This is an intrinsic property of the surface at

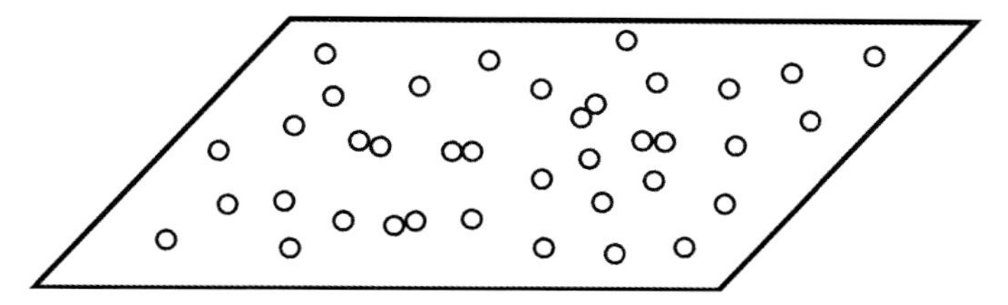

Figure 21.4 There is an equilibrium density of adatoms on a crystal surface.

equilibrium which depends on the nature of the two phases separated by the interface and on the atomic structure of the surface.

This is illustrated in Figure 21.4, and can be analyzed using a simple two-dimensional model for the equilibrium structure of a crystal surface or interface [6].

The change in the free energy, ΔF_S, of an initially flat surface containing N sites due to the random addition of N_a atoms can be written as:

$$\Delta F_S = -N_a\eta_0\phi - N_a\eta_1\frac{N_a}{N}\frac{\phi}{2} + N_a T\Delta S + kT\ln\left[\frac{N!}{N_a!(N-N_a)!}\right] \qquad (21.3)$$

where η_o and η_1 are the number of nearest neighbor sites in the substrate layer and in the surface layer respectively, as illustrated in Figure 21.5, so that $2\eta_o + \eta_1 = Z$, where Z is the total number of nearest neighbors. ΔS is the entropy change associated with the transformation.

The first term in Equation 21.3 is the decrease in energy due to the bonds formed with the substrate. The second term is the interaction with neighboring atoms on the surface, which are assumed to be randomly distributed. The third term is the increase in entropy associated with adding the atoms to the crystal, and the last term is the entropy associated with the random distribution of atoms on the surface.

For $T = T_M$, defining $N_a/N = \theta$, and using the Stirling expansion for the factorials gives:

$$\frac{\Delta F_S}{NkT_M} = -\theta\frac{2\eta_0}{Z}\frac{L}{kT_M} - \frac{\eta_1}{Z}\theta^2\frac{L}{kT_M} + \theta\frac{L}{kT_M} + \theta\ln\theta + (1-\theta)\ln(1-\theta) \qquad (21.4)$$

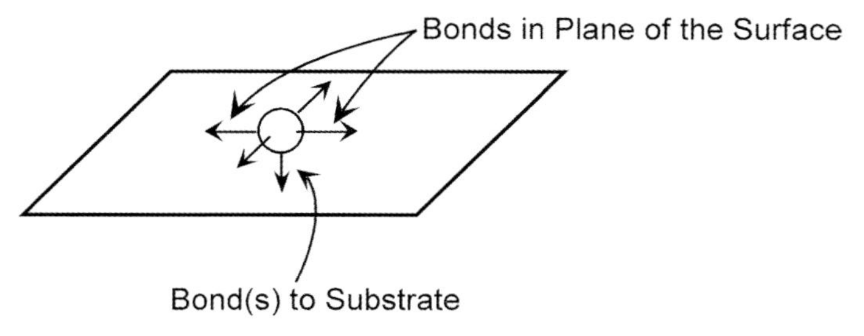

Figure 21.5 η_o is the number of nearest neighbor sites in the layer below, and η_1 is the number of nearest neighbor sites in the same layer as the adatom.

which can be written as:

$$\frac{\Delta F_S}{NkT_M} = \alpha(\theta-\theta^2) + \theta \ln\theta + (1-\theta)\ln(1-\theta) \tag{21.5}$$

Here

$$\alpha = \frac{L}{kT_M}\frac{\eta_1}{Z} \tag{21.6}$$

which is known as the Jackson α-factor. Equation 21.5 is plotted in Figure 21.6.

The equilibrium population of adatoms is defined by the minima in these plots, which is given by:

$$\frac{d}{d\theta}\left(\frac{\Delta F_S}{Nl_e T_M}\right) = 0 = \alpha(1-2\theta) + \ln\theta - \ln(1-\theta)$$

or (21.7)

$$\frac{\theta}{1-\theta} = \exp[-\alpha(1-2\theta)]$$

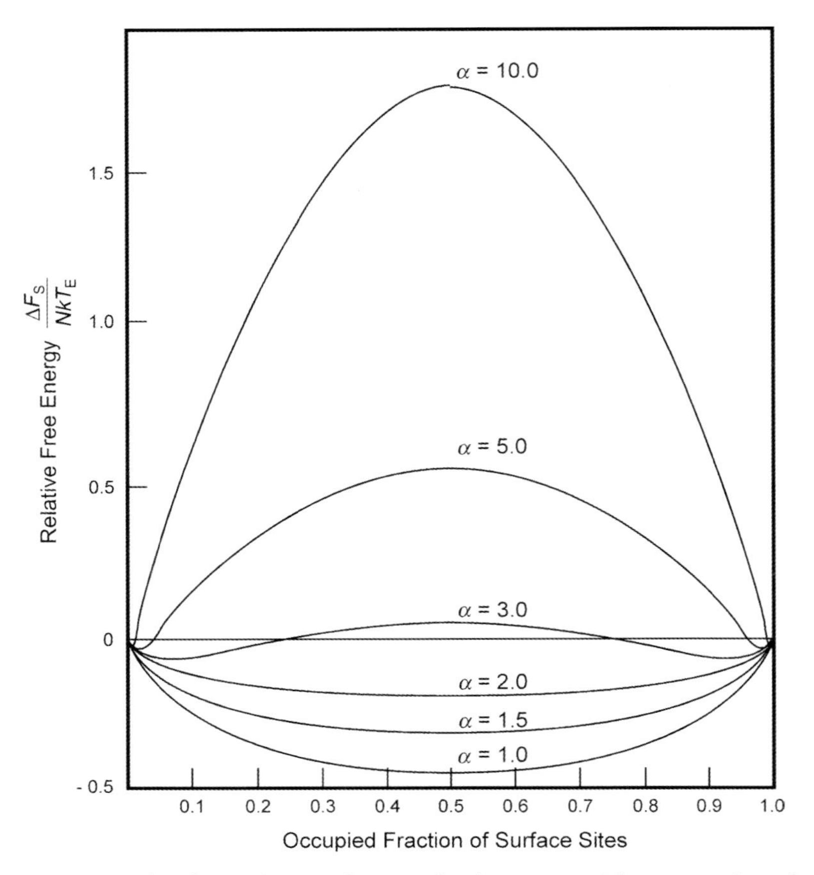

Figure 21.6 Plot of $\Delta F_S/NkT_M$ as a function of surface coverage θ for various values of α.

For large α, there is a minimum at approximately:

$$\theta \approx \exp(-\alpha) \tag{21.8}$$

which implies that the surface has few adatoms, and so is fairly smooth on an atomic scale.

For small α, there is a minimum at $\theta = 1/2$, which is a surface whose sites are half filled with adatoms, which suggests that the surface is rough. Indeed, it is likely to be rougher than the one atomic layer of roughness permitted in the model.

For large α, there is a maximum at $\theta = {}^1/_2$, and the transition between the maximum and minimum defines a critical value of α, which occurs when the curvature changes sign at $\theta = {}^1/_2$.

$$\begin{aligned} \frac{d^2}{d\theta^2}\left(\frac{\Delta F_S}{NkT_M}\right) &= -2\alpha + \frac{1}{\theta} + \frac{1}{1-\theta} \\ &= -2\alpha + 4, \quad \text{for} \quad \theta = \frac{1}{2} \end{aligned} \tag{21.9}$$

So the critical value of α is $\alpha_C = 2$.

The free energy has two minima for $\alpha > 2$, and only one, at $\theta = 1/2$ for $\alpha < 2$.

21.5.2 Application of Surface Roughening to Materials

This value for the surface roughening transition, $\alpha_C = 2$, is in very good agreement with the observed behavior of melt-growing crystals. That is, for surfaces with a small α-factor, the surface is rough and the crystals can grow readily without surface nucleation. Surfaces with a large α-factor are smooth, and there is a nucleation barrier to the growth of each layer.

The α-factor has two components, one of which depends on the change in entropy of the transformation, $\Delta S = L/T_M$, and the other, η_1/Z, on the geometry of the crystal face. The factor η_1/Z is largest for the closest packed planes and smaller for the less closely packed planes of the same crystal.

For materials with a small entropy change on melting, all the crystal faces will be rough. These materials will exhibit relatively isotropic growth, and will grow rapidly at small interface undercoolings. For materials with a large entropy change on melting, the closest packed faces will be smooth and so there is a nucleation barrier to growth on those faces. The less closely packed planes of the same crystal will be rough and there will be no nucleation barrier to growth on them. This is illustrated for a two-dimensional square lattice in Figure 21.7.

The density of growth sites is large for the rough surfaces, and it is also large for the (11) edge shown in Figure 21.7c. The growth rate will be very rapid and relatively isotropic for low entropy change, because the growth site density is similar for both faces. But the growth rate will be much slower on the (10) edge than on the (11) edge

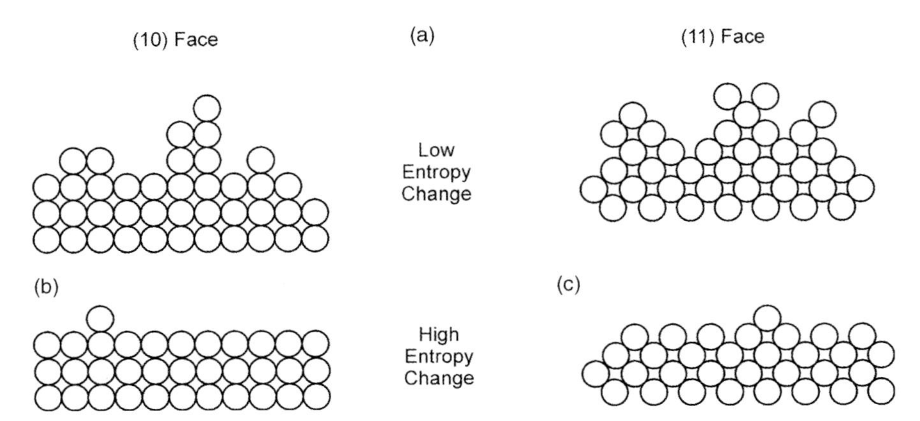

Figure 21.7 There are many growth sites on both rough surfaces (a), few growth sites on the low index edge (b), but many growth sites on the low index edge (c) because of the topology of the lattice.

Table 21.1 η_1/Z for face centered cubic crystals.

(111)	1/2
(100)	1/3
(110)	1/6

for high entropy change, because of the large difference in growth site density on the two edges.

Table 21.1 presents the value of the factor η_1/Z for three faces of the face-centered-cubic structure.

Materials with large entropy changes exhibit very anisotropic growth rates, and the closest packed planes require significant undercooling for reasonable growth rates.

Table 21.2 contains a list of entropies of fusion for various classes of materials.

The α-factor for a particular face of a crystal is the product of the appropriate numbers from Tables 21.1 and 21.2. One is characteristic of the material, and the other depends on the growth direction. The class of materials with low entropy change includes all the metals growing from the melt, which are said to "solidify" because their growth rates are rapid and isotropic. In general, the higher the α-factor,

Table 21.2 Entropy change on crystallization.

Metals from the melt	1
Si, Ge, Sb, Ga from the melt	3
Many organic compounds	6
Metals from the vapor	10
Complex molecules	20
Polymers	>100

the slower and the more anisotropic is the growth rate. This model correctly predicts the morphology and the general characteristics of growth from the melt. It also describes the characteristics of growth from the vapor phase, but the roughening transition occurs at a different value of α for vapor growth.

For molecules or atoms that behave like spheres in the liquid state, the entropy associated with crystallization involves only a change from the randomness of liquid to the ordered structure of the solid, as for the metals in Table 21.2. Materials such as this usually freeze into one of the three close packed crystal structures: face centered cubic, body centered cubic or close packed hexagonal. The entropy change for each of these materials is very close to 1 in reduced units, that is, in units of the gas constant, R, if you are a chemist, or in units of Boltzmann's constant, k, if you are a physicist.

For more open structures such as ice or silicon and some of the semi-metals such as bismuth or gallium, the entropy of fusion is in the range of 3–4. These materials have some directional bonding in the crystal. For most organic materials, the molecules are free to rotate in the liquid state but have a specific orientation in the crystal. The entropy of fusion for many of these materials is about 6. This means that 5 units of entropy are associated with the change in rotational order and only 1 unit of entropy is associated with the disorder in going from the liquid to the crystal state. One way to think about the origin of the entropy is that only some small fraction of the molecules in the liquid will be in the right orientation to join the crystal. The improperly oriented molecules cannot join the crystal until they rotate. Thus only a fraction of the molecules contribute to the arrival rate of atoms at the crystal at any one time. But this effect is not present in the rate at which molecules leave the crystal, because a molecule can leave the liquid in any orientation. This asymmetry in the rates lowers the equilibrium temperature, so that the entropy associated with the transformation is higher.

Most molecular compounds are randomly oriented in the liquid and their orientation is ordered in the crystalline phase. There are two classes of material which are exceptions to this. In one of these, the molecules are free to rotate in the crystal which forms from the liquid. These materials have entropies of fusion which are similar to those of the metals. They crystallize into the highly symmetrical face centered cubic structure, and undergo a solid state phase tranformation at a lower temperature, where the molecules align to form a crystal structure with much a lower symmetry which is dependent on the shape of the molecule. They behave like the metals during crystallization, so they have been used as transparent analogues to study the crystallization charasteristics of metals. There are photographs of the growth morphology of these materials throughout this book.

The other class of materials in which the molecules do not align during crystallization are the liquid crystals. In these materials, the molecules are already aligned in the liquid state. There is a transformation above the melting point in the liquid phase, above which the molecules are not aligned.

The transition between faceted and non-faceted growth forms for melt growth systems occurs at a specific value of the entropy change per molecule. In some materials the effective molecular weight is uncertain. In sulfur, for example, using latent heat per atom, the entropy of fusion is small. However, in liquid sulfur at its melting point, the atoms form eight-membered rings. At higher temperature in the liquid phase, the sulfur atoms polymerize into linear chains. Associated with this structural change in the liquid is an increase in the viscosity of the liquid above the melting point by two orders of magnitude. Using the molecular weight of an eight-membered ring, sulfur has an entropy change typical of the molecular materials which its crystal growth habit resembles. The state of aggregation is important because large molecules in general have a larger entropy of fusion per mol than small molecules, and this difference is reflected in their crystallization behavior. Formally, this enters the α-factor through the latent heat which is in units of energy per mol, so that the latent heat per gram translates into a larger latent heat per mol if the molecular weight is larger. Sulfur can also be crystallized from the polymeric state of the liquid phase, in which case its growth morphology resembles that of other polymers. In phosphorus the same issue arises, but in this case, the state of aggregation in the liquid is not known. The growth morphology suggests a state of aggregation in liquid phosphorus similar to that of the eight-membered ring in sulphur.

Ice is an interesting case. The entropy change for freezing of water is $2.63R$. The crystal structure of ice is hexagonal, but there is asymmetry in the bond lengths so that it is difficult to asses the geometrical factor, η_1/Z, exactly, but it is about $^3/_4$ for both the basal plane and the growth directions in the basal plane. Ice crystals growing from the melt form facets on the basal plane and grow much more rapidly in the basal plane. At a few degrees below 0 °C, ice crystals will grow as sheets one millimeter or so thick, while the dendrites grow several inches across in the basal plane. The growth rate of the dendrites in pure water is limited primarily by the diffusion of latent heat from the dendrite tips. The water–ice transition has an α-factor closer to two than any other pure material, and so is perhaps a test of the critical value of α for melt growth. Most materials have α-factors for their closest packed faces which are not near the critical value.

21.5.3 Snow Flakes

Snow flakes provide a very interesting example of the role of the surface roughening transition in crystal growth. Ice crystals growing in the atmosphere exhibit a wide variety of growth morphologies. The growth morphology of a snow flake depends on both the temperature and the water vapor content of the air. The growth of the snow flake depends on the diffusion of water molecules through the atmosphere to the crystal, as well as on the intrinsic growth characteristics. Some ice crystals grow as needles, elongated perpendicular to the basal plane of the ice structure. A more common form is a disc. The disc shape occurs when the basal plane is smooth, so that the growth rate normal to the basal plane is very slow, as it is for the growth of ice dendrites in undercooled water. Growth in the directions in the basal plane depends sensitively on the local environment. The planes perpendicular to the basal plane can be either rough or smooth, depending on the temperature and on the amount of water in the air. When the conditions are such that these planes are smooth, the growth rate is limited by the growth kinetics, and the morphology is a hexagonal disc.

Figure 21.8 Snow flakes. (From Bentley and Humphreys [7]).

When these planes are rough, the growth kinetics is relatively rapid, so the growth rate is limited by diffusion of water molecules to the snow flake. The growth morphology in this case is a disc with dendritic growth in the basal plane. Of course the growth rate of an ice dendrite in undercooled water, which is limited by thermal diffusion, is much more rapid that the growth of a snow flake, which is limited by the diffusion of water molecules in air.

A snow flake can experience several different local environments as it falls, so it can switch back and forth between these two growth modes. This gives rise to the great variety of snow flakes which are observed, and to the adage that no two snow flakes are alike. The morphologies of some disc-shaped snow flakes are shown in Figure 21.8.

In the top row are a dendritic and a hexagonal snow flake. The other snow flakes have various morphologies. Some, such as the two on the left in the center row, started as faceted discs, and then became dendritic. The snow flake second from the left in the bottom row started as a dendrite and later grew in faceted mode, so the ends of the dendrite arms have become faceted. Others appear to have switched growth modes more frequently.

The disc is not the only morphology of ice crystals which have grown from the vapor phase in air. As mentioned above, there are also needle-shaped crystals, and there are changes in relative growth rates for different orientations with temperature. All of the complex details of how ice crystals grow in air have not yet been sorted out.

21.5.4
Rate Theory Analysis of Surface Roughness

The roughness of the interface can also be analyzed by comparing the rates at which atoms join and leave the crystal. The surface coverage was analyzed in Chapter 16 for the Langmuir model, where it was assumed that the adsorbed atoms at the interface interact only with the substrate, and do not interact with other adsorbed atoms. For crystal growth, this is not a good assumption. The lateral interactions of the atoms must be taken into account. The rate at which atoms join the crystal can be written as in Equation 16.2 for Langmuir adsorption:

$$R^{+} = (N - N_a)\upsilon^{+} \tag{21.10}$$

where N is the number of surface sites, and N_a is the number of adatoms. It is assumed that atoms can arrive only at the $N - N_a$ unoccupied sites. Langmuir assumed that the rate at which atoms leave the surface was independent of their environment, but here we assume that the rate at which an atom leaves the surface depends on how many nearest neighbors it has.

The rate at which an atom with n nearest neighbors leaves the surface is:

$$\upsilon^{+} \exp\left(\frac{L}{kT_M} - \frac{n\phi}{kT}\right) \tag{21.11}$$

as in Equation 20.30. Here n is the number of occupied nearest neighbor sites, which is the number of bonds which are broken when the atom leaves. These equations imply that the rates of arrival and departure are equal at a repeatable step site, where $n = Z/2$.

The overall rate at which atoms leave the surface can be written as:

$$R^{-} = N_a \upsilon^{+} \exp\left(\frac{L}{kT_M} - \frac{\langle n\rangle\phi}{kT}\right) \tag{21.12}$$

where $<n>$ is the average number of nearest neighbors of the atoms on the surface. At equilibrium the overall rates at which atoms join and leave the surface are equal. Equating Equation 21.9 and 21.11 gives:

$$\frac{N_a}{N - N_a} = \frac{\theta}{1 - \theta} = \exp\left(-\frac{L}{kT_M} + \frac{\langle n\rangle 2L}{ZkT_M}\right) \tag{21.13}$$

where $\theta = N_a/N$, and ϕ has been replaced with $2L/Z$. To reproduce the result of the previous analysis, we again assume that the atoms on the surface are randomly distributed, so that the average number of nearest neighbors $<n>$ depends on the average adatom density, and is given by:

$$\langle n\rangle = \eta_0 + \eta_1\theta \tag{21.14}$$

Using $2\eta_o + \eta_1 = Z$, the exponent in Equation 21.12 can be written as:

$$-\frac{L}{kT_M}\left(\frac{Z - 2\langle n\rangle}{Z}\right) = -\frac{L}{kT_M}\left(\frac{\eta_1}{Z}(1 - 2\theta)\right) \tag{21.15}$$

Using

$$\alpha = (L/kT_M)(\eta_1/Z) \tag{21.16}$$

Equation 21.14 can be written:

$$\frac{\theta}{1-\theta} = \exp[-\alpha(1-2\,\theta)] \tag{21.17}$$

which is the previous result found in Equation 21.7. Equilibrating the rates at which atoms join and leave the surface gives the equilibrium condition directly, which corresponds to the minima in Figure 21.6.

The analysis of surface roughening based on rate equations gives the same result as that derived using equilibrium thermodynamic methods. The rate analysis is the basis for Monte Carlo modeling of crystal growth processes, and so Monte Carlo modeling contains not only the basic properties of the system at equilibrium, including the surface roughening transition, but it also provides information about non-equilibrium configurations of the interface and about growth rates.

21.5.5
Surface Roughness in the Ising Model

The analysis of surface roughening presented above is based on an approximate one-level model of the interface. A more rigorous analysis of the interface, based on a three-dimensional Ising model which allows for a multilayer transition between the two phases, results in a different value for the location of the transition [8].

It is usual in the statistical mechanics literature to indicate critical roughening temperatures as calculated based on the Ising model, which implies a simple cubic lattice. It has become usual to compare surface roughening temperatures with the theoretical value for the (100) face of this model. The reduced critical surface roughening temperature is expressed as kT_R/ϕ, where ϕ is the bond energy of the crystal, given by $2L/Z$. This critical roughening temperature is exhibited in the right-hand column of Table 21.3. For the (100) face of a simple cubic crystal, kT_R/ϕ is equal to $2/\alpha_C$. This notation provides an analogy with bulk melting: the surface is smooth (ordered) below the critical roughening temperature, and rough (disordered) above it. On the other hand, this notation tends to obscure the dependence of the surface roughness on the difference in entropy between the two phases. Furthermore, the numerical value of the roughening transition expressed this way depends on both the crystal structure and on the orientation of the interface.

Table 21.3 Surface roughening transition.

	α_C	kT_R/ϕ
Roughening transition (theory)	3.2	0.62
2D Critical point (theory)	3.5	0.57
2D Critical point (Bragg–Wiliams model)	2.0	1.0
Melt growth (expt)	2.0	1.0
Vapor growth (expt)	8.0	0.25

The value for the 2D theoretical critical point presented in the table is from the exact solution for a 2D Ising crystal (square lattice) obtained by Onsager. It is quite close to the value for the roughening transition which is obtained for the three-dimensional Ising model, even though the surface roughness can extend over many layers of atoms in the latter case. This is because the transition is strongly dependent on what happens in the central layer of atoms in the multilayer interface.

Also exhibited in Table 21.3 are the experimentally observed locations of the surface roughening transition for melt growth and for vapor growth. Since only interfaces which are smooth can form macroscopic facets, the formation of macroscopic facets during growth can be used to locate the surface roughening transition. The transition for melt growth was determined by comparing the growth characteristics of a large number of different materials. The critical roughening transition for melt growth happens to coincide with the value predicted by the two-dimensional Bragg–Williams model presented above.

The surface roughening transition for melt growth is significantly higher than the theoretical value. This difference is attributed to the presence of other liquid atoms which interfere with the formation of small solid clusters on the surface, forcing the onset of surface roughening to higher temperatures. For vapor growth systems, the transition is significantly lower than the theoretical value. This difference is believed to arise because the surface atoms are less tightly bound than atoms in the bulk. This promotes surface roughening so that it occurs at a lower temperature. For the vapor case, surface melting, that is, the formation of a layer of mobile atoms, can also occur.

For vapor phase growth, the surface roughening transition can be observed using vapor transport in a closed tube with a fixed temperature difference between the hot and cold ends. The tube is then held at various average temperatures, and the growth morphology is observed. At low temperatures the crystals are faceted and at high temperatures they are rounded. The transition between these two regimes occurs in a very narrow temperature interval. Observations have been made on a large variety of materials. In these experiments, the equilibrium temperature between the crystal and the vapor changes with the average temperature of the system. Since the entropy difference between the two phases depends on the equilibrium temperature, and the location of the roughening transition depends on the entropy change across the interface, growth from a vapor can be taken through the surface roughening temperature to exhibit growth both above and below the transition.

Doing the same for melt growth would require extremely large pressures in order to change the equilibrium temperature significantly, since modest variations in pressure do not significantly change the melting point. The change in equilibrium temperature with pressure depends on the change in volume for the transformation, which is very large for vaporization, but only a small percentage for melting. The change in melting point with pressure is sufficiently small so that a unique value for each material under usual laboratory conditions is tabulated in handbooks. So, for melt growth there is a unique entropy of fusion for each material at normal pressures, and a corresponding α-factor for each crystal face.

Although the mathematical model for the α-factor presented above is a relatively crude approximation, it does indicate the correct physics, and it provides a valuable rule-of-thumb for predicting the surface roughness of various faces of various materials for melt growth.

21.6
Computer Simulations

The above analyses tell the basic story of surface roughening, but there are many more details that have been worked out, primarily using Monte Carlo computer modeling, but also mathematical analyses of approximations to the Ising model. A full analytical treatment of a multilayer interface has been done for one special case, but there is not a general solution. Multilevel cluster expansion models have been used to demonstrate that there is an analytical singularity in the thermodynamic properties of the surface of an Ising crystal at the surface roughening transition, but it is a very weak singularity. The surface tension and most of the other properties of the surface are continuous through this transition. Figure 21.9 is a plot of the surface roughness [9], which is defined as the number of unsatisfied bonds on the surface.

The surface roughness increases as the α-factor decreases. The surface roughness is similar for the three-dimensional Bragg–Williams (zeroth order cluster expansion),

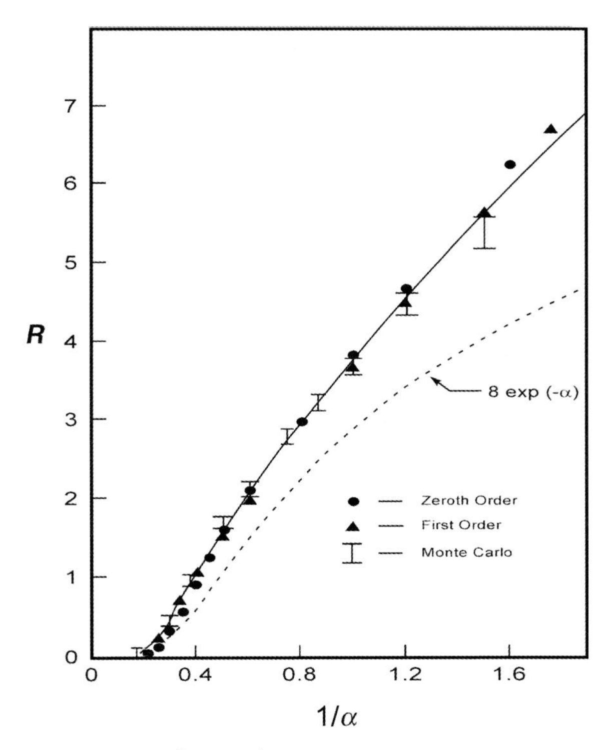

Figure 21.9 Surface roughness versus $1/\alpha$.

for the first order cluster expansion and for Monte Carlo modeling. 8exp($-\alpha$) provides a reasonable approximation. As the surface becomes rougher, the roughness is spread over more atomic layers, but in all the models, nothing much happens to the surface roughness at the roughening transition. The surface roughness increases continuously through the roughening transition.

This is not true for the free energy of a step on the surface. The free energy of a step decreases as the surface roughening transition is approached, and it goes to zero at the surface roughening transition [10], as shown in Figure 21.10.

As the surface roughening transition is approached, a step becomes increasingly jagged. There is a major contribution to the free energy of the step from the entropy associated with this jaggedness. At and above the surface roughening transition, the surface is so rough that steps are lost in the roughness [10], as illustrated in Figure 21.11.

It has been shown that above the surface roughening transition, the interface is like a fluid–fluid interface [11]; the interface ignores the crystal lattice. On the rough side of the surface roughening transition, there is no difficulty in forming new layers. There is no barrier to continuous growth, and so the growth rate is linear with undercooling. On the smooth side of the surface roughening transition, steps on the surface have a finite free energy, and so the growth rate depends on the rate of

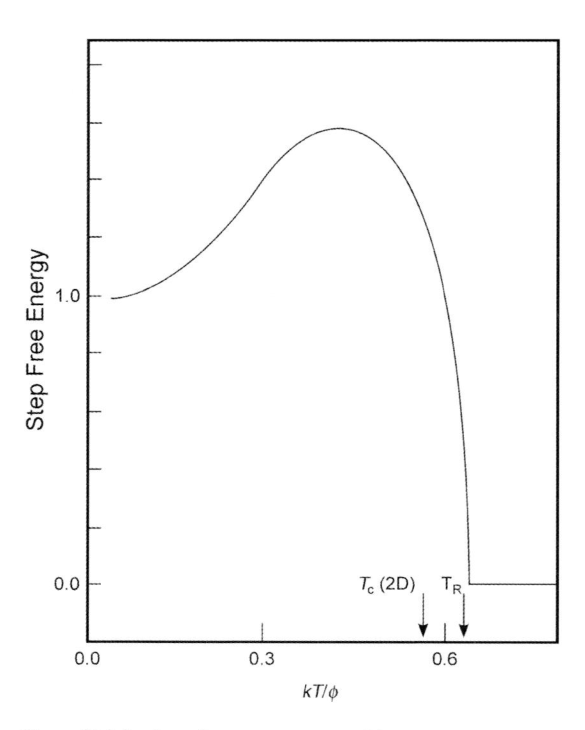

Figure 21.10 Step free energy versus $1/\alpha$.

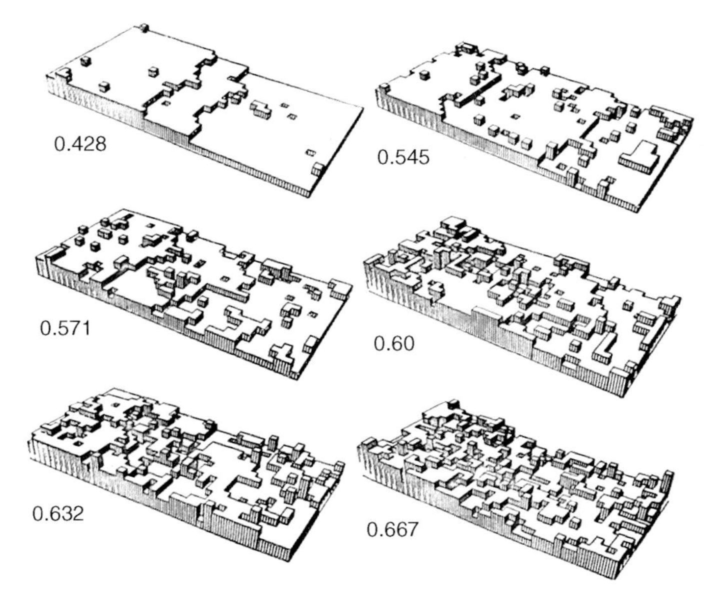

Figure 21.11 Configuration of surfaces below and above the roughening transition, which is at 0.62.

nucleation of new layers. This is illustrated in Figure 21.12, where growth rates from Monte Carlo computer simulation studies are shown. In Figure 21.12 the critical surface roughening transition is at 1.0. The curve labeled 1.08 is just on the rough side of the transition, and the growth rate is linear with undercooling. The other

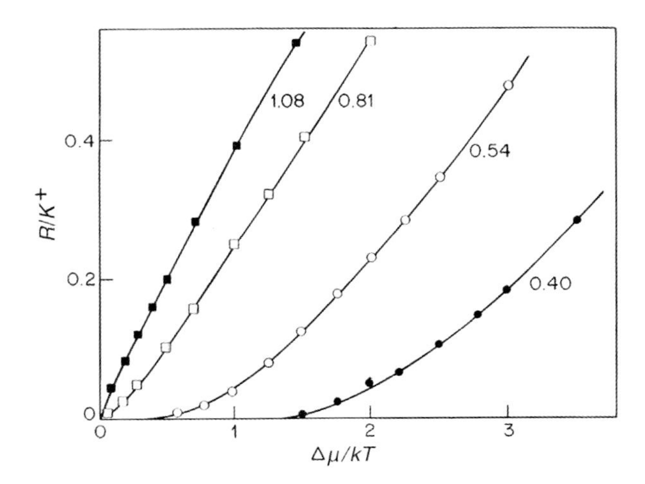

Figure 21.12 Monte Carlo computer simulations results for the growth rate versus undercooling above and below the roughening transition for various values of α_C/α [5].

curves are for smooth surfaces. They come into the origin with zero slope [11], and farther from the roughening transition they are flatter and more "nucleation-like".

As suggested by Figure 21.12, it has been proved as an exact result, using linear response theory, that the growth rate on rough interfaces is linear with undercooling, whereas the growth rate on smooth surfaces approaches equilibrium with zero slope. So there is a discontinuity in the kinetic properties of the interface at the roughening transition.

The free energy of a step depends on the proximity of the surface roughening transition, and so the difficulty in nucleating new layers increases with the distance from the roughening transition. The growth rate for crystals with large entropies of transformation is highly anisotropic.

21.6.1
Determination of the Kinetic Coefficient

The kinetic coefficient can be determined using the fluctuation dissipation theorem as described in Chapter 20. Figure 21.13 shows the kinetic coefficient determined this way from Monte Carlo simulations of the (100) face of a simple cubic crystal.

The interface was at equilibrium in a temperature gradient. The interface fluctuations were analyzed using time correlations, and the kinetic coefficient was determined from Equation 20.45.

The theoretical surface roughening transition for this interface is at $\alpha = 3.2$. There is a sharp change in the kinetic coefficient at the surface roughening transition. Below the surface roughening transition, the growth rate is linear with undercooling, and so a valid kinetic coefficient is obtained. The growth rate is not linear with undercooling

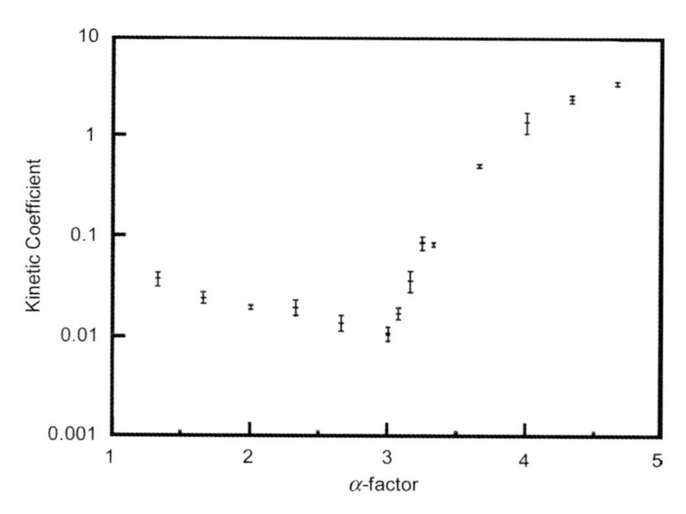

Figure 21.13 Kinetic coefficient as a function of the α-factor from Monte Carlo simulations of the (100) face of a simple cubic crystal [12].

above the surface roughening transition, and so the data do not correspond to a kinetic coefficient. Instead, the time correlation function is picking up very rapid changes in adatom density on the smooth surface.

As illustrated in Figures 21.9 and 21.11 many of the properties of the surface are continuous through the surface roughening transition. It is difficult to locate the transition in these figures. The surface tension is continuous through the transition, but the kinetics of the interface motion are discontinuous at the surface roughening transition, as demonstrated dramatically by Figure 21.13.

The break in the data in Figure 21.13 corresponds to the transition from the region where the growth rate is linear with undercooling in Figure 21.12, to where the growth rate approaches the melting point with zero slope.

21.6.2
Simulations of Silicon Growth

As an example of this anisotropy, Monte Carlo simulations have been carried out for silicon, with the roughening transition in the simulations scaled to the roughening transition for silicon [13]. The entropy change for melt growth of silicon, $\Delta S/R = 3.6$, so the α-factor for (111) is 2.7, and for (100) is 1.8, and it is lower for all the other faces. The crystallization rate for the (100) face, as determined by both experiment and MD simulations, is approximately $v = 0.12\ \Delta T\,\mathrm{m\,s^{-1}}$. The growth on the (111) face is stepwise for a range of interface temperatures, as is shown in Figure 21.14.

These data exhibit nucleation limited behavior, as is evident from the irregular rate of addition of new layers. When the growth rate is plotted versus $1/\Delta T$, as in Figure 21.15, the data fall on a straight line, which also indicates nucleation limited growth.

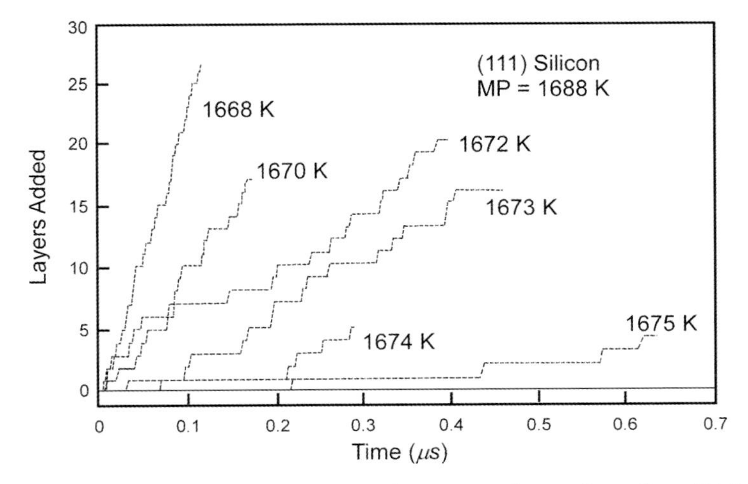

Figure 21.14 Layers added versus time for Monte Carlo growth on silicon (111).

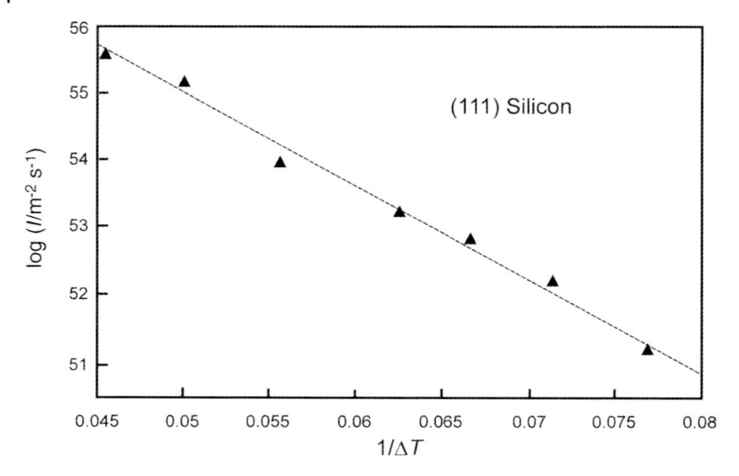

Figure 21.15 Growth rate from the data in Figure 21.14 plotted against$1/\Delta T$.

Adapting the data in Figure 21.15 to a multiple nucleation model, which explicitly takes into account the interaction between nuclei on the surface to extrapolate to the size of a silicon boule, gives the growth rate data shown in Figure 21.16. The Monte Carlo growth rate for the (100) orientation was scaled to $v = 0.12\ \Delta T$, a value derived from experiment and from molecular dynamics modeling. The difference between the growth rates for the two faces comes from the Monte Carlo modeling.

Typical Czochralski growth rates for silicon are between 5×10^{-5} and $10^{-4}\ \mathrm{m\,s^{-1}}$. At these growth rates the undercooling on the (100) face, as well as for all the other growth orientations except for (111), is less than a milli-degree. So these interfaces

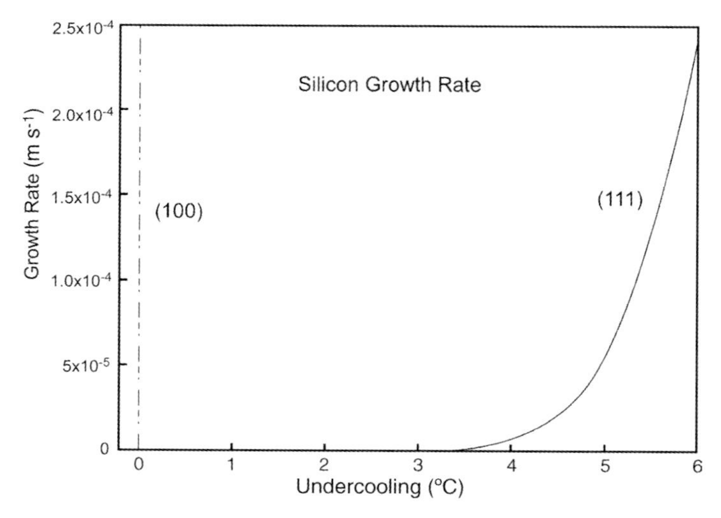

Figure 21.16 Growth rate versus undercooling for silicon (100) and (111).

will be essentially at the melting point isotherm. The undercooling where the steps are nucleating on the (111) face is about 5 °C. This is in accord with experiment and explains the large facet observed on silicon (111) during growth. The undercooling on the two faces for the same growth rate differs by about four orders of magnitude. By fitting the growth rate to ae nucleation model, the edge free energy of the steps on the (111) face is only about 10% of the free energy of the (111) face.

21.7 Growth Morphologies

For a comparison of growth morphologies on smooth and rough surfaces, Figures 11.2, 11.4 and 26.1 show the growth morphologies on rough interfaces, where there is no indication of faceting. This can be compared with Figure 21.17 which shows a dendritic type of growth in tertiary butyl alcohol, for which $\Delta S/R$ is about 3. The dendrite morphology exhibits facets on its side branches.

Figure 21.18 shows faceted growth of benzil, for which $\Delta S/R$ is about 6. This material will not grow dendritically. The morphology is always faceted. The anisotropy in the growth rate suppresses the instabilities that result in side branching.

Figure 21.17 Faceted dendritic growth in *tert*-butyl alcohol.

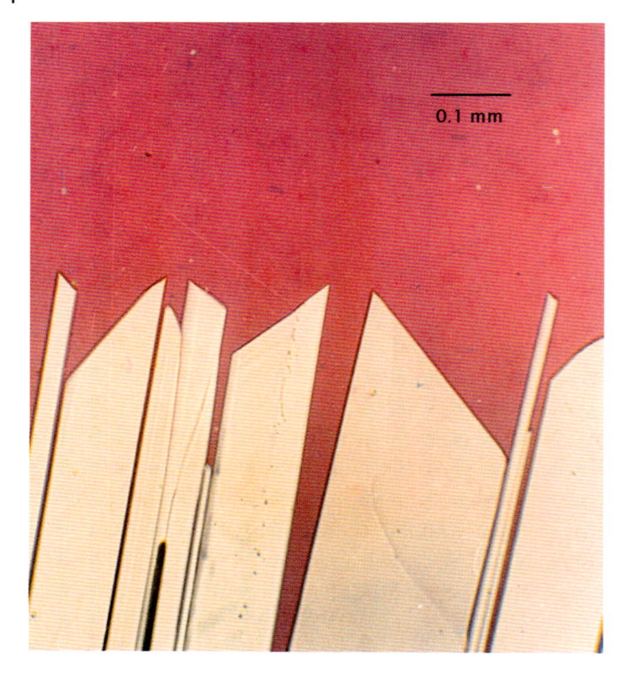

Figure 21.18 Faceted growth in benzil.

21.8 Kinetic Roughening

Experimentally, the growth rate is one of the most anisotropic properties of a crystal. As can be seen in Figure 21.12, at small undercoolings there can be very large anisotropies in the growth rate. At small undercoolings, some faces grow and others do not: the difference in growth rates can be orders of magnitude. On very slowly growing faces, defects can significantly promote growth However, at large undercoolings, when the growth rate lifts off the horizontal axis in Figure 21.18, then the anisotropy in the growth rate is much less. In this regime, a surface which was smooth at equilibrium becomes rough. This has been termed kinetic roughening, and is illustrated in Figure 21.19.

This results in a transition in the growth morphology from faceted or spiky crystals at smaller undercoolings to isotropic growth at large undercoolings, as illustrated in Figure 21.20.

At the lower temperatures, the growth rate has become isotropic.

Because the mobility of molecules in the melt decreases at the lower temperatures, the growth rate of these spherulites is much slower at the lower temperatures. But the normalized difference between the arrival rates at which molecules join and leave the crystal increases with undercooling.

The growth forms shown in the bottom right photograph of Figure 21.20 are known as spherulites, and it is the common form of crystallization in polymers.

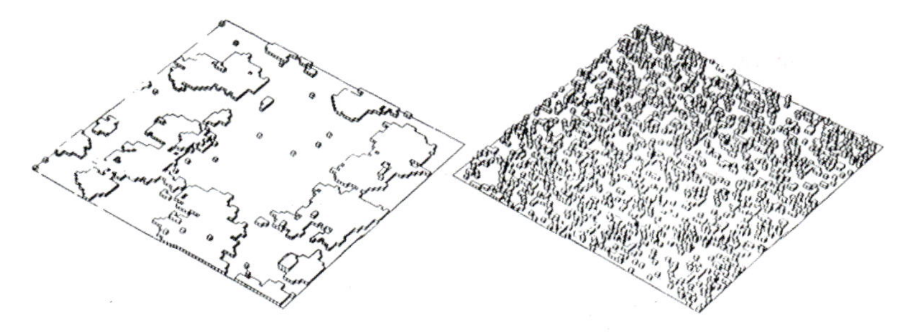

Figure 21.19 Kinetic roughening (From [5]) (a) small undercooling, (b) large undercooling.

The compact morphology illustrated is observed in the growth of minerals such as hematite and malachite, as well as in the iron carbide crystals that grow in the liquid ahead of the interface in nodular cast iron.

Spherulitic growth in polybutene-1 is shown in Figure 21.21. The spherulite starts from a single seed but the result is a spherical polycrystalline mass growing with the same fast growing orientation pointing radially outward everywhere. Initially, the crystal grows and spreads out like a sheaf of wheat in its fast growth direction, due to defect generation, and then faster growing orientations nucleate on the slow growing lateral faces, ultimately resulting in the spherulite.

The growth of large defect-free crystals such as silicon depends on the process being close to the equilibrium condition, so that atoms are joining and leaving the crystal much more rapidly than the net growth rate. The atoms which join the crystal

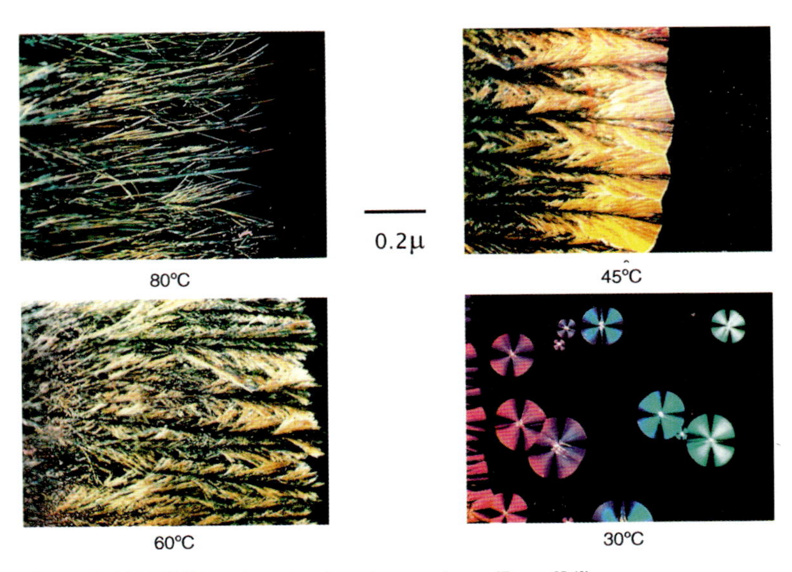

Figure 21.20 DDT growing at various temperatures (From [14]).

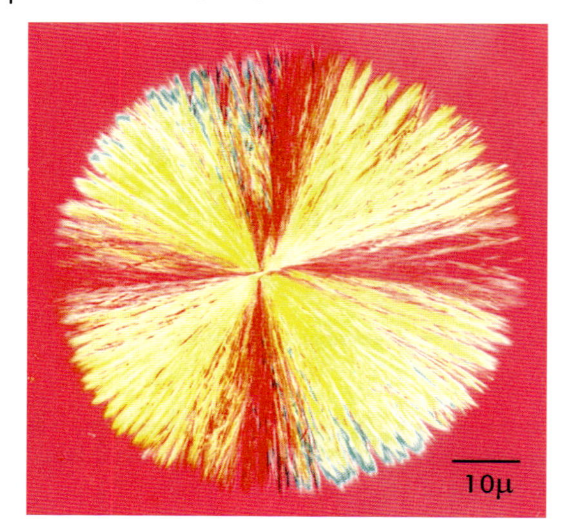

Figure 21.21 Spherulitic growth in polybutene-1.

in wrong positions are even more likely to leave, and so there are many opportunities for defects to be repaired. At large undercooling, far from equilibrium, the defects are much less likely to be repaired.

If the growth is at very large undercooling because the nucleation of new layers is difficult, then defects are not only likely to get built into the crystal, but when they do, the crystal grows faster. At fast enough growth rates there can be a transition to an amorphous solid. But before that stage, a fine grained, highly defective, polycrystal will grow, as in the low temperature growth of DDT in Figure 21.20. Polysilicon has a structure like this, with the scale of the grain structure depending on the substrate temperature during deposition.

The surface roughening transition plays the same role in the deposition of thin films as in growth from the melt. At high substrate temperatures, the equilibrium configuration of the surface is rough. At lower substrate temperatures, the surface becomes smooth, and deposition proceeds by the nucleation and spreading of layers. At very low temperatures, the rate of desorption decreases, and a significant fraction of the incident atoms stick. The critical nucleus size decreases, and the surface becomes kinetically rough.

21.9 Polymer Crystallization

Some polymers can crystallize, some cannot. Among the ones that can, the degree of crystallization, the size of the crystals and the structure of the crystals depend upon a number of parameters, such as whether the crystallization is from the melt or solution, the temperature and time under which the crystallization occurs, and the concentration of the solution and whether any stresses are present to direct the crystallization (mechanical, electrical or magnetic).

Many polymers crystallize by a chain-folding process. The polymer molecule folds back and forth along itself to join the crystal. The size of the crystal is a small fraction of the length of the polymer molecule.

Polymers with highly regular structures are more likely to crystallize than irregular or highly branched structures. In dilute solution, a polymer molecule will tend to chain fold and most of an individual chain can be found in an individual crystal. From the melt, the chains are densely packed, and instead of neat chain folding, bits of chains from different molecules tend to randomly line up to form a crystallite with the rest of the chains extending into the amorphous region, or joining other crystallites.

The degree of crystallization varies widely for different polymers. For example, linear polyethylene can be nearly 95% crystalline, while highly branched polyethylene will only be about 30% crystalline. Polypropylene is usually less than 80% crystalline, PET is usually less than 60%, and nylon 6,6 is usually less than 50%.

Polymers such as polyethylene have a highly regular structure, and highly flexible chains and are easily crystallized. Materials like PET (polyethylene terephthalate) have a benzene ring in the polymer backbone which makes the chain quite stiff and so the melting point of the crystal is high (270 °C). An additional important factor in polymer crystallization is intermolecular forces between adjacent chains. Polar molecules have relatively strong intermolecular forces, and hydrogen bonds formed between chains can be very strong. These can be seen in materials such as polyamides, polyurethanes and polyureas.

Polymers with large side groups such as polystyrene or even polypropylene can crystallize, but the side groups are so bulky that the polymer chains cannot lie flat in the standard planar zig-zag configuration. The chains relieve the strain by twisting and the polymer crystallizes into a helix formation. Many biopolymers crystallize in this helical structure and also have hydrogen bonding that greatly strengthens the helix. For example, DNA forms a two-stranded helix and collagen forms a three-stranded helix.

These cylindrical rods can then pack into a hexagonal parallel array for a super crystalline structure. Polymers such as nylons can form hydrogen bonds between flat planar zig-zag sheets. The C=O from one chain is attracted to the N–H of another.

In dilute solution, polymers tend to form larger, more perfect crystals with the chains folded to fit into the crystal thickness. In polyethylene the crystallite thickness is nearly proportional to the crystallization temperature. At 60 °C the thickness is about 90 Å, at 100 °C about 150 Å and at 130 °C, nearly 200 Å.

Another format seen in polymers crystallized from the melt is called a spherulite, as illustrated in Figure 21.21. These supermolecular structures are built from chain folded laminar ribbons that branch as the spherical structure grows, forming a sphere with constant density with diameter. Spherulites can be hundreds of microns in diameter.

The oligomeric components of the polymeric material as well as impurities tend to segregate the intercrystalline spaces which remain amorphous.

References

1 Stranski, I.N. (1928) *Z. Phys. Chem. (Leipzig)*, **136**, 259.
2 Kossel, W. (1927) *Nachr. Ges. Wiss. Göttingen*, 135.
3 Garrod, C. (1995) *Statistical Mechanics and Thermodynamics*, Oxford University Press, New York, p. 250.
4 Burton, W.K., Cabrerra, N. and Frank, F.C. (1951) *Phil. Trans. R. Soc. London Ser. A*, **243**, 299.
5 Gilmer, G.H. and Jackson, K.A. (1977) *1976 Crystal Growth and Materials* (eds E. Kaldisand H.J. Scheel), North-Holland, Amsterdam, p. 80.
6 Jackson, K.A. (1958) *Liquid Metals and Solidification*, ASM Cleveland, OH, p. 174. Doremus, R.H., Roberts, B.W. and Turnbull, D. (eds) (1958) *Growth and Perfection of Crystals*, Wiley, New York, p. 319;
7 Bentley, W.A. and Humphreys, W.J. (1962) *Snow Crystals*, Dover, New York; Nakaya, U. (1954) *Snow Crystals*, Harvard, Cambridge, MA.
8 Weeks, J.D. and Gilmer, G.H. (1979) *Adv. Chem. Phys.*, **40**, 157.
9 Leamy, H.J., Gilmer, G.H. and Jackson, K.A. (1975) *Surface Physics of Materials* (ed. J.B. Blakeley), Academic Press, New York, p. 121.
10 Leamy, H.J. and Gilmer, G.H. (1974) *J. Cryst. Growth*, **24/25**, 499.
11 Weeks, J.D. and Chui, S.T. (1976) *Phys. Rev. B*, **14**, 4978.
12 Bentz, D.N. and Jackson, K.A. (2003) *Mater. Res. Soc. Symp. Proc.*, **778**, 255.
13 Beatty, K.M. and Jackson, K.A. (2000) *J. Cryst. Growth*, **211**, 13.
14 McCrone, W.C. (1957) *Fusion Methods in Chemical Microscopy*, Interscience, New York.

Problems

21.1. Discuss the surface roughening transition and its importance for crystal growth processes.

21.2. Discuss the similarities and differences between the Knudsen description of vapor deposition, the Langmuir description of adsorption on a surface and the Ising model description of the equilibrium roughening of a surface.

21.3. Solid-on-solid (SOS) simulation
Start with a set of 50 adjacent columns, each of height 25 units.
Add units to columns randomly at a rate k^+.
Remove units from the top of the columns randomly using a probability which depends on the number of nearest neighbors of the unit:

$$k^- = k^+ \exp((2-n)\phi/kT)$$

where n is the number of nearest neighbor sites which are occupied.
Continue until the configuration stops changing significantly.
The information needed for each column can be stored as its height.
Use periodic boundary conditions for the two end columns.
k^+ must be chosen so that k^- is always ≤ 1.
Show the final distribution of column heights for $\phi/kT = 2.5$, 5, and 10.

This scheme is known as the solid-on-solid approximation, since atoms can arrive only on top of other atoms. Overhanging configurations are not permitted, as they are in the Ising model. If the interface is not too rough, the number of overhanging configurations is usually small, and so this is usually quite a good approximation to an Ising model interface.

22
Alloys: Thermodynamics and Kinetics

The thermodynamic properties of alloys are discussed in many texts. A few of these are listed at the end of this chapter [1–4]. There also compilations of phase diagrams, such as [5]. In this chapter, alloys will be discussed in terms of both their thermodynamic properties and the kinetics of their crystallization.

22.1
Crystallization of Alloys

For the crystallization of an alloy, separate equations, such as Equations 13.29 and 13.30 can be written for the rate at which each species makes the transition between the two phases. The chemical potentials, μ_i, of each species, i, should be used instead of the free energy difference as in Equations 13.29 or 13.30. The rate at which species i leaves the liquid to join the crystal can be written:

$$\nu_i^+ = \nu_i^{0+} \exp\left(\frac{\mu_i^L}{kT}\right) \tag{22.1}$$

and the rate at which it leaves the crystal to join the liquid:

$$\nu_i^- = \nu_i^{0-} \exp\left(\frac{\mu_i^S}{kT}\right) \tag{22.2}$$

For an alloy phase with ideal entropy of mixing, the chemical potential is given by:

$$\mu_i = \mu_i^0 + kT \ln C_i \tag{22.3}$$

where $\mu_i^0 = H_i - TS_i$ is the chemical potential for pure i, and C_i is the concentration of species i.

For non-ideal alloys, the same form can be retained, but the concentration is replaced by the "activity" to account for the departures from ideality, although departures from ideality are much more common in the energy than in the entropy of an alloy.

Kinetic Processes: Crystal Growth, Diffusion, and Phase Transitions in Materials. Kenneth A. Jackson

ISBN: 978-3-527-32736-2

The rate at which atoms of species i leave the liquid to join the crystal can thus be written as:

$$\nu_i^+ = C_i^{\mathrm{L}} \nu_i^{0+} \exp\left(\frac{\mu_i^{0\mathrm{L}}}{kT}\right) \tag{22.4}$$

And similarly for the transition from crystal to liquid.

This is the same as the assumptions made in chemical reaction rate theory that the rate of a reaction is proportional to the concentration of the reactants. The strange looking concentration term in Equation 22.3 implies simply that the transformation rate is proportional to the concentration. The concentrations in this case are the concentrations at the interface.

The general form for the growth rate of a crystal, Equation 20.1, will apply for an alloy, but with an individual equation written for each species. The u_{K} in Equation 20.1 becomes:

$$u_{\mathrm{K}}^i = C_i^{\mathrm{L}} - C_i^{\mathrm{S}} \exp\left(\frac{\Delta\mu_i^0}{kT}\right) \tag{22.5}$$

Where $\Delta\mu_{\mathrm{i}}^0 = \mu_{\mathrm{i}}^{0\mathrm{S}} - \mu_{\mathrm{i}}^{0\mathrm{L}}$.

When the chemical potentials of Equation 22.3 are inserted into the standard expression for the free energy of a binary alloy, these $kT\ln C$ terms result in the standard entropy of mixing term:

$$G = C_{\mathrm{A}}\mu_{\mathrm{A}}^0 + C_{\mathrm{B}}\mu_{\mathrm{B}}^0 + kT\,(C_{\mathrm{A}}\ln C_{\mathrm{A}} + C_{\mathrm{B}}\ln C_{\mathrm{B}}) \tag{22.6}$$

So the entropy of mixing term, strange as it looks, has a very simple physical interpretation. It derives from the assumption that the rate at which a component reacts or makes a transition is proportional to its concentration.

In terms of the general form for the growth rate of a crystal, Equation 20.1, ν_{i}^0 incorporates v^+, the rate at which atoms join the crystal at an active growth site, and f, the fraction of interface sites which are active growth sites:

$$\nu_i^0 = (a\mathrm{v}^+ f)_i \tag{22.7}$$

For a binary alloy, the net growth rates can be written:

$$\nu_A = \nu_{\mathrm{A}}^0\left[C_{\mathrm{A}}^{\mathrm{L}} - C_{\mathrm{A}}^{\mathrm{S}} \exp\left(\frac{\Delta\mu_A^0}{kT}\right)\right] \tag{22.8}$$

and:

$$\nu_{\mathrm{B}} = \nu_{\mathrm{B}}^0\left[C_{\mathrm{B}}^{\mathrm{L}} - C_{\mathrm{B}}^{\mathrm{S}} \exp\left(\frac{\Delta\mu_{\mathrm{B}}^0}{kT}\right)\right] \tag{22.9}$$

At equilibrium,

$$\nu_A = \nu_B = 0 \tag{22.10}$$

Equating the rates at which each species of atom joins and leaves the crystal implies that the chemical potentials of each species are the same in both phases.

22.2 Phase Equilibria

For the simple phase diagram in Figure 22.1, when the temperature is between T_M^A and T_M^B, the A atoms are below their melting point, so they freeze faster than they melt, but the B atoms are above their melting point, so they melt faster than they freeze. This implies

$$\nu_A^{L\to S} > \nu_A^{S\to L} \quad \text{and} \quad \nu_B^{L\to S} < \nu_B^{S\to L} \tag{22.11}$$

Given the four rates as illustrated in Figure 22.2, there will be only one set of compositions for the two phases, at any temperature, which satisfy both the relationships:

$$\begin{aligned} C_A^L \nu_A^{L\to S} &= C_A^S \nu_A^{S\to L} \\ C_B^L \nu_B^{L\to S} &= C_B^S \nu_B^{S\to L} \end{aligned} \tag{22.12}$$

together with $C_A^L + C_A^S = 1 = C_B^L + C_B^S$.

Outside the temperature interval between T_M^A and T_M^B, there will not be a set of compositions which satisfy these equations, because above T_M^A both species melt faster, and below T_M^B, both freeze faster. This explanation is based on the simple phase diagram, where the rates are relatively independent of composition, but the conclusion is generally valid, that a two-phase field occurs where the atoms of one species freeze faster than they melt, and the atoms of the other species melt faster than they freeze.

The overall growth rate of an alloy can be written as:

$$\nu = \nu_A + \nu_B \tag{22.13}$$

The instantaneous ratio of the compositions of the two species in the growing crystal is given by the ratio of the net rates at which they enter the crystal:

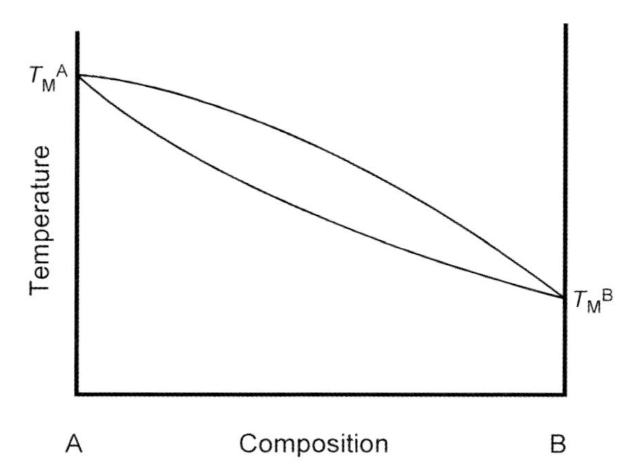

Figure 22.1 Binary alloy phase diagram.

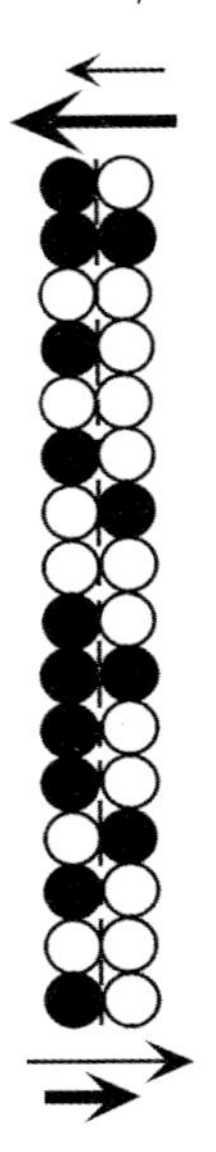

Figure 22.2 Schematic of compositions and transition rates at the interface. The lengths of the heavy and light arrows indicate the transformation rates of the black and white atoms, respectively.

$$\frac{\nu_A}{\nu_B} = \frac{C_A^S}{C_B^S} \tag{22.14}$$

The equilibrium distribution coefficient, k, for the B component is given by the condition $\nu_B = 0$:

$$k \equiv \frac{C_B^S}{C_B^L} = \exp\left(-\frac{\Delta\mu_B^0}{kT}\right) \tag{22.15}$$

22.3
Regular Solution Model

The regular solution model adds a composition-dependent energy term, $\Omega C_A C_B$, to the expression for the free energy of an ideal alloy, Equation 22.6, so the free energy of a regular solution is:

$$G = C_A\mu_A^0 + C_B\mu_B^0 + kT(C_A \ln C_A + C_B \ln C_B) + \Omega C_A C_B \tag{22.16}$$

This regular solution term has a simple interpretation in terms of the bonding between atoms. We will define ϕ_{AA}, ϕ_{AB}, and ϕ_{BB} as the bond energies between neighboring AA, AB, and BB atoms respectively. Z is the number of nearest neighbor sites around each atom in the crystal. The number of A atoms is N_A, the number of B atoms is N_B, and $N_A + N_B = N$, the total number of atoms. The concentrations are given by $C_A = N_A/N$, and $C_B = N_B/N$. The number of AB pairs will be NZX, where X is the probability of finding a B atom on any one of the sites next to an A atom. In general, X will depend on the concentration, and also on how the atoms are arranged in the crystal, whether there is clustering of like atoms, or whether there is ordering.

However, for any configuration of atoms, we can construct Table 22.1.

The first term in the total energy of pairs in the table corresponds to the energy part of the first two terms in Equation 22.16. The second term in the total energy of the pairs in the table corresponds to the regular solution term in Equation 22.16. In the table we have used:

$$W = \frac{\phi_{AA} + \phi_{BB}}{2} - \phi_{AB} \tag{22.17}$$

The total energy of pairs in the table is valid for any configuration of atoms. For the special case of a random distribution, the number of AB pairs is:

$$NX = \frac{N_A N_B}{N} = N\, C_A C_B \tag{22.18}$$

N_A is the number of A atoms, and N_B/N is the fraction of the sites filled with B atoms, so that $X = C_A C_B$. The regular solution parameter, Ω in Equation 22.16, is given by:

$$\Omega = ZW = Z\left[\frac{\phi_{AA} + \phi_{BB}}{2} - \phi_{AB}\right] \tag{22.19}$$

The regular solution model, Equation 22.16 assumes that the two types of atoms are randomly distributed, and, in terms of bond counting, it is equivalent to assuming that the three different bond energies are independent of composition.

For an ideal solution, $W = 0$, so that $(\phi_{AA} + \phi_{BB})/2 = \phi_{AB}$.

For $(\phi_{AA} + \phi_{BB})/2 > \phi_{AB}$, the formation of AA and BB pairs lowers the total energy, so that the alloy will tend to phase separate.

For $(\phi_{AA} + \phi_{BB})/2 < \phi_{AB}$, the formation of AB pairs lowers the total energy, so that the alloy will tend to order.

For a non-ideal alloy, the departure from ideality is usually expressed as an activity, P, which is an "effective concentration" in the chemical potential, in the form of $kT \ln P$. For a regular solution,

$$\mu - \mu^0 = kT \ln P = kT \ln C + \Omega(1 - C)^2 \tag{22.20}$$

Table 22.1 Scheme for counting bonds in an alloy. ZX is the probable number of B atoms next to an A atom.

	Number of Pairs	**Energy of Pairs**
AA	$(Z/2)(N_A - NX)$	$-(Z/2)(N_A - NX)\phi_{AA}$
AB	NZX	$-NZX\phi_{AB}$
BB	$(Z/2)(N_B - NX)$	$-(Z/2)(N_B - NX)\phi_{BB}$
Total	$(Z/2)(N_A + N_B)$	$-(Z/2)(N_A\phi_{AA} + N_B\phi_{BB}) + NZWX$

So that the activity is:

$$P = C \exp\left(\frac{\Omega(1-C)^2}{kT}\right) \tag{22.21}$$

When there are interaction between atoms, the fluxes depend on activity gradients, and not simply on concentration gradients, as in an ideal solution. The interaction energies included in the activity play a major role in determining the fluxes.

22.4 Near Equilibrium Conditions

The equations for the growth rate can be written:

$$v_{\mathrm{A}} = v_{\mathrm{A}}^{0}\left[C_{\mathrm{A}}^{\mathrm{L}} - C_{\mathrm{A}}^{\mathrm{S}} \exp\left(\frac{\Delta H_{\mathrm{A}}}{k}\left(\frac{1}{T_{\mathrm{A}}^{\mathrm{M}}} - \frac{1}{T}\right)\right)\right] \tag{22.22}$$

$$v_{\mathrm{B}} = v_{\mathrm{B}}^{0}\left[C_{\mathrm{B}}^{\mathrm{L}} - C_{\mathrm{B}}^{\mathrm{S}} \exp\left(\frac{\Delta H_{\mathrm{B}}}{k}\left(\frac{1}{T_{\mathrm{B}}^{\mathrm{M}}} - \frac{1}{T}\right)\right)\right] \tag{22.23}$$

ΔH_{A} and ΔH_{B} are independent of concentration for an ideal alloy, but, in general, they depend on the composition of the alloy. Alternatively, but equivalently, the compositions can be replaced with activities. The liquidus temperature, T^{L}, for liquid of composition C^{L} at equilibrium with the solid is given by the simultaneous solution of the two equations for $v=0$:

$$\frac{C_{\mathrm{A}}^{\mathrm{S}}}{C_{\mathrm{A}}^{\mathrm{L}}} = \exp\left(\frac{\Delta H_{\mathrm{A}}}{k}\left(\frac{1}{T_{\mathrm{L}}} - \frac{1}{T_{\mathrm{A}}^{\mathrm{M}}}\right)\right) \tag{22.24}$$

and

$$\frac{C_{\mathrm{B}}^{\mathrm{S}}}{C_{\mathrm{B}}^{\mathrm{L}}} = \exp\left(\frac{\Delta H_{\mathrm{B}}}{k}\left(\frac{1}{T_{\mathrm{L}}} - \frac{1}{T_{\mathrm{B}}^{\mathrm{M}}}\right)\right) = k \tag{22.25}$$

$C_{\mathrm{A}}^{\mathrm{S}}/C_{\mathrm{A}}^{\mathrm{L}}$ and $C_{\mathrm{B}}^{\mathrm{S}}/C_{\mathrm{B}}^{\mathrm{L}}$ ($= k$) do not change significantly for small growth rates, so that

$$v_{\mathrm{A}} = v_{\mathrm{A}}^{0} C_{\mathrm{A}}^{\mathrm{L}}\left[1 - \exp\left(\frac{\Delta H_{\mathrm{A}}}{k}\left(\frac{1}{T_{\mathrm{L}}} - \frac{1}{T}\right)\right)\right] \approx v_{\mathrm{A}}^{0} C_{\mathrm{A}}^{\mathrm{L}} \frac{\Delta H_{\mathrm{A}}}{kT}\left(\frac{\Delta T_{\mathrm{L}}}{T_{\mathrm{L}}}\right) \tag{22.26}$$

and similarly,

$$v_{B} = v_{\mathrm{B}}^{0} C_{\mathrm{B}}^{\mathrm{L}}\left[1 - \exp\left(\frac{\Delta H_{\mathrm{B}}}{k}\left(\frac{1}{T_{\mathrm{L}}} - \frac{1}{T}\right)\right)\right] \approx v_{\mathrm{B}}^{0} C_{\mathrm{B}}^{\mathrm{L}} \frac{\Delta H_{\mathrm{B}}}{kT}\left(\frac{\Delta T_{\mathrm{L}}}{T_{\mathrm{L}}}\right) \tag{22.27}$$

where $\Delta T_{\mathrm{L}} = T_{\mathrm{L}} - T$ is the undercooling below the liquidus.

The overall growth rate can be written as:

$$v = v_A + v_B \approx \left(v_A^0 C_A^L \frac{\Delta H_A}{kT} + v_B^0 C_B^L \frac{\Delta H_B}{kT} \right) \left(\frac{\Delta T_L}{T_L} \right) \tag{22.28}$$

For a constant density of active growth sites, *f*, the growth rate is linear with undercooling below the liquidus line. This equation can also be written as:

$$v \approx \left(v_A^0 C_A^L \frac{\Delta \mu_A}{kT} + v_B^0 C_B^L \frac{\Delta \mu_B}{kT} \right) \tag{22.29}$$

An expression which has been used for the growth rate of an alloy:

$$v = v^0 \exp\left(\frac{\Delta G}{kT} \right) \tag{22.30}$$

is only valid for small concentrations of the second component when $C_B^L \approx 0$, $C_A^L \approx 1$ and $\Delta G \approx \Delta \mu_A$, since, in general, $\Delta G = 0$ does not define the equilibrium condition for an alloy. Equilibrium for an alloy occurs when the chemical potential of each component is the same in both phases, which does not imply $\Delta G = 0$.

For small concentrations of the B component, we can write: $\ln(C_A^S) = \ln(1 - C_B^S) = \ln(1 - kC_B^L) \approx -kC_B^L$, where k is the equilibrium distribution coefficient, and $\ln(C_A^L) = \ln(1 - C_B^L) \approx -C_B^L$ for small C_B^L, so that Equation 22.24 for the A component reduces to:

$$C_B^L = \frac{\Delta H_A}{k(1-k)} \left(\frac{1}{T^L} - \frac{1}{T_A^M} \right) \tag{22.31}$$

This equation can be rewritten as:

$$T^L = T_A^M - mC_B^L \tag{22.32}$$

where

$$m \equiv (k(1-k) T^L T_A^M) / \Delta H_A \tag{22.33}$$

which is the standard thermodynamic expression for the slope of the liquidus line for small concentrations of a second component, also known as the freezing point depression. For small concentrations of B, the undercooling at the interface can be written in familiar form as:

$$\Delta T = T_A^M - T = mC_B^L + \mu v_A \tag{22.34}$$

where $\mu = kT^L T_A^M / \Delta H_A v_A^0$ is termed the kinetic coefficient.

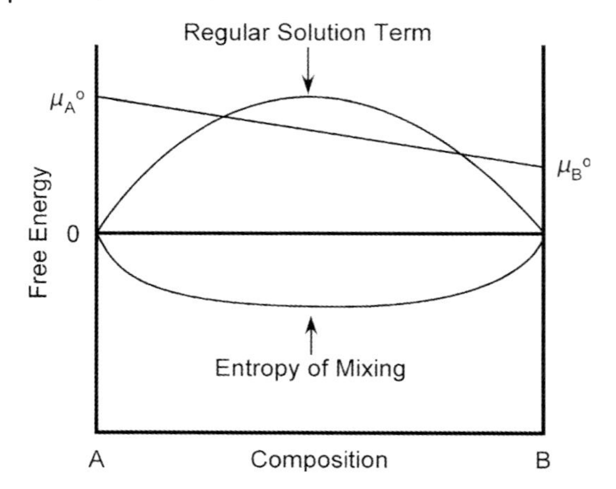

Figure 22.3 Components of the free energy of a regular solution.

22.5 Phase Diagrams

The simplest scheme for constructing phase diagrams is to the fit the liquid to an ideal solution model and the solid to a regular solution model.

$$F_L = (1-C_L)\mu_L^{0A} + C_L\mu_L^{0B} + kT[C_L \ln C_L + (1-C_L)\ln(1-C_L)] \tag{22.35}$$

$$F_S = (1-C_S)\mu_S^{0A} + C_S\mu_S^{0B} + kT[C_S \ln C_S + (1-C_S)\ln(1-C_S)] + \Omega C_S(1-C_S) \tag{22.36}$$

More complex schemes are used for detailed descriptions of alloy phase diagrams. But this model works well for many cases.

The three components of the free energy of a regular solution, Equation 22.36 are shown in Figure 22.3.

The first two terms in Equation 22.36 represent a linear interpolation between the chemical potentials of the pure materials. The second term is the entropy of mixing. The third term is the regular solution term, which adds a simple composition-dependent term to the energy of the alloy. The entropy of mixing term starts off with infinite slope at each end, and so the sum of the three terms always starts off going down at the two ends.

For a case where the regular solution parameter is small, the free energy curves for the solid and liquid at some temperature between the melting points of the two end members will look like Figure 22.4a, and the Gibbs common tangent construction gives the corresponding solidus and liquidus compositions on the phase diagram, Figure 22.4b.

The chemical potentials of the two components of a phase are found graphically by drawing a tangent to the free energy curve at the composition of interest, and extending the tangent line all the way across the diagram. At the left axis, the tangent

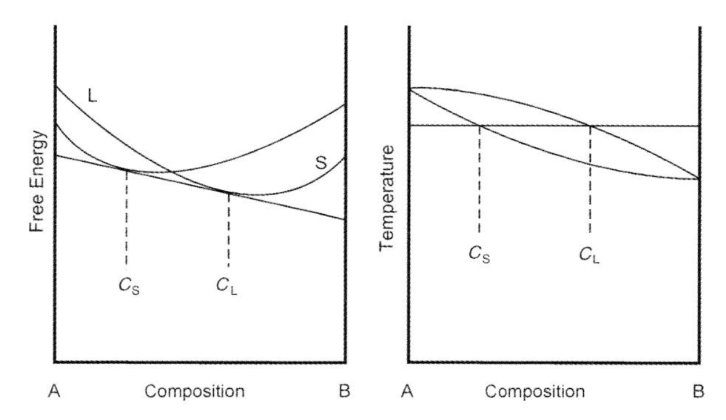

Figure 22.4 (a) The free energy versus composition, and (b) the corresponding phase diagram for a nearly ideal alloy.

line gives the chemical potential of the A component, and at the right axis, it gives the chemical potential for the B component. When both phases are present, drawing the common tangent to both free energy curves ensures that the chemical potentials of both species are equal in the two phases, and so the common tangent points are the equilibrium compositions.

A slightly larger value of the regular solution parameter will flatten out the free energy curve for the solid phase, which can result in the free energy curves sketched in Figure 22.5a, and the corresponding phase diagram, Figure 22.5b.

When the regular solution parameter is larger, the free energy curve for the solid can develop a hump in the middle. This gives rise to a phase separated region or an ordered region on the phase diagram as illustrated in Figure 22.6.

The phase boundary of the two phase region in the lower part of Figure 22.6b is given by:

$$\ln\left(\frac{C}{1-C}\right) = \frac{\Omega}{k_B T}(2C-1) \tag{22.37}$$

The critical point is at $C = \frac{1}{2}$, and $T = T_C$, so that $k_B T_C = \Omega/2$.

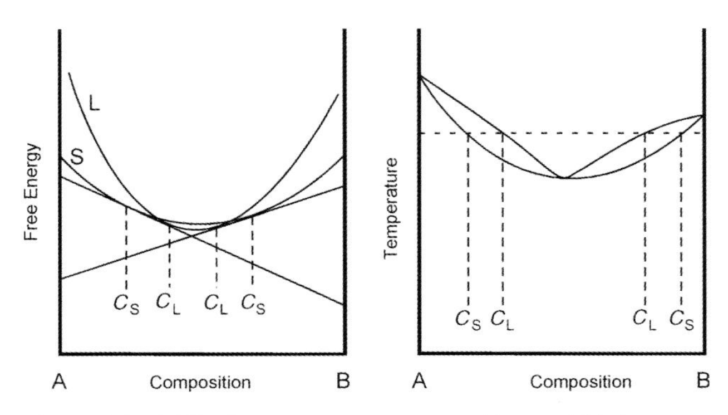

Figure 22.5 (a) The free energy versus composition, and (b) the corresponding phase diagram for a larger value of the regular solution parameter.

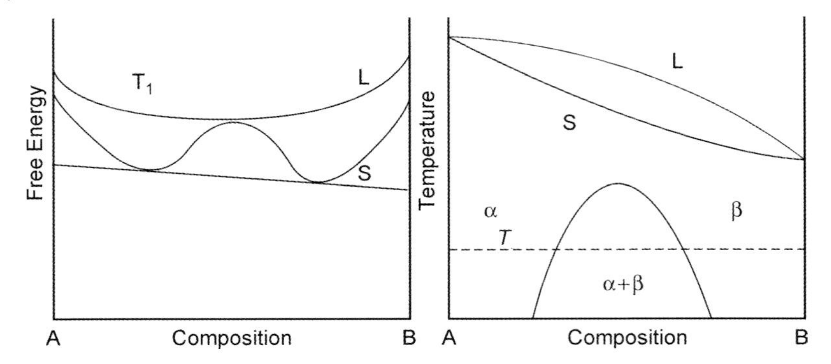

Figure 22.6 (a) The free energy versus composition, and (b) the corresponding phase diagram showing a phase separated region or an ordered region.

Ordering occurs for negative values of Ω. Assuming that the compositions on alternate sites of the ordered structure are given by C and $1-C$, then $X=(C^2+(1-C)^2)/2$. This results in the same equilibrium values of composition as given by Equation 22.37, but in this case, the phase boundary gives the compositions on alternate sites, rather than the region of phase stability.

If the hump in the free energy curve occurs at a temperature where the liquid has a similar free energy, a eutectic phase diagram can result, as illustrated in Figure 22.7.

This occurs when the free energy of the solid can be represented by a regular solution at all compositions. There are many alloys for which the end members have different crystal structures, so that it is not possible to describe the free energy of all the compositions of the solid with a single free energy curve. There are then two different free energy curves, one for each of the two solid phases. The free energy curve for each solid phase will have its own minimum, so there are two different minima to interact with the liquid phase, just as in Figure 22.7a. Even though the free energy diagrams are quite different, the phase diagrams will also be a eutectic.

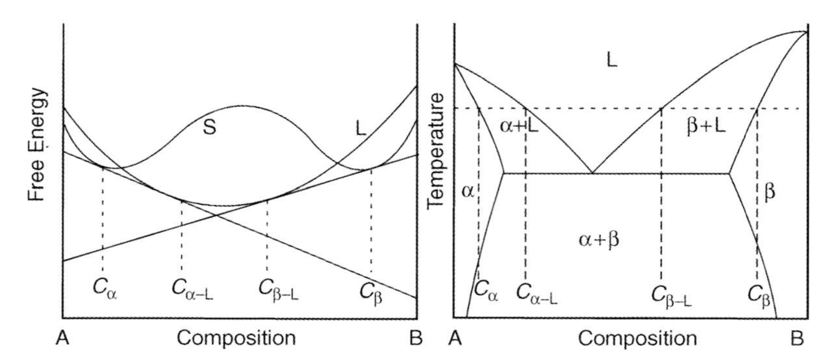

Figure 22.7 (a) The free energy versus composition, and (b) the corresponding eutectic phase diagram.

22.6
The DLP Model

Stringfellow [6] has developed the Delta Lattice Parameter (DLP) model for mixtures of III–V elements. In the DLP model, the regular solution parameter, Ω, is given by:

$$\Omega = 1.15 \times 10^5 \frac{(a_A - a_B)^2}{(a_A + a_B)^{4.5}} \text{ cal mol}^{-1} \quad (22.38)$$

where a_A and a_B are the lattice parameters of the two end components in Angstroms. This scheme works very well, not only for binary alloys, but for pseudo-binary alloys, for example, GaAs–AlAs.

The regular solution parameter cannot be estimated this simply for most alloys, but the thermodynamic properties of many alloys can be described using a regular solution model, especially over a limited composition range.

References

1 Christian, J.W. (1975) *The Theory of Transformations in Metals and Alloys*, Pergamon, Oxford.
2 Hillert, M. (1998) *Phase Equilibria, Phase Diagrams and Phase Transformations*, Cambridge, New York.
3 Porter, D.A. and Easterling, K.E. (1992) *Phase Transformations in Metals and Alloys*, 2nd edn, Chapman and Hall, London.
4 Tiller, W.A. (1991) *The Science of Crystallization*, Cambridge University Press, New York.
5 Okamoto, H. (2000) *Phase Diagrams for Binary Alloys*, ASM, Materials Park, OH.
6 Stringfellow, G.B. (1999) *Organometallic Vapor-Phase Epitaxy*, 2nd edn, Academic Press, Boston, MA, p. 27.

Problems

22.1. What is the difference between an ideal solution and a regular solution?

22.2. Discuss the entropy of mixing.

22.3. Perform a simulation as follows.
Start with a square array of 50 × 100 sites (50 rows by 100 columns).
Fill half of the sites with A atoms and the other half with B atoms (or 1s and 0s) in any pattern you choose.
Use periodic boundary conditions on the top and bottom of the array.
Use periodic boundary conditions on the left and right edges, except use the other type of atom as the nearest neighbor. That is, if the atom at one end of a row is A, then use a B nearest neighbor for the site at the other end of the row.
Pick two sites at random.

Calculate the total energy of the two sites, E_1, due to the bonding energy with their nearest neighbors, using:

$$\phi_{AA}/kT = \phi_{BB}/kT = 5, \quad \phi_{AB}/kT = 2.5$$

(Make sure that a stronger bond *reduces* the energy!)
Now calculate the total energy of the two sites, E_2, which would result if the two atoms were interchanged.
If the interchange would reduce the total energy, that is, if $E_1 > E_2$, perform the interchange.
If the interchange would increase the total energy, that is, if $E_2 > E_1$, perform the interchange if a random number is less than $\exp(E_1 - E_2)/kT)$.
Continue interchanging pairs until the pattern of A and B atoms stops changing.
Display the resulting pattern.
Repeat for:

$$\phi_{AA}/kT = \phi_{BB}/kT = 5, \quad \phi_{AB}/kT = 7.5$$

$$\phi_{AA}/kT = \phi_{BB}/kT = 1, \quad \phi_{AB}/kT = 0.5$$

$$\phi_{AA}/kT = \phi_{BB}/kT = 1, \quad \phi_{AB}/kT = 1.5$$

Discuss the results.

23
Phase Separation and Ordering

Phase separation is a process whereby a homogeneous alloy decomposes into regions of differing compositions. For example, a binary alloy consisting of A and B atoms will form separate A-rich and B-rich regions. This can occur by nucleation or by a spontaneous process known as spinodal decomposition. Ordering is a process whereby the A and B atoms occupy adjacent lattice sites to form an ordered array. These two processes are complementary: one occurs when atoms can lower their free energy by having similar neighbors, and the other occurs when atoms can lower their free energy by having dissimilar neighbors. A scheme based on the interactions between neighbouring atomic sites will be used to discuss these processes, rather than the conventional description, which is based on a continuum model with a correction term. This permits following the time evolution of the microstructures of phase separated and ordered structures.

23.1
Phase Separation versus Ordering

Phase separation or ordering can occur in a regular solution, which is a simple, one-parameter thermodynamic model for non-ideal mixtures. The regular solution parameter, Ω, was expressed in terms of the bond energies in Equation 22.19

$$\Omega = Z\left(\frac{\phi_{AA} + \phi_{BB}}{2} - \phi_{AB}\right) \tag{23.1}$$

When Ω is positive there tends to be phase separation, and when (is negative there tends to be ordering. It is unusual for (to be zero in a solid solution. At first sight it appears that one of these should increase the free energy of the alloy, and the other should decrease it. So how do both ordering and phase separation occur? The form of the regular solution term in Equation 22.16 assumes that the atoms are randomly distributed. This is then the free energy for a random mixture. Both ordering or phase separation decrease the free energy, but do so by changing the number of AB pairs, that is, by changing the value of X in Table 22.1. For an ordered configuration, the number of AB pairs is not given by $N_A N_B / N$.

Kinetic Processes: Crystal Growth, Diffusion, and Phase Transitions in Materials. Kenneth A. Jackson

ISBN: 978-3-527-32736-2

The mathematics for both ordering and phase separation are similar, but the order parameters are defined differently. For phase separation, the order parameter is defined in terms of the probability of a B atom having other B atoms as nearest neighbors. For ordering, the order parameter is defined in terms of the probability of finding a B atom on the B sub-lattice of the ordered structure. It is not unusual to find that the ordered alloy forms a different crystal structure than the disordered solid solutions.

23.2 Phase Separation

For the free energy curve in Figure 22.6a, uniform compositions of the solid phase in the central region of the phase diagram are unstable, as illustrated in Figure 23.1.

Where the curvature of the free energy curve is negative, as illustrated in Figure 23.1, the alloy can reduce its free energy by forming separate regions that are slightly higher and slightly lower than the average composition. The range of compositions where the curvature of the free energy curve is negative is known as a spinodal. The free energy curve of Figure 23.1 is for one temperature. If temperature were plotted as a third dimension the hump in the curve would be part of a ridge on a three-dimensional surface. The ridge resembles a spine, and hence the name.

Phase separation occurs in many alloys and mixtures, for example in aluminum–copper alloys. Phase separation can also occur in a transparent liquid when it is cooled through its critical point. It suddenly becomes opaque. This is known as critical opalescence, and it happens very quickly, because diffusion in liquids is so rapid. Phase separation also occurs in glasses, for example borosilicate glasses, where boron oxide separates from the silica. This process is used commercially as a relatively low-temperature process to make silica. After phase separation, the boron-rich phase is

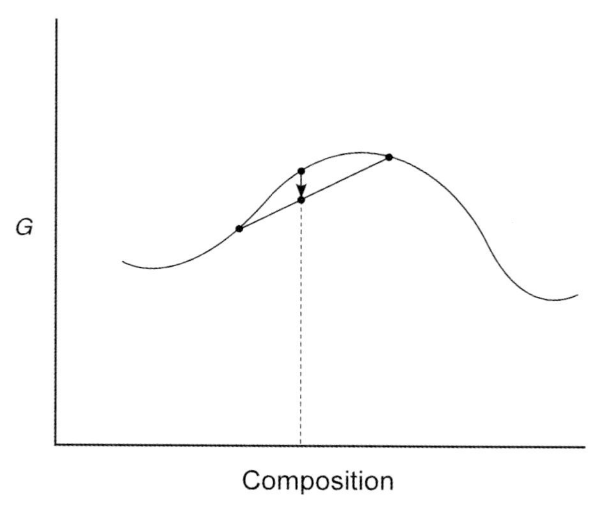

Figure 23.1 The alloy can lower its free energy by separating into two regions, one of lower composition, and the other of higher composition.

etched away, leaving a porous silica structure, which is then densified by annealing. The resulting silica glass is known as VYCOR.

23.3
The Spinodal in a Regular Solution

The free energy and a phase diagram showing the spinodal region for a regular solution are presented in Figure 23.2, as an example.

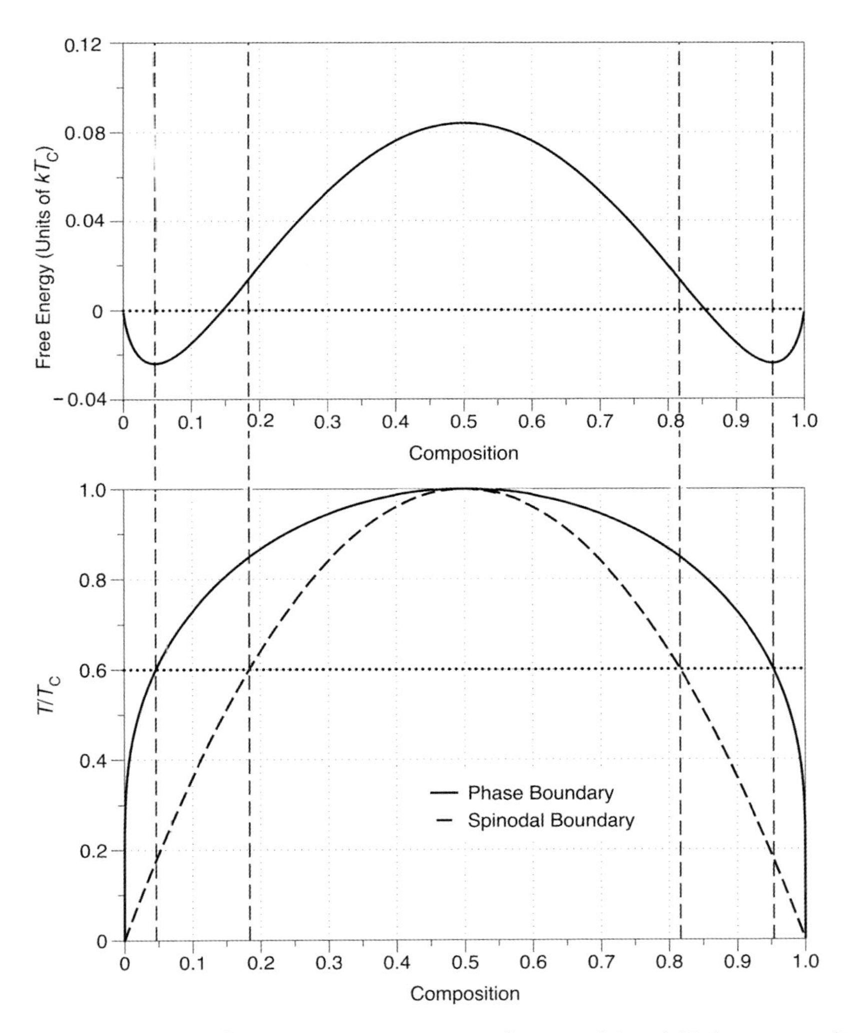

Figure 23.2 (a) The free energy versus composition for $T/T_C = 0.6$, and (b) the corresponding phase diagram.

The free energy of a regular solution is given by Equation 22.16.

$$G = C_A\mu_A^0 + C_B\mu_B^0 + kT(C_A \ln C_A + C_B \ln C_B) + \Omega\, C_A C_B \tag{23.2}$$

The first two terms add a linear term to the free energy, as in Figure 22.3, so only the last two terms in Equation 23.2 for $T/T_C = 0.6$ are plotted in Figure 23.2a).

The equilibrium compositions of the boundaries of the phase-separated region are given by equating the chemical potentials of both components between the two phases. These chemical potentials are given in Equation 22.20. Due to the symmetry of the phase boundary, its equation takes a simple form:

$$kT \ln\left(\frac{C}{1-C}\right) = \Omega(2C-1) \tag{23.3}$$

where $C = C_B = 1 - C_A$. The critical point occurs at the composition $C = \frac{1}{2}$, and at the temperature $T_C = \Omega/2k$. This can be demonstrated by expanding Equation 23.3 using $C = 1/2 + \Delta C$ for small ΔC. Equation 23.3 can be rewritten as:

$$\frac{T}{T_C} = \frac{2\,(2C-1)}{\ln(C/(1-C))} \tag{23.4}$$

This phase boundary is plotted as the solid line in Figure 23.2b.

The spinodal boundary is given by the inflection points in the free energy curve, which are obtained from the second derivative of the free energy. The first derivative of the free energy with respect to composition is:

$$\frac{dG}{dC} = -\mu_A^0 + \mu_B^0 + kT[\ln C - \ln(1-C)] + \Omega(1-2C) \tag{23.5}$$

The curvature of the free energy curve is given by the second derivative:

$$\frac{d^2G}{dC^2} = kT\left(\frac{1}{C} + \frac{1}{1-C}\right) - 2\Omega \tag{23.6}$$

This is negative inside the spinodal, and positive outside it. The spinodal boundary is given by:

$$C(1-C) = \frac{kT}{2\Omega} \tag{23.7}$$

From this equation it is immediately evident that the critical point is at $C = \frac{1}{2}$, $T_C = \Omega/2\,k$.

Equation 23.7 can be rewritten as

$$\frac{T}{T_C} = 4C(1-C) \tag{23.8}$$

which is plotted as the dashed line in Figure 23.2b.

The phase boundary in Figure 23.2a is a plot of Equation 23.2 with $T = 0.6T_C$. The equilibrium compositions of the phase boundaries are given by the Gibbs construction as the common tangent of the two minima. The points on the free energy curve where the curvature changes from positive to negative are also identified, and correspond to the spinodal boundary for $T/T_C = 0.6$.

Inside the spinodal, the alloy is unstable and can lower its free energy by the growth of any small fluctuation in composition. Phase separation occurs by up-hill diffusion, that is, the A atoms diffuse to where the concentration of A atoms is higher, and the B atoms diffuse to where the concentration of B atoms is higher, rather than moving to randomize the distribution of atoms, as is usual in an ideal solution. Outside the spinodal, but inside the phase boundary, the alloy is metastable: phase separation will occur only if there is a sufficiently large fluctuation in composition to result in nucleation of the second phase.

For ordering, the atoms are not randomly distributed, so the value of X in Equation 23.2 is replaced by $(C_A^2 + C_B^2)/2$, which is the probability of an A atom on the A sub-lattice having a B nearest neighbor plus the probability of an B atom on the B sub-lattice having an A nearest neighbor. In this case, the phase boundary gives the compositions on alternate sites, rather than the region of phase stability.

23.4 Analytical Model for Diffusion during Spinodal Decomposition

Difference equations which describe the diffusion in a regular solution will be developed in this section, based on recent work of the author [1]. These equations incorporate the nearest neighbor interactions that were discussed in Chapter 22. There it was shown that the free energy can be evaluated if ZX, the number of AB pairs per atom is known. The number of AB pairs for site at (x,y) is given by the concentration of A on the site multiplied by the sum of the B concentrations of its neighbors, plus the concentration of B on the site multiplied by the sum of the A concentrations of its neighbors.

In order to evaluate the chemical potentials, Equation 23.2 for the free energy should be rewritten in terms on the number of atoms:

$$G = -\frac{Z}{2}(N_A\phi_{AA} + N_B\phi_{BB}) + k_BT\left(N_A\ln\left(\frac{N_A}{N}\right) + N_B\ln\left(\frac{N_B}{N}\right)\right) + \Omega\, N X \tag{23.9}$$

23.4.1 Chemical Potentials in Two Dimensions

As an example, difference equations for diffusion which incorporate nearest neighbor interactions will be derived for a two-dimensional square lattice. Similar difference equations can be readily derived for one-dimensional planar composition variations, or for various other lattices in two or three dimensions. The free energy of each lattice site can be derived from the occupancy of the site and the occupancy of its nearest neighbors. The free energy of a site in a 2D square array is given by:

$$\begin{aligned} G(x,y) = &-\left(\frac{Z}{2}\right)(N_A(x,y)\phi_{AA} + N_B(x,y)\phi_{BB}) \\ &+ k_BT\left(N_A(x,y)\ln\left(\frac{N_A(x,y)}{N_A(x,y)+N_B(x,y}\right) + N_B(x,y)\ln\left(\frac{N_B(x,y)}{N_A(x,y)+N_B(x,y}\right)\right) \\ &+ \Omega(N_A(x,y)+N_B(x,y))X(x,y) \end{aligned} \tag{23.10}$$

where $C(x,y) = \dfrac{N_B(x,y)}{N_A(x,y)+N_B(x,y)} = 1 - \dfrac{N_A(x,y)}{N_A(x,y)+N_B(x,y)}$.

Here $N_A(x,y)$ and $N_B(x,y)$ are the occupancies of the site (x,y). The last term in Equation 23.10 is the probability of finding an A atom on site x,y times the sum of the probabilities of finding B atoms on the neighboring sites, plus the probability of finding an B atom on site x,y times the sum of the probabilities of finding A atoms on the neighboring sites:

$$\Omega(N_A(x,y)+N_B(x,y))X(x,y)$$

$$= \frac{W}{2}\left[\begin{array}{l} N_A(x,y)\dfrac{(N_B(x+d,y)+N_B(x-d,y)+N_B(x,y+d)+N_B(x,y-d))}{N_A(x,y)+N_B(x,y)} \\ \quad + N_B(x,y)\dfrac{(N_A(x+d,y)+N_A(x-d,y)+N_A(x,y+d)+N_A(x,y-d))}{N_A(x,y)+N_B(x,y)} \end{array}\right] \tag{23.11}$$

Here $W = \Omega/Z$. The factor of $\frac{1}{2}$ is necessary because only half of the energy of each bond goes with the site. For a 2D square lattice, Z=4.

Subtracting $4WN_A(x,y)\ N_B(x,y)/N$ from Equation 23.11 turns it into a second difference, and adding this term to Equation 23.10 gives, together with the first two terms, the standard form for the free energy of a regular solution. So Equation 23.10 contains the standard free energy for a regular solution, plus a second difference term for the variations in composition.

The chemical potential for the B component is given by the partial derivative of the free energy with respect to the occupancy of B atoms, N_B, keeping the temperature and the occupancy of A atoms, N_A, constant. The change in free energy of the nearest neighbors due to the change in the number of B atoms on site x,y must also be included.

The corresponding chemical potentials are:

$$\mu_B = -\frac{Z}{2}\phi_{BB} + k_B T \ln C(x,y) + \frac{W}{2}(8-4C(x,y)+(2C(x,y)-3)\Sigma_{2D}C(x,y)) \tag{23.12}$$

$$\mu_A = -\frac{Z}{2}\phi_{AA} + k_B T \ln(1-C(x,y)) + \frac{W}{2}(-4C(x,y)+(2C(x,y)+1)\Sigma_{2D}C(x,y)) \tag{23.13}$$

where

$$\Sigma_{2D}C(x,y) = C(x+d,y)+C(x-d,y)+C(x,y+d)+C(x,y-d) \tag{23.14}$$

is the sum of the compositions of the neighbors of site x,y.

23.4.2 Difference Equations for Diffusion

The diffusion process depends on the fluxes between volume elements. In an ideal solution, the fluxes depend on the gradient of the concentration. In a non-ideal solution, the fluxes depend on gradients of the chemical activities. These are effective concentrations, given by the concentration multiplied by an activity coefficient, as in Equation 22.21. The chemical activities can be derived directly from the chemical potentials in Equations 23.12 and 23.13.

When an alloy is not at equilibrium, the chemical potentials of the various species are different, and so their fluxes are different. Darkin [2] developed equations for diffusion where the fluxes of the two species were different by introducing vacancies as a third component. An effective diffusivity was obtained by averaging the activity coefficients of the two species. Definitive experiments by Reynolds *et al.* [3] compared self-diffusion and inter-diffusion in gold–nickel alloys in the single phase field above the critical temperature. They concluded that Darkin's equations correctly described diffusion. In the present treatment, the fluxes between sites will be averaged, rather than averaging the activity coefficients. These two averages are equivalent in a single phase field when composition variations on an atomic scale are small, but they are quite different in a two-phase field where phase separation occurs. Averaging the activity coefficients does not result in the correct equilibrium phase-separated compositions, but averaging the fluxes does. Darkin did not consider the possibility of negative effective diffusivities that would lead to phase separation.

Below the critical point, a homogeneous alloy is unstable, but a problem arises with a continuum description: the shortest wavelength grows fastest. Cahn and Hilliard's classical treatment of spinodal decomposition [4, 5] introduced a "gradient energy" term into the free energy to limit the exponential growth of very short wavelengths. This term was added as part of an expansion of the free energy for small composition variations. Hillert [6] realized that the problem could be addressed with difference equations, but he applied his scheme only to interfaces between regions having large composition differences. Cahn [7] described diffusion with a continuum model, using an effective diffusivity based on the activity of only one component. But, since the flux of only one species was included, the final compositions cannot converge to the equilibrium phase boundary compositions. Cook *et al.* [8] recognized the importance of the crystal lattice, but maintained the gradient energy term, rather than introducing interactions between atomic sites. As will be seen below, Cook *et al.* [8] derived expressions for the growth of small amplitude sinusoidal perturbations, that are similar to those derived below, but this depends on the special properties of small amplitude sinusoidal perturbations. In general, the results are quite different.

In the analysis outlined here, limiting the size of a volume element to the atomic volume limits the shortest possible wavelength of a perturbation, and incorporating the nearest neighbor interactions introduces the thermodynamic properties of the alloy. This scheme results in fourth order difference equations that can only be solved numerically. This is not a significant drawback today, since the results of complex systems of equations are usually evaluated numerically.

The flux of atoms of species j out of a site is given by the standard expression from chemical rate theory (Chapter 13): a frequency times a Boltzmann factor containing the difference between the chemical potential, μ_j^*, of the activated state between sites, and the chemical potential, μ_j, of the j component on the site.

$$J_j = -\Gamma_0 \exp\left(-\frac{\mu_j^* - \mu_j}{kT}\right) = -\Gamma_0 \exp(((\mu_j^* - \mu_j^0) - (\mu_j - \mu_j^0))/kT) = -\Gamma_j P_j \quad (23.15)$$

where $\Gamma_j = \Gamma_0 \exp((\mu_j^* - \mu_j^0)/kT)$, and $P_j = \exp((\mu_j - \mu_j^0)/kT)$ is the activity of the j component. Γ_0 is usually taken to be the Debye frequency, and departures from this frequency are lumped into the entropy and energy of the activated state. μ_j^0 is the chemical potential of pure j, which is given by the ϕ_{jj} terms in Equations 23.12 and 23.13. These are incorporated in Γ_j to give the activation energy for diffusion as the difference between the energy in the activated state and the equilibrium state. The activity, P_j contains the entropy terms and the interaction energy terms from Equations 23.12 and 23.13. For an ideal solution, $P_j = C_j$, and so Equation 23.15 reduces to the normal concentration driven flux. The net flux into a site is given by the difference between the total flux out of the site to all of its neighbors, and the total flux into the site from the neighbors.

If the system is not at equilibrium, the fluxes for the two species A and B, J_A and J_B will be different because the chemical potential for each of the two species on each site is different. Since the rate at which an A atom changes place with a B atom must be the same as the rate which a B atom changes place with an A atom, we will also assume that Γ is the same for both species.

We will use the average of the fluxes for the two species in the difference equation for diffusion. The flux of A atoms is in the opposite direction to the flux of B atoms, so the fluxes are summed by reversing the sign on the A atom flux. The difference equation for diffusion is:

$$C(x, y, t_{i+1}) = C(x, y, t_i) + \frac{D\Delta t}{d^2}(\Delta^2 P_B(x, y, t_i) - \Delta^2 P_A(x, y, t_i))/2 \quad (23.16)$$

where

$$\Delta^2 P_j(x, y, t_i) = \left[\begin{array}{l} P_j(x-1, y, t_i) + P_j(x+1, y, t_i) \\ \quad + P_j(x, y-1, t_i) + P_j(x, y+1, t_i) - 4P_j(x, y, t_i) \end{array}\right] \Big/ 4 \quad (23.17)$$

Here $D = \Gamma d^2$ is the diffusivity in units of length^2 time^{-1}, d is the inter-planar spacing, and Δt is the time increment between iterations. After n iterations, the elapsed time is given by nd^2/D. Equations for diffusion in one or three dimensions are similar.

A dimensionless time τ, can be used for calculations, where τ is given by

$$\tau = n\Delta\tau = n\frac{D\Delta t}{d^2} \quad (23.18)$$

For a particular case, multiplying τ by d^2/D will convert to real time.

This scheme is difficult to employ analytically, but it is simple to facilitate with difference equations.

Evaluation of the above difference equations for diffusion can be carried out in the same way as evaluation of standard difference equations for diffusion. A set of initial boundary conditions is required, and then the equations provide the development in time of the composition distributions. It is usual to apply periodic boundary conditions in these calculations. A region for the calculations is decided, and then the compositions next to sites at the periphery are taken to be the compositions at the opposite edge of the region. In this case, because the free energies of the nearest neighbors depend on their nearest neighbors, it is necessary to include the compositions of two layers of atoms at the opposite edge of the region.

23.4.3 Growth of a Sinusoidal Perturbation

The instability in the spinodal region can be explored by examining what happens to small amplitude sinusoidal perturbations. For positive values of the regular solution parameter, there are a range of wavelengths inside the spinodal for which the amplitude of the perturbation initially grows exponentially, as $\exp(\alpha\tau)$, as predicted by Cahn [7] and Cook *et al.* [8].

In two dimensions sinusoidal perturbations can be either planar or two-dimensional:

$$\begin{aligned} C(x,y) &= C_0 + A_0\cos(kx) \\ C(x,y) &= C_0 + A_0\cos(kx)\cos(ky) \end{aligned} \tag{23.19}$$

When inserted into Equation 23.16, the amplitude of the perturbation initially grows as $A = A_0 \exp(\alpha\tau)$. Figure 23.3 shows the initial value of the growth exponent

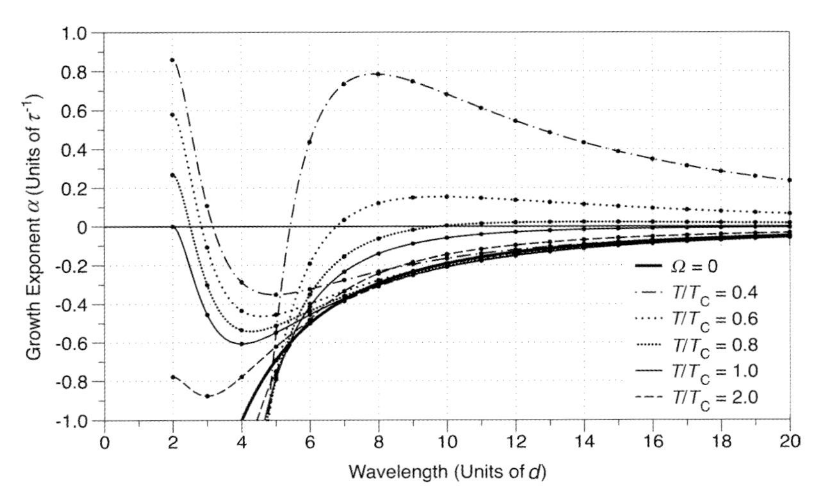

Figure 23.3 The initial growth exponent α, as a function of wavelength in 2D for $C_0 = 0.5$ at various temperatures.

α, found by starting with the 2D initial perturbation from Equation 23.19, for $C_0 = 0.5$. The growth exponent is plotted as a function of the wavelength of the initial sinusoidal perturbation, for five different growth temperatures, and for $C_0 = 0.5$. For simplicity, the curves are labeled with T/T_C, where T_C is the critical temperature: the regular solution parameter Ω has been replaced by Boltzmann's constant times the critical temperature: $\Omega = k_B T_C/2$, as in Equation 23.7. Data for both positive and negative values of the regular solution parameter are included. The data for negative values of Ω have a maximum on the left at $d = 2$.

The points in Figure 23.3 are from simulations, and the curves are from Equation 23.20 for the growth exponent α.

$$\alpha = \left(1 - \frac{T_C}{T}\cos(2\pi/\lambda)\right)(\cos(2\pi/\lambda) - 1)\exp\left(\frac{T_C}{2T}\right) \tag{23.20}$$

This equation was derived by approximating Equation 23.16 for small A_0 at $C_0 = 0.5$. A similar equation was derived by Cook *et al.* [8]. The term in the first bracket is from the chemical potential, and the term in the second bracket is from the diffusion equation. For positive Ω and for $T/T_C < 1$, the critical wavelength where the amplification factor changes sign is given by the zero of the first bracket. The maximum in α is at approximately the square root of 2 times this wavelength. The magnitude of the maximum growth exponents increases with decreasing temperature because of the temperature dependence in the activity. However, the temperature dependence of the diffusivity is not included in these curves, so, in experiment, perturbations will actually grow more slowly at lower temperatures, but at a rate faster than predicted by the temperature dependence of the diffusivity. Exponential growth of the initial perturbation can cover decades for wavelengths near the maximum, until the composition reaches the phase boundary composition. However, for an initial wavelength which is far from the fastest growing wavelength, the initial wavelength grows for a while, and then a faster growing wavelength takes over.

Above the critical point, for $T/T_C > 1$, the growth exponents approach those for $\Omega = 0$, which is ideal solution behavior. For $\Omega = 0$, perturbations or fluctuations of all wavelengths decay, as described by the usual diffusion equations. Above the critical point, there are no positive values of α. For temperatures approaching the critical point from above, the magnitude of the negative growth exponents decreases, indicating that perturbations or fluctuations decay more slowly. This is known as critical slowing down.

The data for negative values of Ω were derived using the same equations, with only the sign of the regular solution parameter changed. As expected, the fastest growing wavelength was found to be twice the lattice spacing, which leads to ordering. For large values of the negative regular solution parameters, that is, far above the critical temperature, the curves also approach the curve for $\Omega = 0$, which corresponds to standard diffusion.

23.4.4 Phase Boundary Widths

The composition profiles through the (10) phase boundaries in 2D for various temperatures in are shown in Figure 23.4. Far from the boundary, the compositions

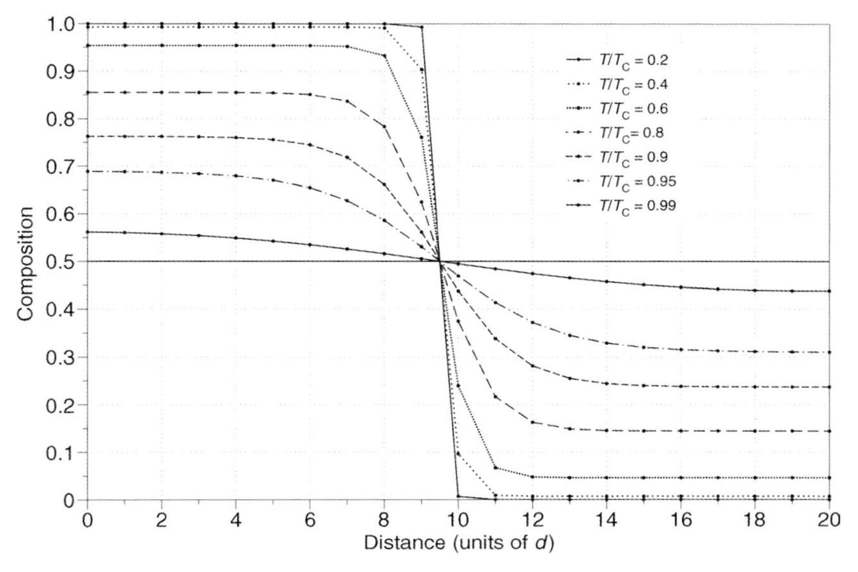

Figure 23.4 Composition profiles of (10) phase boundaries in 2D.

are at the equilibrium concentrations. The boundaries are very broad near T_C, and become very narrow at low temperatures.

The widths of anti-phase boundaries in ordered structures are similar, but somewhat different than those of phase separated boundaries.

Figure 23.5 compares the morphologies of particles at $T/Tc = 0.3$, with $C_0 = 0.1$, and at $T/Tc = 0.9$, with $C_0 = 0.3$. The interface on the left is quite sharp, and the figure is almost square, suggesting significant anisotropy in the surface free energy. The figure on the right is almost circular indicating that the surface free energy is more or less isotropic. The interface is quite diffuse, as indicated in Figure 23.4. This is reminiscent of the surface roughening in first order phase transformations discussed in Chapter 21.

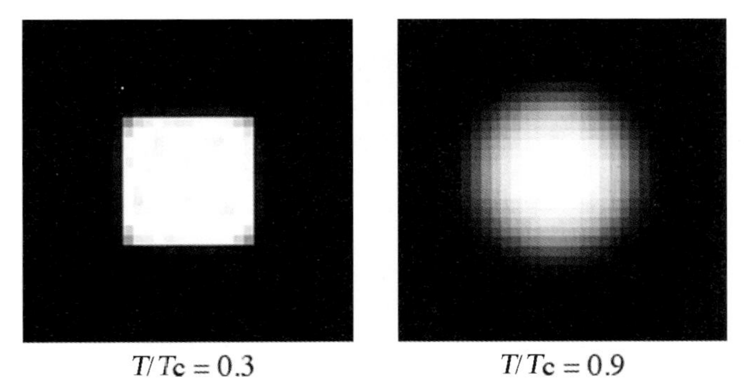

Figure 23.5 Comparison of the growth morphologies of precipitates at $T/T_C = 0.3$ and at $T/T_C = 0.9$. The grey level indicates the composition of the site.

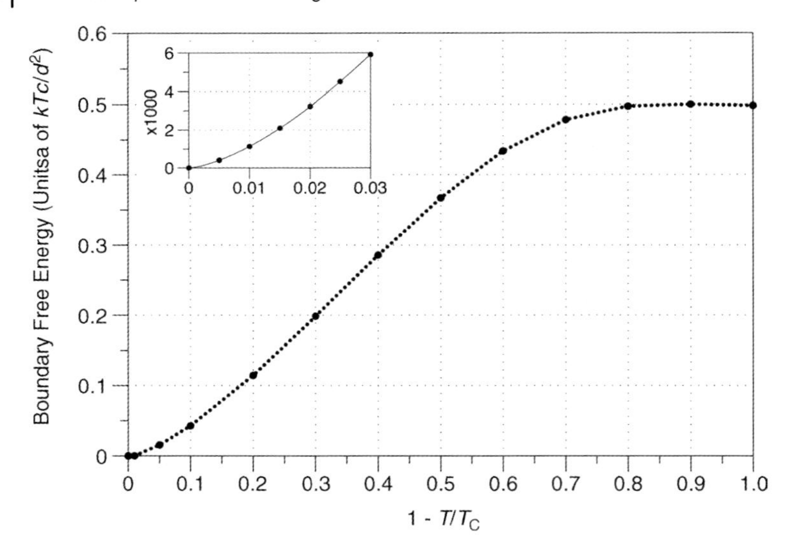

Figure 23.6 Free energies of 2D boundaries in both phase separated and ordered structures as a function of $1 - T/T_C$. The inset shows the free energies of 3D boundaries near T_C.

23.4.5
Phase Boundary Free Energies

The free energies of the 2D boundaries are plotted in Figure 23.6. The free energies of boundaries in phase separated structures were found to be identical to those of anti-phase boundaries in ordered structures. The boundary free energy at low temperatures, where entropy is not important, is the energy of an abrupt boundary, $1/2\ kT_C/d^2$.

Inset is the free energy of 3D boundaries for temperatures very close to T_C. The data fit to $(1 - T/T_C)^{1.5}$, which is the limiting behavior near T_C predicted by van der Waals [9] for the liquid–vapor transition, and later by Cahn and Hilliard [4] for the spinodal.

23.5
Microstructure Development

The development of microstructures can be followed using calculations based on Equation 23.16.

23.5.1
Phase Separation

The development of a typical composition distribution in 2D for $C_0 = 0.5$ at $T/T_C = 0.6$ is shown in Figure 23.7. The two phases are interconnected, and the structure coarsens with time.

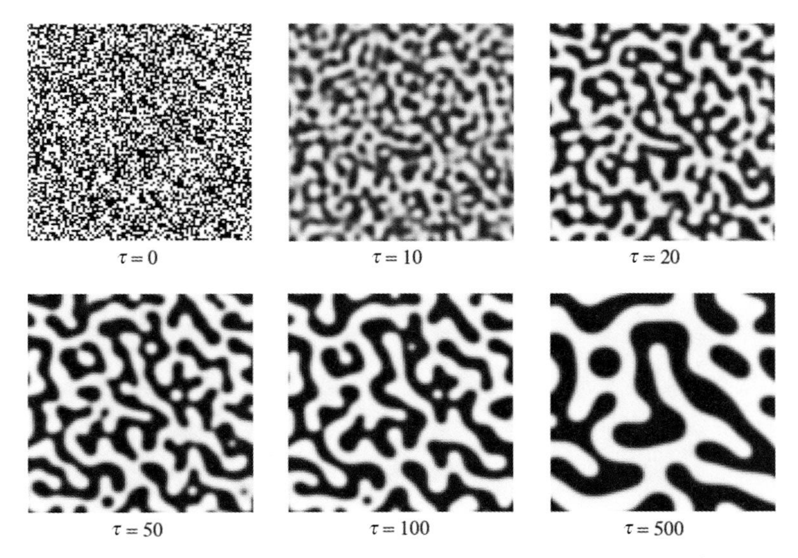

Figure 23.7 Development of phase separation in 2D at $C_0 = 0.5$ starting with a 50/50 random distribution of As and Bs at $T/T_C = 0.6$.

Figure 23.8 is a micrograph of a typical phase separated structure in a glass, which is similar to the structure in Figure 23.7.

Figure 23.9 shows the development of the composition distribution at $T/T_C = 0.6$, starting from a random distribution of 25% B and 75% A atoms, so that $C_0 = 0.25$.

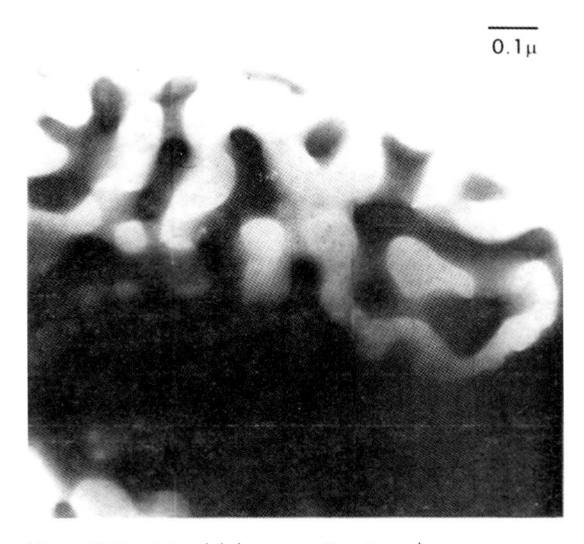

Figure 23.8 Spinodal decomposition in a glass.

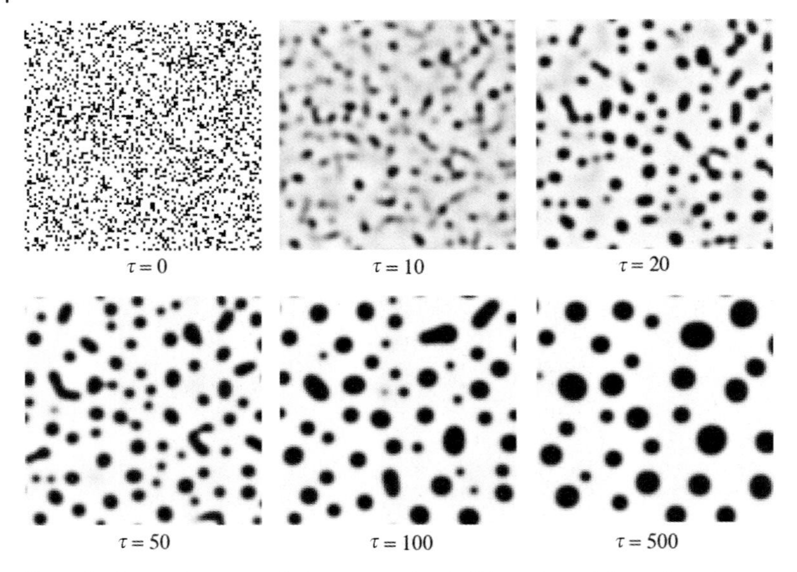

Figure 23.9 Development of phase separation in 2D at $T/T_C = 0.6$, starting with a random distribution of 25% B and 75% A atoms, which gives $C_0 = 0.25$.

The structure which develops consists of islands which are rich in one component in a matrix which is rich in the other component.

At $C_0 = 0.5$, the microstructure consists of two interpenetrating phases, whereas at $C_0 = 0.25$, the microstructure consists of islands of one component in a matrix of the other component. This is consistent with the simple analysis of the difference between the total inter-phase boundary area of the two structures. For comparable center-to-center distances between the two phases, there is less total surface area when the structure consists of two continuous, intertwined phases if:

$$\pi > \frac{V_A}{V_B} > \frac{1}{\pi} \tag{23.21}$$

Otherwise, the structure consists of separate spheres that are rich in the minor component in a matrix of the major component, as in Figure 23.9.

23.5.2
Ordering

Ordering occurs for negative values of the regular solution parameter Ω. Figure 23.10 shows the development of an ordered configuration, starting from a random distribution of A and B atoms. Anti-phase boundaries can be seen coarsening in the last two images of this sequence. The concentrations on the lattice sites in

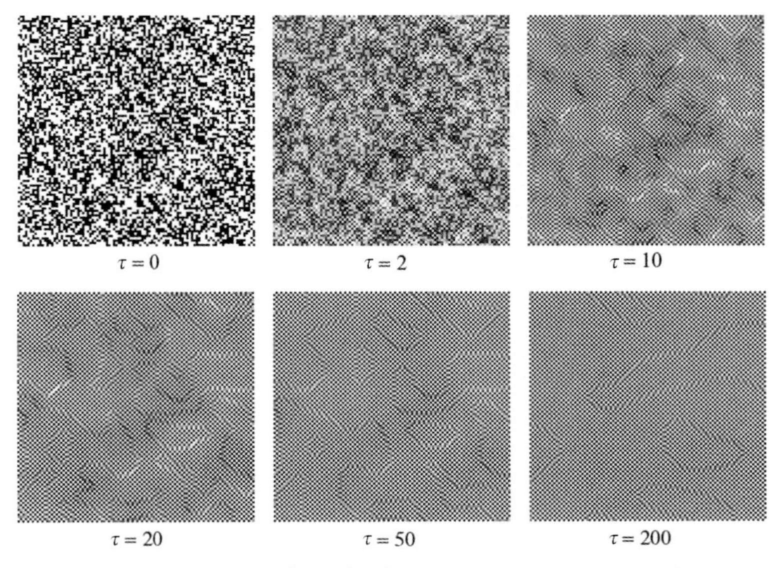

Figure 23.10 Development of an ordered microstructure in 2D at $T/T_C = 0.6$.

the ordered state are at the phase boundaries. Only compositions within a small percentage of the 50–50 composition produce stable ordered structures: small departures from 50–50 can be accommodated in the anti-phase boundaries. For large departures from 50–50 compositions, the diffusion equations do not converge to a reasonable result.

23.5.3 Precipitates

The growth of a precipitate is a classical Stefan problem, discussed in Chapter 8, where the boundary conditions for diffusion are applied on a moving boundary. The difference equations for diffusion presented in this chapter automatically include the boundary conditions on moving the boundary, as well as the departures from the equilibrium values of the concentrations inside and outside the precipitate due to curvature of the interface, as described in Section 8.5.

Figure 23.11 shows the development of precipitate particles in 2D which started with a random distribution of 7.5% B, at $T/T_C = 0.3$. This is inside the two-phase field, but outside the spinodal region. There is a large undercooling below the phase boundary, so there are many nucleation events.

Similar behavior was observed at the same temperature, starting with 10% B, which is inside the spinodal region. Starting with an initial uniform composition of 10% B, with a superimposed random perturbation of 0.5% produces a final distribution similar to that shown in Figure 23.11. However, starting with this same perturbation superimposed on a background composition of 7.5% B, which is

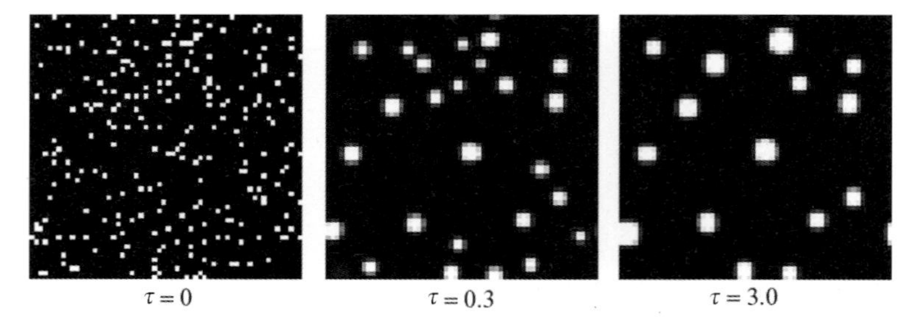

Figure 23.11 Precipitation starting from an initial random distribution of 7.5% B at $T/T_C = 0.3$.

outside the spinodal but inside the two-phase field, results in the decay of the perturbations to a uniform composition of 7.5%.

23.5.4
Growth of Precipitate Particles

The next several figures illustrate the growth of precipitate particles in 2D, starting with a small seed, which grows into a matrix of uniform composition. The development of the composition distribution depends on the perturbation provided by the initial configuration.

Figure 23.12 shows the growth initiated by a 5×5 particle of composition 0.9 in a 100×100 matrix of composition 0.1, at $T/T_c = 0.3$. At this composition, the usual

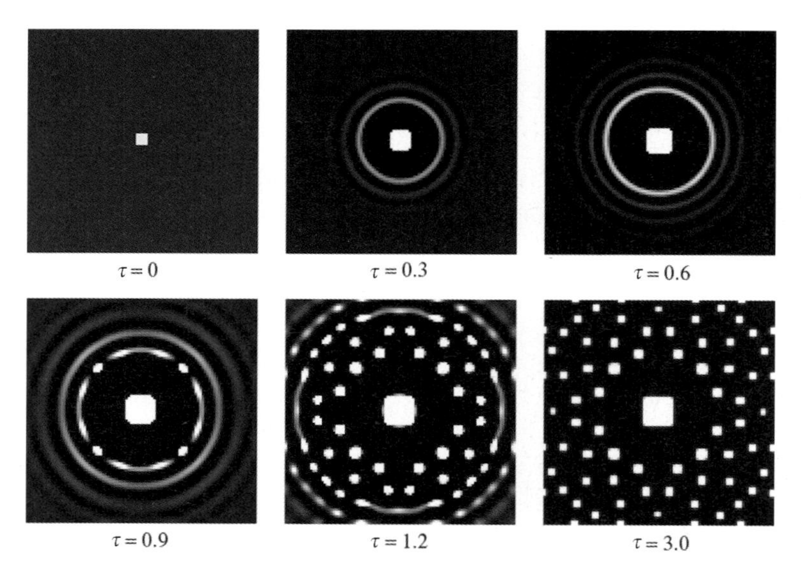

Figure 23.12 Growth generated by a single particle in a background concentration of $C_0 = 0.1$ at $T/T_C = 0.3$.

microstructure is particles of the minor component in a matrix of the major component. Here, rings of higher concentration form around the initial particle. The rings expand outward, and, at $\tau = 0.9$, the inner ring begins to break up into particles. The subsequent development of particles from the rings is evident in the last two images. The symmetry of the precipitate pattern derives initially from the symmetry of the precipitate particle, as is evident at $t = 0.9$. Later, the symmetry is also influenced by the shape and size of the simulation.

The rings form by uphill diffusion in the concentration gradient around the initial particle. The initial ring moves outward at first, and then stops.

23.5.5
Growth of Rings at C = 0.5

Figure 23.13 shows a similar growth process but with a background concentration of 0.5, starting from an 8×8 particle with composition 1.0, at $T/T_C = 0.9$. The usual structure at this composition is two interpenetrating, interconnected phases, as in Figure 23.7.

During growth concentric rings develop. The first rings to form initially move outward, but after a few rings form, the expansion of the inner rings slows, and their spacing becomes the same, and does not change significantly as the number of rings increases. The ring structure then continues to expand by adding rings with a constant spacing. Rings such as this form at all temperatures for $C_0 = 0.5$.

At longer times, the ring structure coarsens slowly by the shrinking of the initial precipitate and the inner rings.

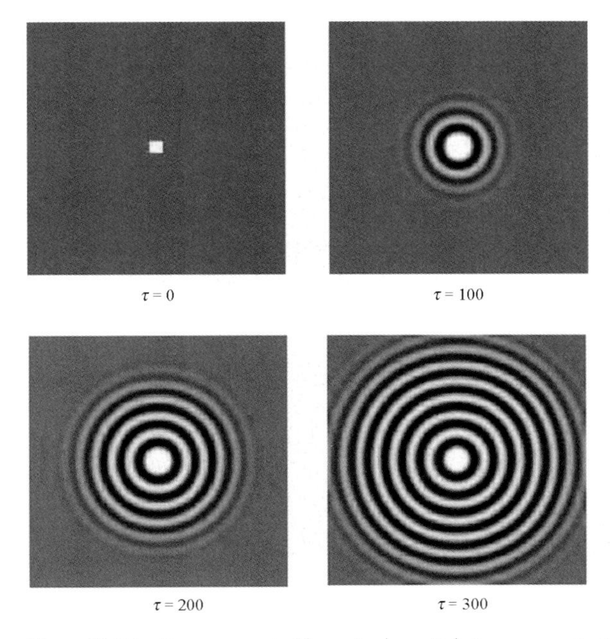

Figure 23.13 Rings generated by a single particle in a concentration of $C_0 = 0.5$ at $T/T_C = 0.9$.

Table 23.1 Comparison of the measured ring spacings with λ max for $C_0 = 0.5$ at various temperatures in 2D, and of the measured propagation velocity (in units of d/T) of the instability front with 4.4 λmax α max.

T/TC max	Ring Spacing	λ max	α max	Velocity	4.4 λ max α
0.2	4.5	4.6	9.75	191	197
0.3	4.8	5.0	2.16	43	48
0.4	5.3	5.3	0.785	18	18
0.5	6.0	5.8	0.340	9.3	8.7
0.6	6.7	6.8	0.153	4.3	4.6
0.7	7.5	7.9	0.066	2.1	2.3
0.8	9.4	9.8	0.0234	1.03	1.01
0.9	13.6	13.9	0.0048	0.30	0.29

The spacing of the rings turns out to be λ_{max}, as shown in Table 23.1. λ_{max} is the wavelength that grows most rapidly, as in Figure 23.3. λ_{max} depends only on the composition and temperature, as in Equation 23.20, and not on the diffusivity. Rings were found to add to each pattern at a constant rate that was about 4.4 times α_{max}, the value of the growth exponent for λ_{max} in Figure 23.3. $1/\alpha_{max}$ is a measure of the time it takes for a sinusoidal perturbation of wavelength λ_{max} to grow from a given initial amplitude to the phase boundary composition. The measured velocity at which the ring system expands is 4.4 λ_{max} α_{max}, in units of d/τ. This rate depends on the diffusivity, which is included in the definition of τ. This rate can be compared with the characteristic diffusion distance $(Dt)^{1/2}$, which does not increase linearly with time. The rings expand much more rapidly than atoms can move by diffusion.

This is illustrated dramatically in Figure 23.14 where growth inside the spinodal is compared with the growth outside the spinodal but inside the two-phase field at a low temperature, $T/T_c = 0.2$. The growth patterns are completely different. The image on the left shows precipitation inside the spinodal, in a matrix with a composition 0.1, where growth is by propagation of an instability front.

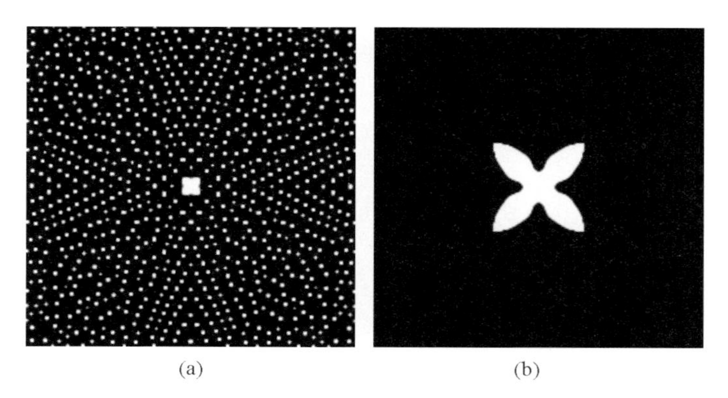

Figure 23.14 Comparison of the growth generated by a single particle at $T/T_C = 0.2$. (a) Inside the spinodal, at $C_0 = 0.1$. (b) Outside the spinodal but in the two-phase field, at $C_0 = 0.05$.

The image on the right shows growth into a matrix with a composition 0.05, at the same temperature. The elapsed time was over 10 times longer than for the image on the left. Diffusion processes dominate the growth of the precipitate in Figure 23.14b, and dendrite-like instabilities have developed at the corners of the initial particle.

23.6 Modeling of Phase Separation and Ordering

There has been a massive amount of work devoted to the study of cooperative phenomena. There is a long history of analyzing second order λ transitions using Classical models, starting with van der Waals [9], and including Bragg-Williams, Bethe, Landau [4–8], as well as current phase field models. The description of phase separation and ordering presented here is also a Classical model. In these models, the composition is defined as the probable occupancy of lattice sites or even of volume elements. These models describe the transition behavior properly, but get the details wrong. In the real world, a B atom in mostly A-rich region will be surrounded mostly by A atoms, and not by sites that are part A and part B. So the energy of the B atom will be higher, and so there will be fewer of them in the A-rich region than predicted by a Classical model. The Ising model captures this behavior, and it provides a better fit to experimental data. The average compositions in the phase-separated regions in the Ising model are much closer to the pure material compositions than in Classical models. The corresponding phase diagram for the two-phase region is much flatter on top than in Figure 23.2b. The phase boundary composition near the critical point varies approximately as $(1\text{-}T/T_C)^{1/3}$, rather than $(1\text{-}T/T_C)^{1/2}$, as in Figure 23.2b.

Cyril Domb [10] provides a splendid overview of the great effort that has been expended on the study of critical point phenomena and the development of our understanding of these processes. The treatment presented here suffers from the limitations of all Classical models, but it does provide an insightful description of the development of microstructures.

References

1 Jackson, K.A. (2010) *J. Non-Cryst. Solids*, in press.
2 Darken, L.S. (1948) *Trans. AIME*, **175**, 184.
3 Reynolds, J.E., Averbach, B.L., and Cohen, M. (1957) *Acta Metall.*, **5**, 29.
4 Cahn, J.W. and Hilliard, J.E., (1958) *J. Chem. Phys.*, **28**, 258.
5 Cahn, J.W. and Hilliard, J.E. (1959) *J. Chem. Phys.*, **31**, 688.
6 Hillert, M. (1961) *Acta Metall.*, **9**, 525.
7 Cahn, J.W. (1968) *Trans. Met. Soc. AIME*, **242**, 166.
8 Cook, H.E., de Fontaine, D., and Hilliard, J.E. (1969) *Acta Metall.*, **178**, 765.
9 van der Waals, J.D. (1893) Sect. 1, vol. 1, Verhandel. Kon. Acad. Weten, Amsterdam, p. 56.
10 Domb, C. (1996) The Critical Point, Taylor and Francis.

Problems

23.1. Derive the equation for the growth exponent α for an initial perturbation of the form $C(x, y) = C_0 + A_0 \cos(kx)$. Plot the result as in Figure 23.3.

23.2. Derive Equations 23.12 and 23.13 for the chemical potentials in 2D.

23.3. Write a computer program based on Equation 23.16 to generate one (or all) of Figures 23.3–23.14.
(The ImageJ software made available by W. S. Rasband of the U. S. National Institutes of Health, Bethesda, Maryland, USA http://rsb.info.nih.gov/ij/ can be used to convert numerical data into gray scale images.)

24
Non-Equilibrium Crystallization of Alloys

24.1
Non Equilibrium Crystallization

In this chapter we will consider what happens when the rate of advance of the interface becomes comparable to the rate at which atoms can move by diffusion. In this regime, the quasi-equilibrium treatment based on thermodynamics, which is presented in Chapter 22 will be modified by kinetic effects.

24.2
Experiment

The compositions, structure and properties of multicomponent materials produced by phase transformations which occur under conditions which are far from equilibrium are often quite different from those predicted by equilibrium thermodynamics. The first extensive observations of this were made by Duwez [1] using a technique he called splat quenching. A molten drop of the alloy was propelled onto a curved copper sheet. The copper sheet was held at an angle so that the droplet spread out along it into a thin layer, typically a few microns thick. The samples crystallized very quickly. The samples were analyzed using X-rays and transmission electron microscopy (TEM). It was found that many metastable solid solutions could be formed. That is, rather than the solid solubility which is found at equilibrium, and reported on a phase diagram, solids with compositions in the two-phase field of the phase diagram, containing much more of the second component than the equilibrium value, were obtained. In some systems, solid solutions were obtained for all compositions of the alloy.

The alloys which form metallic glasses are usually a mixture of metallic elements and semi-metals or semiconductor elements, which crystallize into a complicated crystal structure. And so these materials have large entropies of fusion. Since these alloys crystallize slowly, they can be quenched into a glassy state where the atom mobility is very small. For pure metals, the crystallization rate is so fast even at very low temperatures that this cannot happen.

Kinetic Processes: Crystal Growth, Diffusion, and Phase Transitions in Materials. Kenneth A. Jackson

ISBN: 978-3-527-32736-2

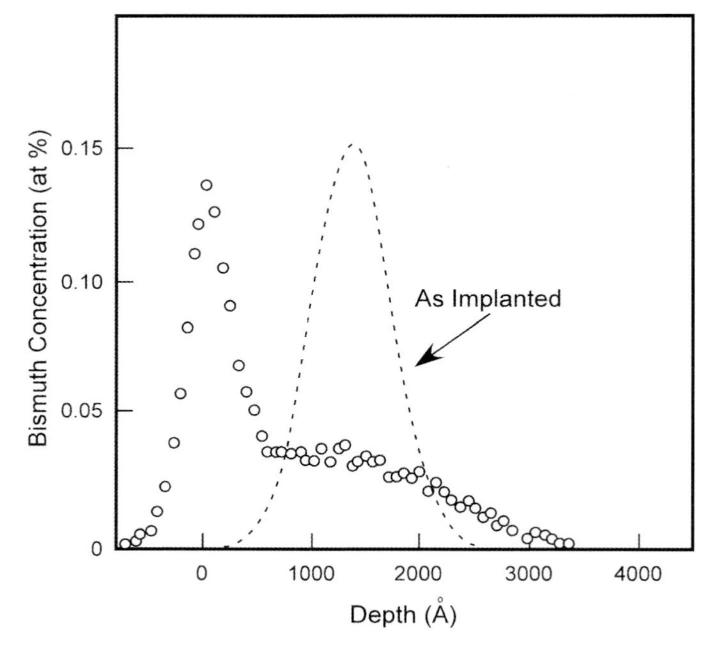

Figure 24.1 The as-implanted distribution of bismuth in silicon, and the distribution after a surface layer of the crystal was melted and rapidly recrystallized.

The composition of the solid which forms during crystallization is given by the distribution coefficient, also known as the k-value, which is defined as $k \equiv C_B^S / C_B^L$.

Quantitative data are available for the dependence of the k-value on the growth rate during the very rapid recrystallization of laser-melted, ion-implanted silicon [2, 3]. These data were obtained by first ion implanting a dopant into a silicon wafer. Then a very short, high power laser pulse, which is a centimeter or so in diameter, is directed at the surface. The total energy in the pulse is enough to melt a 1–2 μm thick layer of the surface of the wafer. The wafer is at room temperature, so the bulk of the wafer acts as an efficient heat sink. The liquid layer recrystallizes in around 1 ms, with growth rates in the range of $\mathrm{m\,s^{-1}}$. Typical data for the observed concentration profiles are shown in Figure 24.1.

The as-implanted dopant distribution is shown, together with the final distribution of the dopant after laser melting of the surface. If the equilibrium distribution coefficient, 7×10^{-4}, had applied, all of the dopant would have been pushed to the surface by the crystallization process. The data can be fitted only by using a k-value of 0.1, which is over 100 times the equilibrium value. Similar data have been collected for aluminum alloys.

The growth rate dependence of the k-value is also responsible for the so-called "facet effect" observed during the slow growth of semiconductor crystals. An increased incorporation of most dopants at a faceted region of the interface during growth at normal laboratory or production growth rates is observed.

24.3 Computer Modeling

Computer modeling has reproduced all of the main features of these observations, including the orientation dependence of the *k*-value. In addition the modeling has provided a definitive explanation for the orientation dependence. This modeling does not rely on any special properties of the atoms, it assumes the same interactions between atoms which are responsible for the equilibrium properties of the alloy. The modeling does incorporate the non-equilibrium effects which occur when the rate of advance of the interface becomes comparable to the rate at which atoms can move by diffusion.

Simulations of "diffusionless" transformations have been carried out, where the transformation takes place by the motion of the interface, but the atoms do not move: their position is fixed on lattice sites. These simulations correspond, for example, to a "shear" or martensite-type transformation where the interface moves very rapidly, at rates approaching the speed of sound. The atoms only have sufficient time to shift their positions a small amount to conform to the new structure as the transformation front passes. Growth and melting of the alloy in these simulations was observed below and above the T_0 line, respectively, where T_0 is the locus of the temperatures on a phase diagram where the free energy of the solid alloy is equal to the free energy of the liquid alloy with the same composition. This is clearly the expected behavior for a diffusionless transformation since the kinetics of a diffusionless transformation should depend on the difference between the free energies of the two phases, rather than on the difference between the chemical potentials in the two phases of the species present. For the diffusionless case, freezing or melting should depend on which phase has the lower free energy, and this occurs above and below T_0, as is observed in the simulations.

Data from Monte Carlo simulations of alloys which have been accumulated for a variety of different growth temperatures, growth rates and diffusion coefficients, both above and below the roughening transition [4–7] are presented in Figure 24.2. All of these data fall on a single curve when they are plotted against the dimensionless parameter β, as in Figure 24.2.

$$\beta = \sqrt{\frac{avu_{\mathrm{k}}}{D}} \tag{24.1}$$

where a is the a lattice dimension, v is the growth rate, u_{k} is the net rate at which atoms join the crystal at active growth sites, and D is the diffusion coefficient in the liquid.

Near equilibrium, where both v and u_{k} are small, β is small, and the *k*-value approaches the equilibrium value. When diffusion is relatively slow, and the growth rate is relatively large, β becomes large, and the *k* -value approaches one.

β was first identified by Temkin [8] in his analytical modeling of alloy crystallization. He suggested that β depends on the time it takes a fluctuating interface to pass an atom, compared to the diffusion jump time.

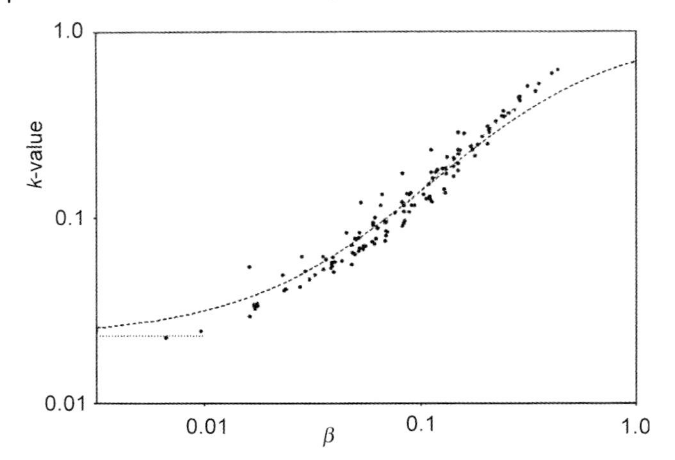

Figure 24.2 The distribution coefficient (k-value) data from Monte Carlo computer simulations for a wide variety of growth conditions, plotted against β.

An alternate explanation is based on whether a B atom at the interface is likely to be engulfed during the time it spends at the interface. How far the interface moves during the average time that a liquid B atom spends at the interface can be compared with the distance that an atom can diffuse during that time. If the interface moves farther than the B atom can diffuse, the probability that the B atom will be incorporated into the crystal will be greatly increased.

Recall that the growth rate of a crystal can be written in general form as in Equation 20.1:

$$\mathrm{v} = a\upsilon^{+} u_{\mathrm{k}} f \tag{24.2}$$

Here a is the cube root of the atomic volume, f is the probability of finding an active growth site in an area a^2 of the interface, ν^{+} is the rate at which atoms join the crystal at the active growth sites, and u_{k} is the normalized net rate of growth.

The rate at which a liquid atom at the interface joins the crystal is $\nu^{+}f$, which is the probability that the atom is at an active growth site, times the rate at which it joins the crystal when it is at an active growth site. The average time that a liquid atom spends at the interface before it joins the crystal is the reciprocal of this rate, $1/\nu^{+}f$.

The distance which the interface advances during this time is $v/\nu^{+}f$. The distance which an atom can move by diffusion during this time is $(D/\nu^{+}f)^{1/2}$. The ratio of these two distances is β:

$$\frac{\dfrac{v}{\nu^{+}f}}{\sqrt{\dfrac{D}{\nu^{+}f}}} = \sqrt{\frac{a v u_{\mathrm{k}}}{D}} = \beta \tag{24.3}$$

Near equilibrium, β is small. For very fast growth or very slow diffusion, β approaches infinity.

24.4 Analytical Model

For near-equilibrium growth, the growth rate for the *i*th component of an alloy can be written as in Equation 22.8 [4, 9]:

$$\nu_i = a\nu^+ f\left[C_i^L - C_i^S \exp\left(\frac{\Delta\mu_i^0}{kT}\right)\right] \tag{24.4}$$

Where the term in the square brackets becomes u_k for each species. Here $\Delta\mu_i$ is the chemical potential difference for each species, *i*, between the two phases. The overall growth rate is then given by the sum of the ν_i, as in Equation 22.13. This formulation is valid when the atoms in the liquid move around rapidly compared to the rate at which the interface passes. Each growth site then samples the average composition of the liquid. Under these conditions, each atom effectively acts independently on being incorporated into the crystal, as is assumed implicitly in Equation 24.4. As a result, this formulation predicts that the distribution coefficient does not change significantly with growth rate, and so it does not include the phenomena known as solute trapping.

When the growth rate becomes comparable to the rate at which atoms can move around in the parent phase, the equations must be modified to take into account that the atoms can no longer act independently. Equation 24.4 can be modified to take into account these interactions by introducing a parameter *P*, which is related to β. *P* is zero when β is zero, and *P* is one when β is infinite.:

$$1 - P = \frac{1}{1 + A\beta} \tag{24.5}$$

A is a constant. When $P = 0$, the atoms act independently; when $P = 1$, all species present in the liquid at the interface are incorporated at the same rate. In the modified formulation, the Equation 24.4 becomes:

$$\nu_i = a\nu^+ f\left[C_i^L - C_i^S \exp\left(\frac{(1-P)\Delta\mu_i^0 + P\Delta G^0}{kT}\right)\right] \tag{24.6}$$

where ΔG^0 is the free energy difference between the crystal and the liquid.

For small ν, when near equilibrium conditions prevail at the interface, *P* is small, and Equation 24.6 reduces to the quasi-equilibrium equation for the growth rate, Equation 24.4. When β is very large, that is when the diffusion rate is small compared to the growth rate, *P* approaches 1. The growth rates of all the species depend on the free energies of the alloy in the two phases, rather than on their individual chemical potentials. Because the atoms cannot move as the interface passes, they must enter the crystal cooperatively. The composition of the solid will be the same as the composition of the liquid.

A simple approximate expression for the distribution coefficient for a small concentration of a second component, B, can be obtained from Equation 24.6 by noting, as in Equation 22.15, that the equilibrium k-value, k_e is given by:

$$k_e = \exp\left(-\frac{\Delta\mu_B^0}{kT}\right) \tag{24.7}$$

Since ν_B/av^+f is usually small, and ΔF^0 is small compared to $\Delta\mu_i^0$ for small concentrations of B, we have:

$$k \approx k_e^{1\text{-}P} = k_e^{1/(1+A\beta)} \tag{24.8}$$

For near equilibrium growth, β is small, so $k \approx k_e$. For large β, k approaches 1.

For growth on a rough interface, the growth rate, ν, is proportional to u_k. In this case, β is proportional to ν, and so Equation 24.8 can be written as:

$$k \approx k_e^{1/(1+A'\nu)} \tag{24.9}$$

Where A' is a constant which can be determined from the constant A.

24.5
Comparison with Experiment

The critical experiments to explore the solute trapping phenomenon were performed by laser melting of a thin layer on the surface of a silicon single crystal which had been ion implanted with a dopant as described above. In Figure 24.3, experimental data are presented for the growth rate dependence of the k-value for silicon implanted with bismuth [10–14], tin [15] and germanium [16].

Monte Carlo computer simulations have been performed using the crystal structure of silicon, inputting the equilibrium k -values for silicon doped with germanium, tin and bismuth. Data from these simulations are also presented in Figure 24.3.

The lines in Figure 24.3 are Equation 24.8 using $A = 8$. The agreement between Equation 24.8 and the simulation results is quite good. There is also quite good agreement with the experimental data for germanium, and there is good agreement with some of the experimental data for bismuth. There is not good agreement with some of the experimental data for bismuth, or with the experimental data for tin, which are from the same laboratory.

It is well known that the k -value of dopants in silicon, as well as in compound semiconductors, depends on the orientation of the growth front. In order to investigate this, an extensive set of simulations were carried out in three dimensions using the crystal structure of silicon, with various orientations of the interface. The growth rates for the simulations using the crystal structure of silicon were correlated to experimental growth rates, as outlined above. The data agree well with experimental measurements of the orientation dependence of bismuth incorporation into laser-melted silicon wafers [17]. The k-value does not depend on orientation when plotted as a function of β rather than as a function of the growth rate. The magnitude

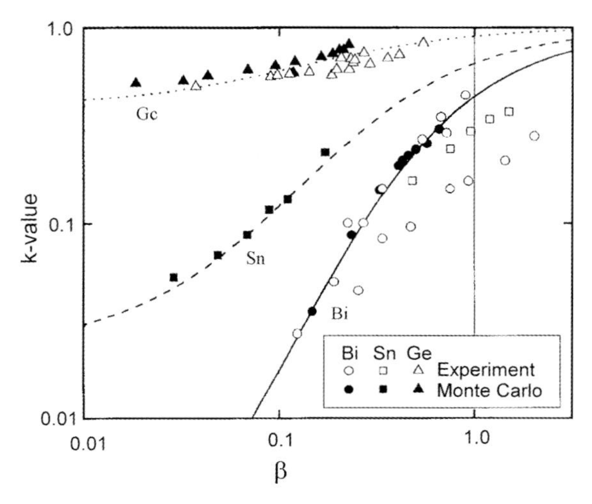

Figure 24.3 Simulation data and experimental data for the k -value for silicon doped with germanium, tin and bismuth.

of u_k, the undercooling required for growth, is quite different on and off a facet for the same growth rate. This difference accounts for the anisotropy in the incorporation of dopants on different orientations during growth at the same rate. In physical terms, this means that in order to keep up with the growth rate on a rough surface, the steps on a smooth surface must move very rapidly, and this increases the incorporation of the dopant on the facet.

Equation 24.8 is also in reasonable agreement with experimental results for the growth rate dependence of the k-value for aluminum alloys and for a nickel alloy, all using the same value of A.

Equation 24.6 can be used to calculate kinetic phase diagrams, such as the illustration in Figure 24.4. As the growth rate increases, the solidus and liquidus lines both collapse towards the T_0 line given by the equality of the free energies of the two phases.

For growth under conditions where these non-equilibrium effects are present, the kinetic phase diagram and the associated k-value, rather than the equilibrium values, should be used to describe the growth conditions and segregation effects.

24.6
Crystallization of Glasses

Glasses usually crystallize at reasonable rates only far below their equilibrium melting points. However, in glasses, unlike in liquids, the diffusion rate of dopants or second components is often much faster than the diffusion rate of the components of the glass matrix. The crystallization rate depends on the mobility of the major components of the glass matrix. Dopants which can move by diffusion much faster than the growth front moves will not be trapped into the crystal. The segregation

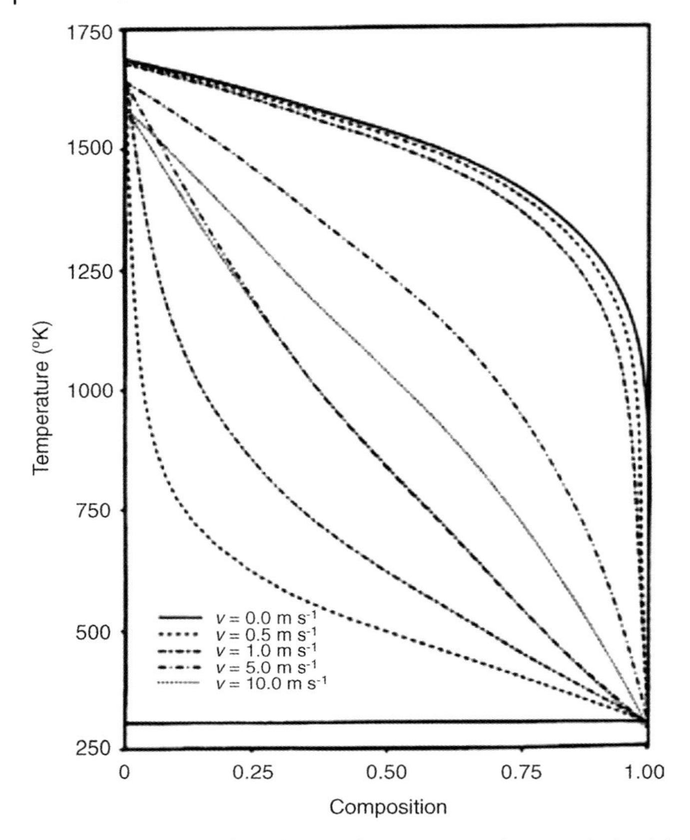

Figure 24.4 Kinetic phase diagram for various growth rates calculated from Equation 24.6.

coefficient for these dopants will be very close to the equilibrium value, even though the crystallization is taking place far from equilibrium.

References

1. Duwez, P. (1967) *ASM Trans. Quart.*, **60**, 606;*Prog. Solid State Chem.*, **3**, 377;Duwez, P.(ed.) (1978) *Metallic Glasses*, ASM, Metals Park, OH.
2. Baeri, P., Poate, J.M., Campisano, S.U., Foti, G., Rimini, E., and Cullis, A.G. (1980) *Appl. Phys. Lett.*, **37**, 912.
3. White, C.W., Wilson, S.R., Appleton, B.R., and Young, F.W. Jr (1980) *J. Appl. Phys.*, **51**, 738.
4. Jackson, K.A. (2002) *Interface Sci.*, **10**, 159.
5. Jackson, K.A., Gilmer, G.H., and Temkin, D.E. (1995) *Phys. Rev. Lett.*, **75**, 2530.
6. Jackson, K.A., Gilmer, G.H., Temkin, D.E., and Beatty, K.M. (1996) *J. Cryst. Growth*, **163**, 461.
7. Jackson, K.A. and Beatty, K.M. (2004) *J. Cryst. Growth*, **271**, 495.
8. Temkin, D.E. (1972) *Sov. Phys. Cryst.*, **17**, 405.
9. Jackson, K.A., Beatty, K.M., and Gudgel, K.A. (2004) *J. Cryst. Growth*, **271**, 481.

10 White, C.W., Wilson, S.R., Appleton, B.R., and Young, F.W. Jr (1980) *J. Appl. Phys.*, **51**, 738.
11 White, C.W., Appleton, B.R., Stritzker, B., Zehner, D.M., and Wilson, S.R. (1981) *Mater. Res. Soc. Symp. Proc.*, **1**, 59.
12 Baeri, P., Foti, G., Poate, J.M., Campisano, S.U., and Cullis, A.G. (1981) *Appl. Phys. Lett.*, **38**, 800.
13 Baeri, P., Foti, G., Poate, J.M., Campisano, S.U., Rimini, E., and Cullis, A.G. (1981) *Mater. Res. Soc. Symp. Proc.*, **1**, 67.
14 Aziz, M.J., Tsao, J.Y., Thompson, M.O., Peercy, P.S., White, C.W., and Christie, W.H. (1985) *Mater. Res. Soc. Symp. Proc.*, **35**, 153.
15 Hoglund, D.E., Aziz, M.J., Stiffler, S.R., Thompson, M.O., Tsao, J.Y., and Peercy, P.S. (1991) *J. Cryst. Growth*, **109**, 107.
16 Brunco, D.P., Thompson, M.O., Hogland, D.E., Aziz, M.J., and Grossman, H.J. (1995) *J. Appl. Phys.*, **78**, 1575.
17 Beatty, K.M. and Jackson, K.A. (1997) *J. Cryst. Growth*, **174**, 28.

Problems

24.1. At what growth rate will one tenth of the dopant atoms be incorporated into the crystal for $k_e = 0.01$ and a liquid diffusion coefficient of $D = 10^{-8}\,\mathrm{m^2\,s^{-1}}$ At what growth rate will half be incorporated? Use $A = 8$, $L/kT_M = 1$, $a = 3 \times 10^{-10}\,\mathrm{m}$, and $v = 0.1\,\Delta T\,\mathrm{m\,s^{-1}}$ in Equations 24.1 and 24.8.

24.2. A massive phase transformation is the name given to solid state phase transformations which take place at low temperatures in alloys. Reports in the literature are uncertain about whether a massive transformation occurs above or below the solidus line on the phase diagram. Discuss.

25 Coarsening, Ripening

25.1 Coarsening

A collection of particles of different sizes will tend to coarsen, that is the average particle size will grow, in order to reduce the total surface area of the collection of particles. This process is illustrated in Figures 23.7, 23.9 and 23.10. This coarsening proceeds only when there is enough mobility of the atoms. This process is also known as Ostwald ripening [1].

25.2 Free Energy of a Small Particle

The size dependence of the free energy of a particle was discussed in Chapter 15.

The total free energy to form a particle depends on the volume free energy and on the surface free energy of the particle. There is a critical radius r^* at which the particle can reduce its free energy either by growing or shrinking. The free energy of a particle of radius r can be written:

$$\Delta G = -\frac{4}{3}\pi r^3 \Delta G_V + 4\pi r^2 \sigma \tag{25.1}$$

The critical radius is:

$$r^* = \frac{2\sigma}{\Delta G_V} \tag{25.2}$$

This is illustrated in Figure 25.1. Atoms join and leave a cluster of atoms of radius r^* at equal rates, so it is in equilibrium, but it is an unstable equilibrium. It can reduce its free energy either by growing or by shrinking. If there is a distribution of particles of various sizes, as illustrated in Figure 25.2, the large ones, with radius greater then r_C will tend to grow, and the small ones will tend to shrink. On average, the particle size increases with time, and this is coarsening. The total surface area of the coarse

Kinetic Processes: Crystal Growth, Diffusion, and Phase Transitions in Materials. Kenneth A. Jackson

ISBN: 978-3-527-32736-2

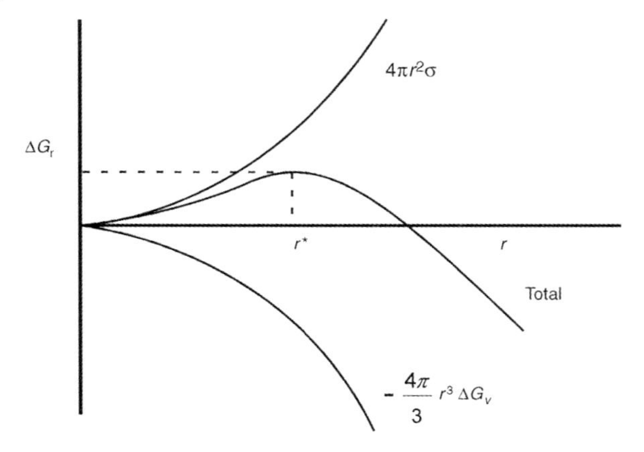

Figure 25.1 Free energy of a particle.

structure is less than the total surface area of the same volume of material with a finer structure.

In a solution, either solid, liquid or gaseous, containing particles, there will be some average concentration, $\overline{C}$, of the species which form the particles. This concentration depends on the average size of the particles present. There is a size of particle which is in unstable equilibrium at this concentration. Particles smaller than this size will shrink, and larger particles will grow.

In order to describe this process, we would like to know how many particles there are of various sizes, how the average concentration, $\overline{C}$, changes with time, and how the particle size distribution evolves over time. Since the large particles in the distribution are growing, and the smaller ones are shrinking, there is always a wide range of particle sizes. As time goes on, the average particle size increases, the average concentration, $\overline{C}$, decreases towards the equilibrium concentration, and since the critical radius increases, the size of particles which will shrink increases. This is a very complex mathematical problem. The problem has been simplified by looking for particle size distribution, $N(n)$, of the form [2, 3]:

$$N(n, t) = P(n) \cdot T(t) \tag{25.3}$$

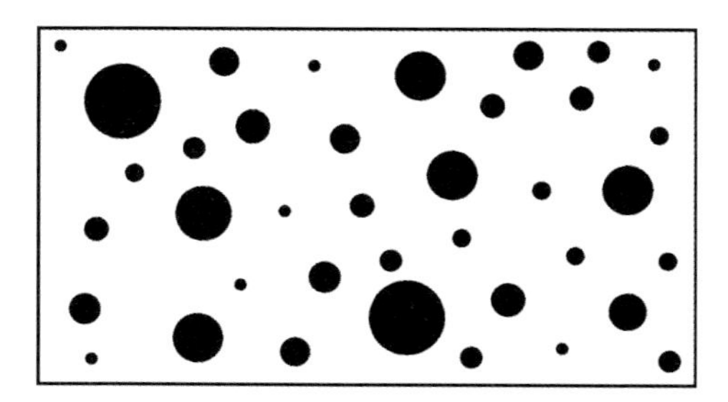

Figure 25.2 The larger particles will tend to grow, and the smaller particles will tend to shrink.

This form implies that the cluster distribution has the same functional form, with the amplitude evolving in time. Even these mathematical descriptions are complex. Instead, we will look at a simple model which suggests how the system evolves.

25.3 Coarsening in a Solution

The concentration of solution C_r at the surface of a particle of radius r is given by the unstable equilibrium condition:

$$C_r - C_0 \approx C_0 \frac{a}{r} \tag{25.4}$$

Here C_0 is the equilibrium concentration for a very large particle, and a is a parameter which depends on the surface tension, σ:

$$a = \frac{2\sigma\Omega}{kT} \tag{25.5}$$

Ω is the atomic volume. The average concentration of the solution, $\overline{C}$, will depend on the distribution of sizes of the particles. The growth rate of a spherical particle can be written as:

$$(C_P - C_0)\frac{dr}{dt} = D\frac{\overline{C} - C_r}{\delta} \tag{25.6}$$

where δ is the thickness of the composition boundary layer around the particle. If the particles are not isolated, and the effective boundary layer thickness depends on the proximity of other particles. Combining Equations 25.6 and 25.4, the growth rate of a particle radius is:

$$\frac{dr}{dt} = D\frac{\overline{C} - C_0\left(1 + \frac{a}{r}\right)}{\delta(C_P - C_0)} \tag{25.7}$$

which can be either positive or negative, depending on the size of r, as illustrated in Figure 25.3.

The radius critical size of a particle, r_C, where $dr/dt = 0$, is:

$$r_C = \frac{a\,C_0}{\overline{C} - C_0} \tag{25.8}$$

When $\overline{C} \approx C_0$, Equation 25.7 can be integrated to give the change in the radius with time:

$$r^2 \approx \frac{2DaC_0}{\delta C_P}t \tag{25.9}$$

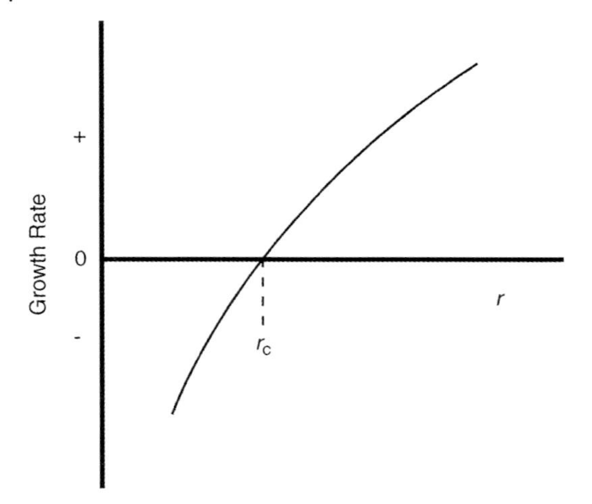

Figure 25.3 Growth rate for various particle sizes.

But this assumes that δ does not change in time, so this growth rate is approximately correct when the clusters are close together and their spacing determines δ. It also applies when the coarsening process is limited by the surface reaction rate. For well-separated particles, δ will be approximately the particle radius, the value obtained from the steady state diffusion solution. In this case:

$$\frac{dr}{dt} = D\frac{\left(\overline{C} - C_0 - \frac{C_0 a}{r}\right)}{rC_p} \tag{25.10}$$

So that:

$$r^3 \sim \frac{3DaC_0}{C_p}t \tag{25.11}$$

The coarsening of the structure in Figure 23.7 can be fitted with

$$l = l_0 + \alpha\tau^{1/2} \tag{25.12}$$

The data extrapolate back a value of l_0 that is approximately the fastest growing wavelength for a small amplitude sinusoidal perturbation, as in Figure 23.20. Coarsening of the structure shown in Figure 23.9 proceeds with the average radius increasing as $t^{1/6}$. Coarsening of the ordered structure shown in Figure 23.10 proceeds as $t^{1/2}$.

Extensive Monte Carlo simulations of the coarsening of phase separated structures for $C = 0.5$ [4, 5] give a time dependence of $t^{1/3}$. Mathematical modeling [6] of phase separated structures, based on the evolution of the structure factor, give a time dependence of $t^{1/3}$, going to $t^{1/4}$ or $t^{1/6}$ at long times.

25.4
Coarsening of Dendritic Structures

Dendritic structures coarsen as the dendrites grow. This can be seen from a detailed examination of the growing dendrite shown in Figure 26.1. The spacing of the ripples near the tip is much smaller than the spacing of the dendrite arms farther back from the dendrite tip. Only some of the bumps develop into branches, and as the growth proceeds, the smaller branches shrink and disappear, while the larger branches continue to grow.

The dendritic structure continues to coarsen after growth, as illustrated in Figure 25.4.

Both positive and negative curvatures of the surface are present in these structures.

The coarsening process continues as long as there is still liquid in the interdendritic spaces.

The coarsening of dendritic microstructures has been studied extensively in copper, aluminum, zinc, and so on alloys, by Flemings and his coworkers [7]. Their experimental results are illustrated schematically in Figure 25.5. The dendrite arm spacing was found to depend on the square root of the time, as suggested by Equation 25.9. The data cover times varying by six orders of magnitude, with the corresponding spacing changing by a factor of 1000.

The time on the horizontal axis in Figure 25.5 is the interval between when the sample passes through the liquidus temperature, which is when the dendrites grow, and when the temperature falls below the solidus line, which is when everything in the sample is solid, as illustrated in Figure 25.6.

The rate of coarsening is very rapid for fine structures, but slows significantly as the structure coarsens. The initial scale of the dendrite structure depends on diffusion processes, and so it is much finer for rapid growth rates. The dendrite arm spacing in a casting does not depend on how rapidly the dendrites grow, since the as-grown structure coarsens very quickly. The final dendrite arm spacing depends entirely on the coarsening process.

Figure 25.4 Coarsening of a dendritic structure in an alloy. The gross features remain constant, but the fine features disappear. The pictures were taken 1, 3 and 5 min after the dendrites had grown.

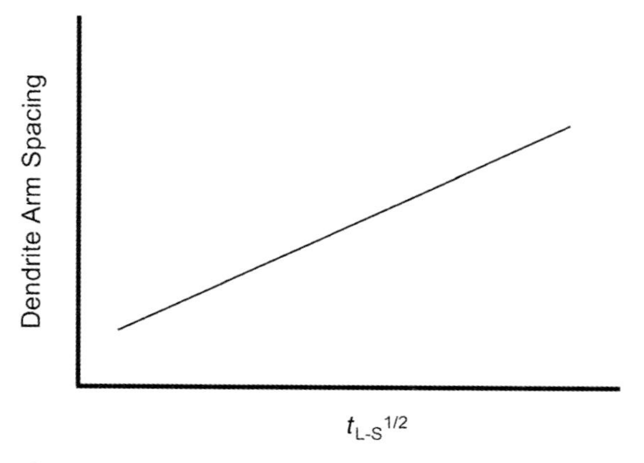

Figure 25.5 Dendrite arm spacing plotted against the square root of the time taken for the sample to cool from the liquidus temperature to the solidus temperature of the alloy.

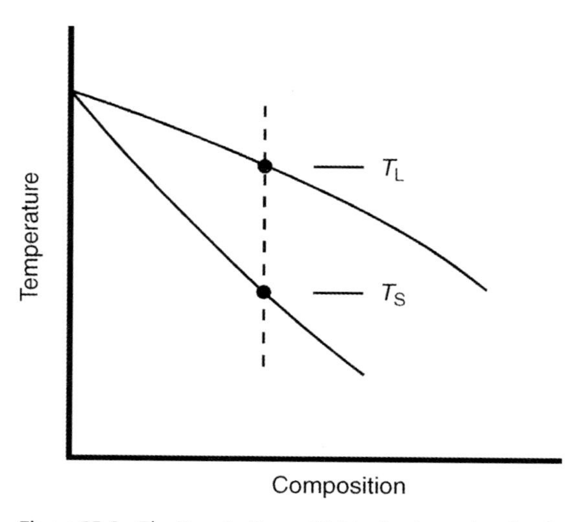

Figure 25.6 The time in Figure 25.5 is the time taken for the sample to cool from the liquidus temperature to the solidus temperature of the alloy.

Glicksman studied the isothermal coarsening of dendrites, and found that the coarsening proceeded as $\tau^{1/3}$.

25.5 Sintering

The sintering process is driven by the reduction in the surface area of the sintering particles. This is an important process for making ceramic objects. The melting point of many ceramic materials is so high that melt processing is prohibitive, and so

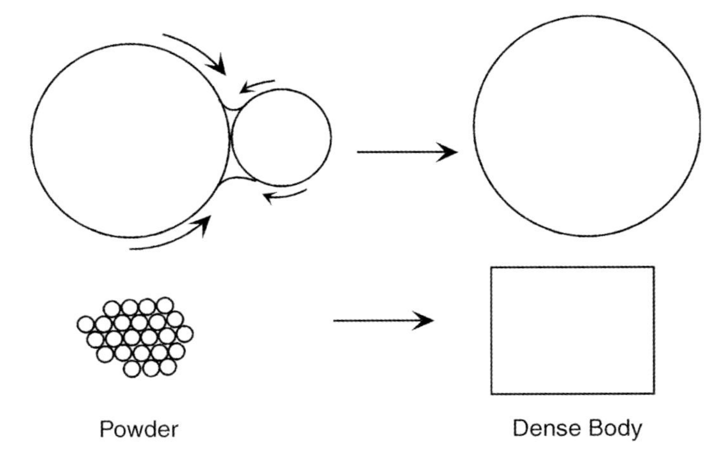

Figure 25.7 Sintering of particles.

objects are fabricated by sintering of powders far below the melting point. The powder particles agglomerate and merge, as illustrated in Figure 25.7.

Where the powder particles are in contact, they form a neck. The surface curvature is negative in the neck region and this drives diffusion processes which tend to merge two particles into a single sphere, and to merge a powder compact into a solid body. The process is driven by surface tension, but there must be mobility of the atoms. This can be bulk diffusion, surface diffusion, or even vapor transport. Usually the sintering is carried out at elevated temperatures to promote the mobility. Often a second component with increased mobility, such as a glass with a low softening temperature, is added.

An interesting process was developed by Coble to make the quartz sheaths which are used for high power halide light bulbs. The problem was that gas became trapped in the pockets which formed in the spaces between the particles, during sintering. The diffusion of the gas through the bulk was too slow to get rid of the pockets, and the pockets became hot spots, limiting the operating temperature. Coble found that the gas pockets shrank more rapidly when the gas could diffuse out along grain boundaries. When grain growth occurred so that the grain boundary left a pocket it shank much more slowly. So Coble added a component which pinned the grain boundaries. Bubble free "Lucolux" quartz sheathes are made by this process.

25.6
Bubbles

Soap bubbles blown into the air pop because the water evaporates. But in a closed container, a froth of bubbles coarsens by the migration of the soap films, not by popping. The air inside the closed container is saturated with water, but there is local vapor transport from surfaces of positive curvature to surfaces with negative

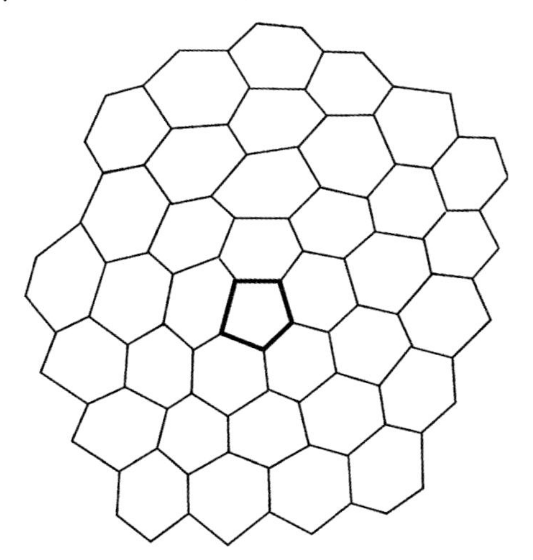

Figure 25.8 Soap film in 2D.

curvature. Locally, the junction between soap films forms at 120°, as illustrated in the two-dimensional drawing in Figure 25.8.

In two dimensions, an array of hexagonal shapes is stable. The junctions where the hexagons join can all form angles of 120°. Cells with other than six sides, such as the pentagon in Figure 25.8, must have curved sides in order to form 120° angles at the junctions. The sides of the pentagon will bow out, as illustrated in Figure 25.9.

Evaporation occurs relatively faster from concave surfaces than from convex surfaces This asymmetry in evaporation rate means that a soap film will move towards its center of curvature. So the pentagon in Figure 25.8 will shrink. A cell with more than six sides will expand. A pattern which consists only of six-sided cells will

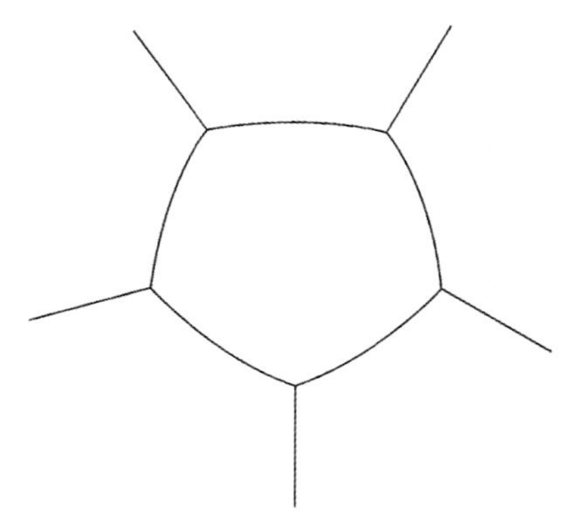

Figure 25.9 The sides of a pentagon must bow out if the corner angles are maintained at 120°.

have all straight walls, and so will be relatively stable. The coarsening action comes from the cells which do not have six sides.

An interesting topological fact is that a five-sided cell in a sea of six-sided cells does not disappear when it collapses to a point. When it disappears, another five-sided figure is created. The topological defect persists.

Coarsening has been illustrated above with a two-dimensional drawing, but similar considerations apply to three-dimensional soap bubble arrays.

25.7 Grain Boundaries

Grain boundaries in metals tend to behave just like soap bubbles except that their motion is a lot more complicated. There is much more known about this subject than the brief outline which will be presented here.

First, grain boundaries can have a variety of energies, although in many metals, high angle grain boundaries all have relatively similar energies. The energy of these high angle boundaries is about the same as the energy of one or two atomic layers of liquid. Small angle tilt boundaries are made up of dislocations. Their energies are much lower, and depend on the misorientation between the grains. There are also twin boundaries, where the atom sites in the boundary plane are common to both lattices. In a coherent twin boundary, all the nearest neighbor atoms are in the right place, but the second nearest neighbors are on wrong sites across the twin boundary. There are also coincidence site boundaries where some of the atoms in the boundary share sites with both lattices, as illustrated in Figure 25.10.

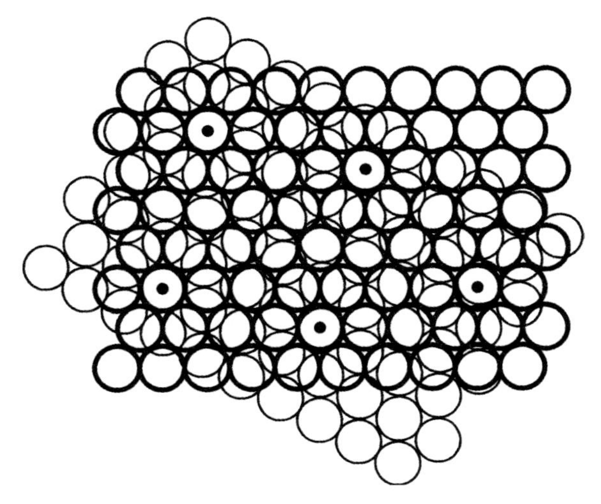

Figure 25.10 Coincidence site boundary. The upper layer with light circles has been rotated with respect to the heavy circles. The dots mark the lattice sites that are common to both layers.

High angle boundaries will split into a coincidence site boundary and small angle boundary if that lowers the total energy. The angle which grain boundaries form at their junctions depends on their relative energies.

The energy of a twin boundary in some crystals is very small. For example, it is very low in copper, so many twin boundaries form during grain growth in copper. It is also very low in silicon and many compound semiconductors, so that the formation of twins during crystal growth can be a problem. The twin boundaries are relatively immobile. In general, coincidence site boundaries are less mobile than high angle boundaries.

25.8 Scratch Smoothing

The effective mobility of surface atoms can be determined by measuring the rate at which corrugations on a surface smooth out [8, 9]. This is also a process which occurs to reduce the surface area. Surfaces can be made corrugated either by scratching them, or by using photolithography. Atoms move from the hills to the valleys to reduce the surface area. In general the smoothing can occur by both bulk and surface diffusion, as described by:

$$Z = A_0 \exp[-(B\omega^4 + S\omega^3)t]\sin \omega x \qquad (25.13)$$

Here A_0 is the initial amplitude of the sinusoidal corrugation of the surface, which has a wavelength $2\pi/\omega$. The decay rate of corrugation due to bulk diffusion is given by:

$$B = D_B \frac{(\gamma_0 + \gamma''_0)}{kT}\Omega \qquad (25.14)$$

and the decay rate due to surface diffusion is given by:

$$S = n\,D_S \frac{(\gamma_0 + \gamma''_0)}{kT}\Omega \qquad (25.15)$$

Here Ω is the atomic volume, D_B is the bulk diffusion coefficient, γ_0 and γ''_0 are the surface tension and the second derivative of the surface tension with respect to surface orientation (the curvature of the Wulff plot), n is the density of mobile adatoms on the surface, and D_S is the diffusion coefficient of the adatoms on the surface. The surface diffusion process can be distinguished from bulk diffusion because the two have different wavelength dependences. Both the number of adatoms and their mobility change with temperature, and the analysis gives only the product of the two, nD_S. This product is the effective surface mobility.

The experiment can be carried out by diffracting a laser beam from the corrugations. The intensity of the diffracted beam depends on the amplitude of the corrugation.

The surface mobility of individual atoms has also been determined using scanning tunneling microscopy (STM), by following the motion of individual atoms. This requires many measurements in order to gather statistical information about the motion. There is also the issue of whether stresses from the presence of the STM tip have an effect on the measurement.

References

1 Ostwald, W. (1897) *Z. Phys. Chem.*, **22**, 289.
2 Wagner, C. (1961) *Z. Electrochem.*, **65**, 581.
3 Zener, C. (1949) *J. Appl. Phys.*, **20**, 950.
4 Amar, J.G., Sullivan, F.E., and Mountain, R.D. (1988) *Phys. Rev. B*, **37**, 196.
5 Yaldram, K. and Binder, K. (1991) *Acta Metal. Mater.*, **39**, 707.
6 Langer, J.S., Bar-on, M., and Miller, H.D. (1975) *Phys. Rev. A*, **11**, 1417.
7 Flemings, M.C. (1974) *Solidification Processing*, McGraw-Hill, New York, NY.
8 Mullins, W.W. (1956) *J. Appl. Phys.*, **27**, 900.
9 King, R.T. and Mullins, W.W. (1962) *Acta Metall. Mater.*, **10**, 601.

Problems

25.1. (a) Plot the critical radius size as a function of time using Equation 25.11, with $a = 3 \times 10^{-10}$ m, $D = 10^{-8}\,\text{m}^2\,\text{s}^{-1}$, and $C_0/C_P = 0.1$.
(b) How long does it take for the critical size to grow from 100 nm to 1 μm?
(c) How long does it take for the critical size to grow from 1 mm to 1 cm?

25.2. For $D_B = 10^{-10}\,\text{m}^2\,\text{s}^{-1}$, and $nD_S = 10^{-4}\,\text{m}^2\,\text{s}^{-1}$, at what wavelength will the bulk and surface contributions to scratch smoothing be equal?

26
Dendrites

26.1
Dendritic Growth

The word dendrite derives from the Greek word for tree, and means “tree-like”. Dendritic growth occurs only when a diffusion process dominates the rate at which the phase transformation proceeds. An extensive study of this mode of growth was carried out by Papapetrou [1]. Dendrites grow into a metastable phase which is either supercooled or supersaturated. The supercooling can be uniform throughout the sample, or it can be a region of constitutional supercooling ahead of an advancing interface in an alloy.

26.2
Conditions for Dendritic Growth

Dendritic growth occurs when the interface kinetic processes are rapid, so that a planar growth front is unstable. Essentially what happens is that the growth front subdivides the metastable phase into regions which are small enough so that diffusion processes can remove the remaining instability.

The diffusion process responsible for dendritic growth can be thermal, compositional, or both. In a pure material, where dendrites grow into a supercooled liquid, the dendrites are a result of thermal diffusion. An example is shown in Figure 26.1.

Experimentally, a liquid can be supercooled, and then, when the solid nucleates, the growth is dendritic. Most liquids cannot be supercooled below about 80% of their melting points. But the latent heat in most materials is large enough to heat them up through about 30% of their melting points. So in most materials, the latent heat of the freezing process is more than enough to heat an undercooled sample up to the melting point before it is all frozen.

Dendritic growth can be very rapid: dendrites have been observed to grow at a rate of $40\,\mathrm{m\,s^{-1}}$ into supercooled pure nickel.

Dendrites can also grow as a result of constitutional supercooling. In this case they grow in an array, as shown in Figure 11.2. The growth proceeds at a rate determined

Kinetic Processes: Crystal Growth, Diffusion, and Phase Transitions in Materials. Kenneth A. Jackson

ISBN: 978-3-527-32736-2

Figure 26.1 Dendritic growth into an undercooled melt of pure succinonitrile.

by the rate at which heat is extracted from the sample. This is the usual growth mode in the columnar zone in a casting, where the dendrites grow inwards from the mold wall as heat is extracted. Dendrites can also grow into a supercooled alloy, in which case both thermal and compositional diffusion are important. Since thermal diffusivities are usually much larger than compositional diffusivities, there are two different diffusion length scales involved, which makes for complex growth patterns and behavior.

Dendrites grow in crystallographic directions. For face centered and body centered cubic crystals, the growth axis is $\langle 1\,0\,0 \rangle$. The simple rule for this is to make a closed figure out of the closest packed planes of the crystal, and the dendrites will grow out of the corners. For example, the closest packed planes of the face centered cubic structure are the (1 1 1) planes, and these form an octahedron, as illustrated in Figure 26.2. The corners of this figure are in the $\langle 1\,0\,0 \rangle$ directions.

The dendrite stem and the braches in Figure 26.1 are growing in $\langle 1\,0\,0 \rangle$ directions.

The elongated shape of the dendrites is a result of a non-linear amplification of the growth rates at the corners. This process is illustrated in Figure 26.3. A small anisotropy in the kinetic coefficient or in the surface tension is amplified by the diffusion process, because the diffusion process is faster where the interface is more convex. This turns a small bump into a rapidly growing tip.

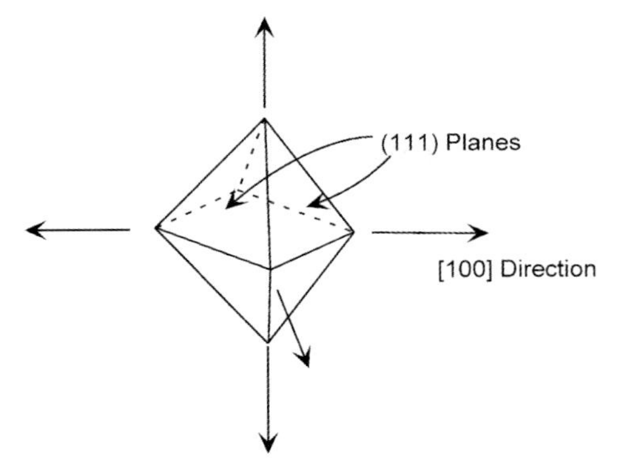

Figure 26.2 Dendrites grow in the directions of the corners.

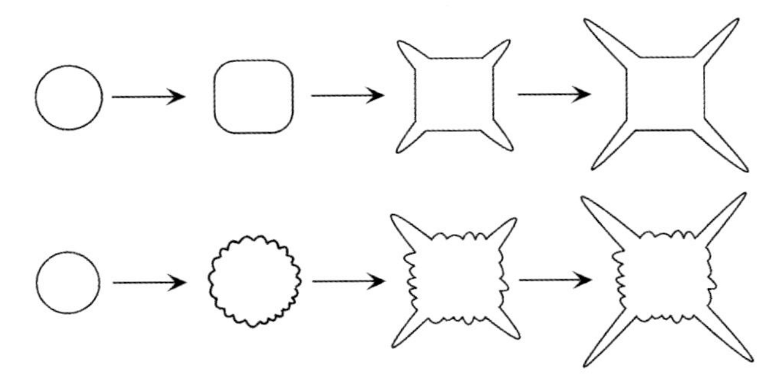

Figure 26.3 Diffusion is faster away from the surface where the radius of curvature is smaller.

The anisotropy can create corners where growth is more rapid and these corners develop into dendrites, as in the top sequence. Alternatively, the interface can become unstable and develop an array of perturbations, and then the perturbations in the fast growing direction will take off, as in the lower sequence.

26.3 Simple Dendrite Model

The analysis of dendritic growth is a complex Stefan problem. A moving interface is a moving source of heat and/or solute. The strength of the source depends on the rate of motion of the interface. The position and rate of motion of the interface depend on the local fields and gradients. These are lovely non-linear mathematical problems, and the observed dendrite interface shapes are very complex.

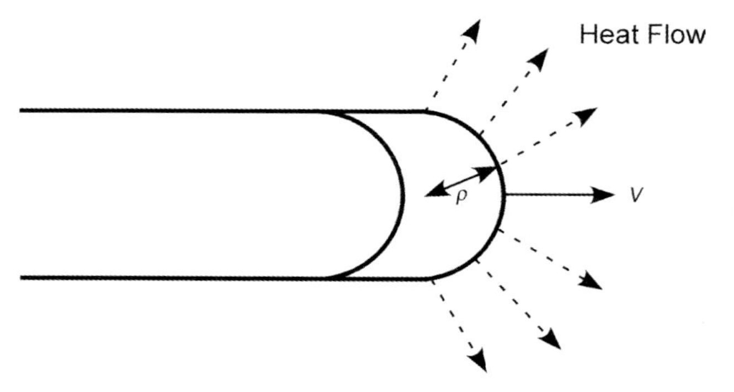

Figure 26.4 Simple model for a dendrite.

In this section, an oversimplified model of a thermal dendrite growing in a pure material will be presented. It includes the major components of the problem, and illustrates the inherent dilemma found in analyzing dendritic growth.

The dendrite will be modeled as a cylinder of radius ϱ, with a hemispherical cap on the end, also of radius ϱ, as illustrated in Figure 26.4.

We will assume that the dendrite is growing at a constant rate v, into a supercooled liquid, and that the thermal field around the tip is given by the steady state solution for diffusion from a sphere, as derived previously for a composition field, Equation 8.25. The steady state thermal field around the hemispherical tip of radius ϱ is:

$$T = T_\infty + \left(T_\mathrm{I} - T_\infty \frac{\varrho}{r}\right) \qquad (26.1)$$

The temperature at the surface of the hemisphere is T_I, and the far-field temperature is T_∞.

The total heat conducted away from the hemisphere is given by its surface area, times the local heat flux:

$$-2\pi\varrho^2 K \left(\frac{\mathrm{d}T}{\mathrm{d}r}\right)_{r=\varrho} = 2\pi\, K(T_\mathrm{I} - T_\infty)\varrho \qquad (26.2)$$

where K is the thermal conductivity. The thermal gradient at the surface was derived from the temperature distribution given by Equation 26.1. We can take into account the heating of the liquid along the side of the cylinder as the tip passes by limiting the angle at the tip through which the heat is conducted away. We will assume that the effective area through which heat can be conducted is reduced by a factor g^2.

The rate at which the cylinder increases in volume is given by its cross-sectional area times the growth rate, v. Heat is produced by the growing dendrite at a rate given by the rate of increase in volume times the latent heat, L:

$$\pi\varrho^2 v\, L \qquad (26.3)$$

The latent heat generated is carried away by the thermal gradient at steady state, so the heat flux in Equation 26.2 is equal to the heat generated, Equation 26.3:

$$(T_I - T_\infty) = \frac{Lv\varrho}{2Kg^2} \tag{26.4}$$

The undercooling depends on a product of the radius and the growth rate. The dimensionless quantity $v\varrho/\varkappa$, where $\varkappa$ is the thermal diffusivity, is known as the Peclet number. Our analysis of the heat flow has not provided a specific growth rate for the dendrite. Instead, the solution is in terms of the Peclet number. A dendrite with a small tip radius growing rapidly or a dendrite with a large tip radius growing slowly is equally valid. This is a common feature of all the analytical solutions for the diffusion fields around a dendrite tip.

A similar analysis to the above can be carried out for solute diffusion away from a growing dendrite tip.

Papapetrou [1] suggested that a dendrite tip should be parabolic, rather than the cylinder which we used. He noted that the electric field gradient around a parabolic tip was such that, if the surface was displaced normal to itself everywhere by an amount proportional to the gradient, then the parabola would be displaced along its axis, without changing shape. So this should be the steady state shape for a dendrite tip.

Many years later, Ivantsov [2] proved that the parabola (or a parabaloid of revolution, which has a circular cross-section) provides a valid steady solution to the heat flow from a dendrite, provided that the interface is isothermal. His expression for the undercooling at the interface is now known as the Ivantsov function:

$$\Delta T = Iv\left(\frac{v\varrho}{2\varkappa}\right) = -\frac{v\varrho L}{2K}\exp\left(\frac{v\varrho}{2\varkappa}\right)E_i\left(-\frac{v\varrho}{2\varkappa}\right) \tag{26.5}$$

where $\varkappa = K/C$ is the thermal diffusivity, C is the specific heat per unit volume, and $E_i(x)$ is the integral error function. He solved the problem by using the equation for thermal diffusion as a boundary condition for the equation for the interface shape. This is now known as Ivantsov's method. But notice that the thermal field is expressed in terms of the dimensionless Peclet number, $v\varrho/\varkappa$.

Some other condition must be applied to determine the relationship between the tip radius and the growth rate. The thermal analysis does not take into account either kinetic undercooling at the interface or the undercooling due to surface curvature. These can be incorporated formally by expressing the difference between the interface temperature and the temperature far from the interface as the sum of a heat flow term, ΔT_H, a kinetic undercooling term ΔT_K, and undercooling due to surface curvature, ΔT_σ:

$$T_M - T_\infty = \Delta T_H + \Delta T_K + T_\sigma \tag{26.6}$$

For our simple model this becomes:

$$T_M - T_\infty = \frac{Lv\varrho}{2Kg^2} + v/\mu + \frac{2\sigma T_M}{\varrho L} \tag{26.7}$$

In the kinetic term, the growth rate is assumed to be linear with undercooling and μ is known as the kinetic coefficient. The surface tension term is the standard expression for the undercooling of an interface which has a radius of curvature ϱ, as in Equation 15.3.

With the curvature term, the dendrite will grow more slowly if the radius of curvature is too small, so this suggests that the dendrite should have a tip radius such that it grows as fast as possible at a given undercooling. This is given by maximizing the above expression, which gives:

$$\varrho^2 = \frac{4\sigma\, T_M K g^2}{L^2 v} \tag{26.8}$$

Inserting this into the equation for the undercooling:

$$\Delta T = \Delta T_K + 2\sqrt{\frac{\sigma\, T_M v}{K g^2}} \tag{26.9}$$

So that for small values of ΔT_K we have:

$$v \,\alpha\, \Delta T^2 \tag{26.10}$$

Which is approximately the relationship which is observed for dependence of the growth rates of dendrites on the thermal undercooling.

Data for dendritic growth in nickel in Figure 26.5 were obtained by undercooling liquid nickel, then initiating crystallization, and measuring the rate at which the dendrites grow. The line in Figure 26.5 is $v \,\alpha\, \Delta T^2$, which fits the data reasonably well using a value of g in Equation 26.9 which is about 0.5.

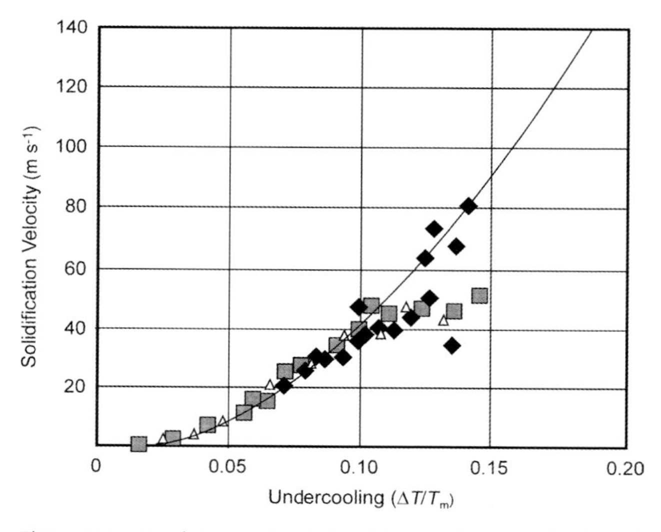

Figure 26.5 Dendrite growth rate for nickel as a function of melt undercooling [3].

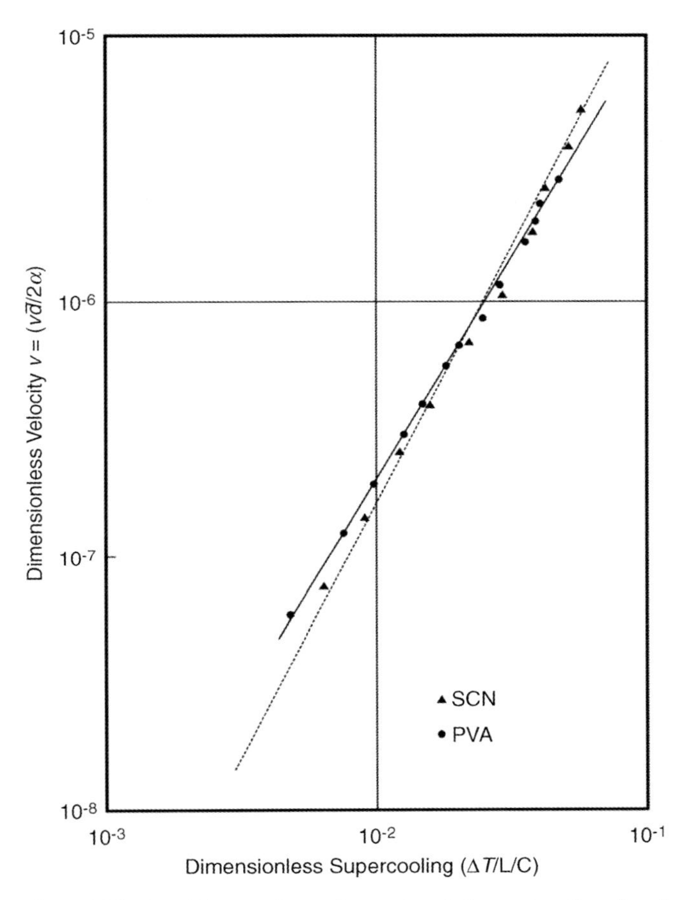

Figure 26.6 Dendrite growth rate for succinonitrile as a function of melt undercooling [4–6].

Data for dendritic growth in succinonitrile are presented in Figure 26.6. The dotted line is $v \alpha \Delta T^2$, and the solid line is a least squares fit to the data. The parabola fits the data reasonably well, but again, the prefactor is not given correctly by Equation 26.9.

The curvature term has been added to the Ivantsov expression for the thermal field, using the undercooling corresponding to the curvature at the tip. This also does not fit the data. So maximizing the growth rate is not the answer. Furthermore, the curvature varies along a parabolic dendrite, and so the undercooling along the dendrite is not constant because of the curvature. The Ivantsov solution is not valid for this case.

The issue of what determines the tip radius is still not altogether settled. There have been two schools of thought, one of which is known as the Stability Condition and the other as the Solvability Condition [7–13].

Those who favor the Stability Condition have argued that the dendrite becomes unstable behind the tip, and develops ripples which grow into the side branches. They argue that the instability process causes the tip radius to oscillate slightly about an average value, and so the dendrite tip radius is controlled by the development of the perturbations.

The Solvability Condition derives from numerical modeling of the dendrite tip. In our analysis above, we assumed that the tip was a hemisphere, so it had a constant radius of curvature everywhere on the surface. We also assumed that the kinetic undercooling term was the same everywhere on the surface. But neither of these assumptions is correct for a parabolic dendrite tip. Both the curvature of the surface and the growth rate vary along the parabolic surface of the dendrite tip. The modeling suggests that there is no way to adjust the shape of a dendrite tip to obtain stable growth if both the surface tension and the kinetic coefficient are isotropic. But both the surface tension and the kinetic coefficient are anisotropic. It seems that the surface of the dendrite can assume a shape which results in stable steady state growth by adjusting to these anisotropies. The anisotropy compensates in some way for the variation in curvature and growth rate along the parabola. The modeling indicates that there are solutions, which depend on the anisotropy, for a limited number of tip sizes and shapes. And there are specific growth rates corresponding to these stable configurations.

There are now serious attempts to sort this out, using a combination of experimental measurements and computer simulations to obtain the appropriate parameters. But these are clearly not simple calculations. The modeling in the past has focused on the anisotropy in the surface tension. Molecular dynamics simulations suggest that the kinetic coefficient is both larger than had been used for calculations, and is also more anisotropic. So the kinetic coefficient may well prove to be the key to dendritic growth.

26.4 Phase Field Modeling

Phase field modeling is a computer modeling method for describing processes such as dendritic growth [14–16]. It is used to provide a bridge between the diverse length scales involved in dendritic growth. There are equations for the thermal field, and equations for the compositional field. The phase field model adds a parameter which varies in space, and depends on which phase is present. The boundary between the phases is assumed to have some finite thickness, and the properties vary continuously from one phase to the other phase on going through the boundary. When the interface moves, it generates heat and rejects solute over its thickness, and its motion depends on its curvature. The thermal field, the composition field and the phase field are all coupled. The thickness of the boundary is arbitrary, but it must be small compared to the diffusion lengths involved in the problem, and it must be small compared to the radius of curvature of the interface. The mesh size for computations in the interface region must be small compared to the interface thickness. And so the computations are not as simple and rapid as was hoped, but this is an extremely powerful method, which works very well for many cases where the input parameters are known [17].

26.5 Faceted Growth

A small amount of anisotropy is important to determine the relationship between the interface curvature and the dendrite growth rate. But the models assume that the growth rate is linearly proportional to the local interface undercooling. This is valid for metals growing from a melt, where all the surfaces are rough. There will, in general, be a different kinetic coefficient for each growth direction, even though all the growth rates are linear with undercooling. But for smooth surfaces, the growth rate is not linear with undercooling. Moreover, where the growth rate is very anisotropic, growth occurs by the rapid motion of steps across facets. The growth rate depends on the formation of the steps, which occurs where there are defects, or on the part of the facet where the undercooling is large enough for the nucleation of steps. So the motion of the interface does not depend on the local conditions. There is not a satisfactory analysis of the overall interface motion for such cases.

26.6 Distribution Coefficient

The distribution coefficient can increase from its equilibrium value at very rapid growth rates, a process which is discussed in Chapter 24. This can result in a jump in the growth rate of dendrites at some undercooling [18], as illustrated in Figure 26.7.

As the growth rate increases, the distribution coefficient increases. At a critical growth rate the segregation coefficient starts to increase significantly. This decreases the segregation, so the dendrite grows faster, and the segregation decreases more,

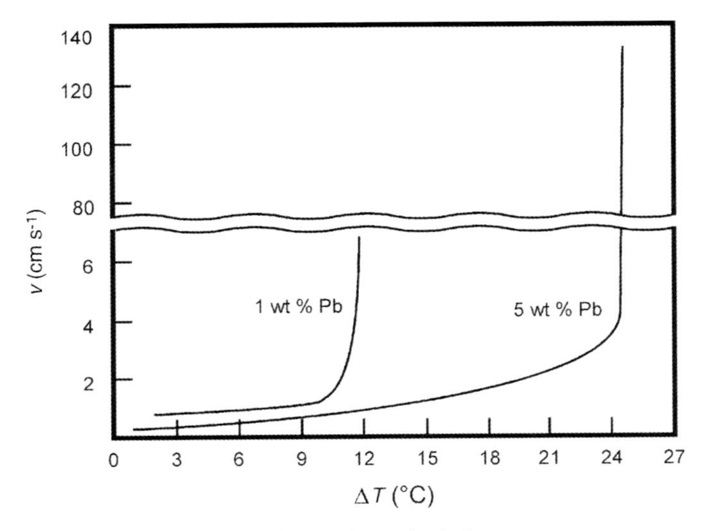

Figure 26.7 Dendrite growth rates for tin–lead alloys.

and so on. In the figure, the dendrite growth rate for a 5% lead in tin alloy jumps from a diffusion-limited rate at 4 cm s^{-1} to a thermally-limited growth rate of about 130 cm s^{-1} at an undercooling of 25 °C. Similar phenomena have been observed in other alloys [19, 20].

References

1 Papapetrou, A. (1935) *Z. Kristallogr.*, **92**, 89.
2 Ivantsov, G.P. (1947) *Dokl. Akad. Nauk. SSSR*, **58**, 567.
3 Walker, J.L. (1961) Phys. Chem of Proc. Met. II (ed. G.R. St. Pierre), p. 845; (1963) Trans. 6th Vac. Met. Conf., p. 3.
4 Glicksman, M.E., Schaefer, R.J., and Ayers, J.D. (1976) *Metal. Trans. A*, **7**, 1747.
5 Huang, S.-C.and Glicksman, M.E. (1981) *Acta Metall.*, **29**, 701.
6 Glicksman, M.E. and Singh, N.B. (1989) *J. Cryst. Growth*, **98**, 277.
7 Ben-Jacob, E., Goldenfeld, N.D., Kotliar, B.G., and Langer, J.S. (1984) *Phys. Rev. Lett.*, **53**, 2110.
8 Kessler, D.A., Koplik, J., and Levine, H. (1985) *Phys. Rev. A*, **31**, 1712.
9 Langer, J.S. (1986) *Phys. Rev. A*, **33**, 435.
10 Kessler, D.A. and Levine, H. (1986) *Phys. Rev. B*, **33**, 7687.
11 Kessler, D., Koplik, J., and Levine, H. (1988) *Adv. Phys.*, **37**, 255.
12 Kurz, W. and Fisher, D.J. (1989) *Fundamentals of Solidification*, Trans Tech., Aedermannsdorf.
13 Trivedi, R., and Karma, A. (1998) *Encyclopedia of Applied Physics*, vol. 23, VCH, New York, p. 441.
14 Warren, J.A. and Boettinger, W.J. (1995) *Acta Metall. Mater.*, **43**, 689.
15 Wang, S.-L. and Sekerka, R.F. (1996) *Phys. Rev. E*, **53**, 3760.
16 Karma, A. and Rappel, W.-J. (1996) *Phys. Rev. E*, **53**, R3017.
17 Bragard, J., Karma, A., Lee, Y.H., and Plapp, M. (2002) *Interface Sci.*, **10**, 121.
18 Nikonova, V.V. and Temkin, D.E. (1968) *Growth and Perfection of Metallic Crystals* (ed. D.E. Ovsienko), transl. by Consultants Bureau, Plenum, New York, USA, p. 43.
19 Eckler, K., Cochrane, R.F., Herlach, D.M., and Feuerbacher, B. (1992) *Phys. Rev. B*, **45**, 5019.
20 Eckler, K., Herlach, D.M., and Aziz, M.J. (1994) *Acta Metall. Mater.*, **42**, 975.

Problems

26.1. For $\sigma = 0.3\,\mathrm{J\,m^{-2}}$, $\varkappa = 10^{-5}\,\mathrm{m^2\,s^{-1}}$, $C = 5\times10^{-6}\,\mathrm{J\,m^{-1}\,°C^{-1}}$, $m = 0.5\,\mathrm{m\,s^{-1}\,°C^{-1}}$, and $T_M = 1700\,°\mathrm{C}$, what value of g in Equation 26.9 will fit the data in Figure 26.5?

27
Eutectics

27.1
Eutectic Phase Diagram

Eutectic refers to a specific type of phase diagram, or part of a phase diagram, as illustrated in Figure 27.1, where a homogeneous liquid solidifies into two different solid phases.

If a diffusion couple is formed between pure samples of the two solid phases at a temperature below the eutectic temperature, the composition profile after some time will be as illustrated in Figure 27.2.

The composition at the spatial interface jumps from the value at one phase boundary on the phase diagram to the other. The composition in the single phase regions varies from the value far from the interface to the equilibrium value at the interface.

27.2
Classes of Eutectic Microstructures

There are three classes of eutectic microstructures, and these depend on the solidification characteristics of the two primary phases [1].

I. Both phases have small α-factors.
II. One phase has a small, and the other a large α-factor.
III. Both phases have large α-factors.

The first two of these microstructure classes were identified before the α-factor was developed, and the classification was based on the relative magnitudes of the latent heats of the end components.

Kinetic Processes: Crystal Growth, Diffusion, and Phase Transitions in Materials. Kenneth A. Jackson

ISBN: 978-3-527-32736-2

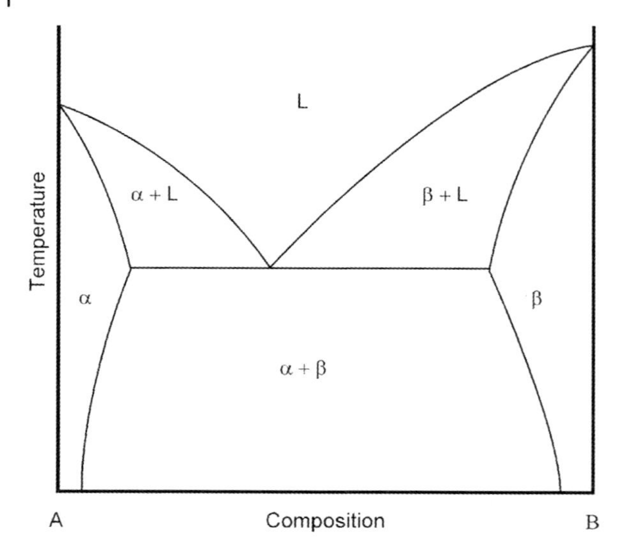

Figure 27.1 Eutectic phase diagram.

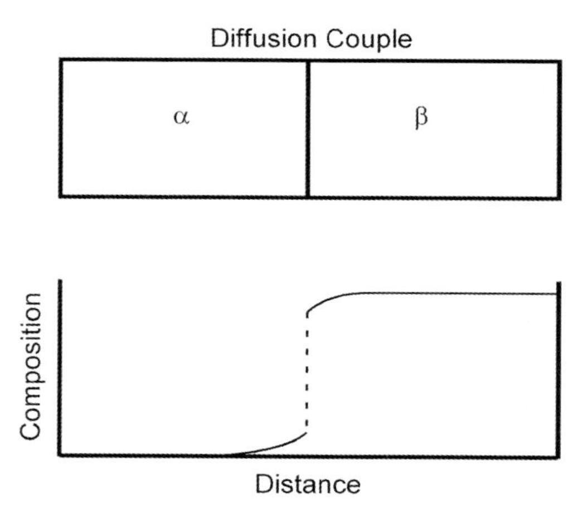

Figure 27.2 Composition profile for a diffusion couple at a temperature below the eutectic temperature.

27.2.1
Class I Eutectics

Most eutectic alloys between metals are in Class I. The most common example is lead–tin solder. These alloys form lamellar or rod microstructures. Figure 27.3 shows a lamellar microstructure in a cadmium–tin alloy.

The growth front of a lamellar eutectic is illustrated in Figure 27.4.

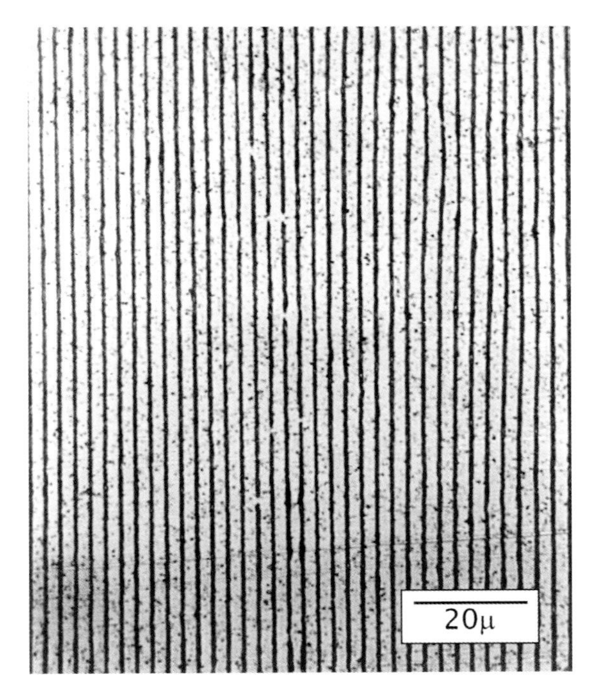

Figure 27.3 Lamellar microstructure in a cadmium–tin alloy.

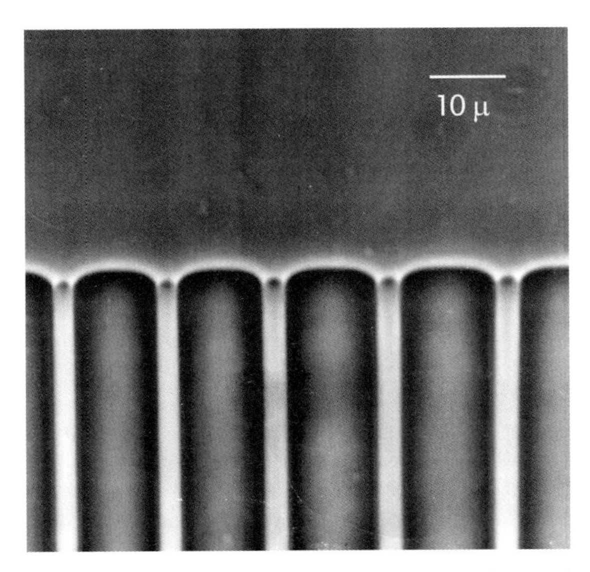

Figure 27.4 Growth front of a lamellar eutectic, carbon tetrabromide-hexachloroethane.

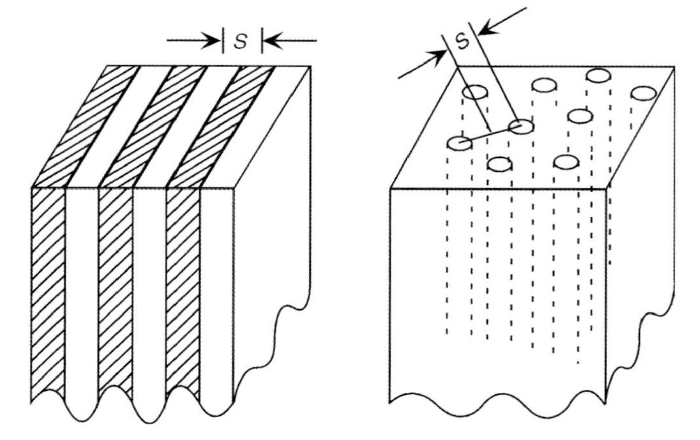

Figure 27.5 Surface area in rod and lamellar structures.

Each phase rejects the other component, and the rejected components interdiffuse in the liquid ahead of the interface. This happens more readily if the lamellar spacing is small. On the other hand, making the lamellar spacing small introduces more boundaries into the solid. Accordingly, the formation of these microstructures depends on a balance between the segregation process necessary to form the two separate phases and the energy required to introduce more interphase boundaries. For rough interfaces, typical of metals, the crystallization kinetics is sufficiently rapid that they do not play a role. The formation of this type of microstructure will be treated in more detail below.

Whether the microstructure is lamellar or rod depends on the relative volume fraction of the two phases. For the same distance, S, between the centers of the two phases, as illustrated in Figure 27.5, the total interface area is less for the lamellar structure if $\pi > V_\alpha/V > 1/\pi$.

For $\pi < V_\alpha/V < 1/\pi$, rod microstructures will have smaller total interface area. So lamellar structures are observed when the eutectic composition is close to the middle between the compositions of the solid phases. Rods form when the phase diagram is very asymmetric, with the eutectic composition much closer to the composition of one of the primary solid phases.

A very common example of a lamellar structure is low carbon steel, which is a common structural steel used for bridges and buildings. In this case, the lamellar structure, known as pearlite, forms by a solid state reaction, on the decomposition of austenite into alpha iron and iron carbide. This is called a eutectoid reaction because the high temperature phase is a solid. But carbon is interstitial in austenite, and its diffusion coefficient is like that of a liquid, so the lamellar structure forms on a scale which is similar to that of lamellar eutectics. The transformation rate for a thermally activated transformation is shown schematically in Figure 27.6a. This can be converted this into a transformation time plot, known as a TTT plot, as in Figure 27.6b.

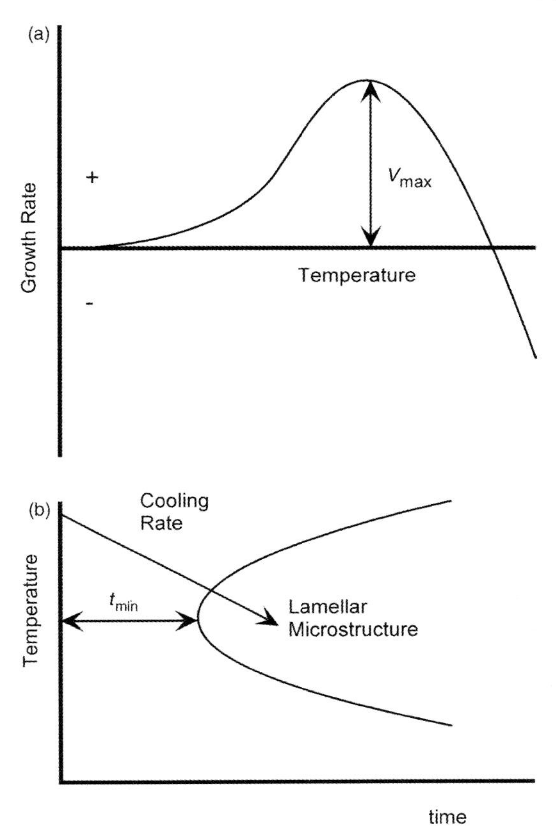

Figure 27.6 (a) Growth rate versus undercooling, and (b) the corresponding transformation time.

The fastest rate corresponds to the shortest transformation time. Since this is a solid state reaction which depends on the mobility of the iron atoms in the crystal, it is relatively slow. It is possible to quench a sample rapidly enough so that the cooling path misses the nose of the TTT curve, in which case the lamellar structure will not form, and so, at a lower temperature, a distorted, highly stressed, iron carbide structure forms. These TTT diagrams are available for many steel alloys.

27.2.2
Class II Eutectics

The second class of eutectics includes one phase which only grows with significant kinetic undercooling. Eutectics between semiconductors and metals, and between intermetallic compounds and a primary metal phase usually fall into this class. Figure 27.7 is an aluminum–silicon alloy, of the kind used for airframes in airplanes.

The silicon phase provides significant strengthening of the aluminum matrix. The growth of such a structure is illustrated in Figure 27.8.

Figure 27.7 Aluminum–silicon alloy.

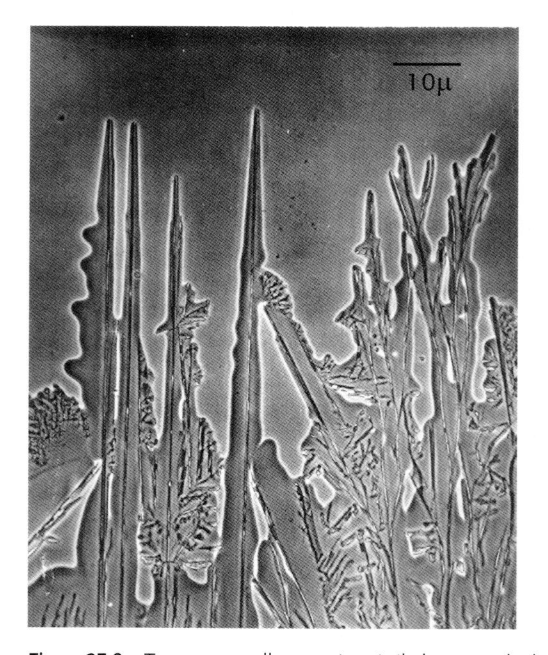

Figure 27.8 Transparent alloy, succinonitrile-borneo, which resembles the aluminum–silicon alloy.

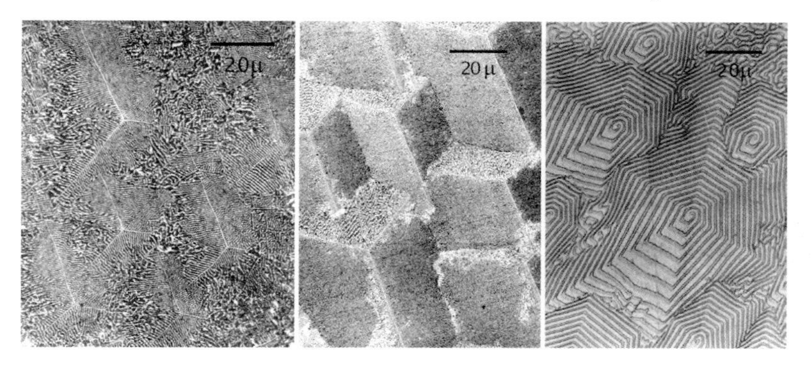

Figure 27.9 Microstructures of eutectic alloys with one primary metal phase and one high alpha factor phase, such as bismuth, or an intermetallic compound. (a) Sn–Bi; (b): Zn-Mg_2Zn_{11}; (c) Zn-$MgZn_2$.

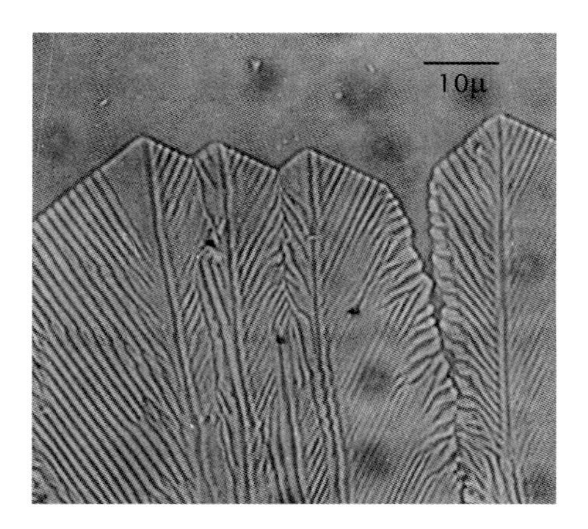

Figure 27.10 Faceted interface in cyclohexane–camphene, a Class II eutectic alloy.

The silicon grows as a rod into the liquid, and the aluminum grows preferentially into the enriched regions around these rods.

Other examples of Class II eutectics are shown in Figure 27.9.

The microstructure depends on the shape of the solid–liquid interface during growth, as illustrated in Figure 27.10.

27.2.3 Class III Eutectics

The growth of a Class III eutectic alloy is illustrated in Figure 27.11.

In both Class I and Class II the two phases can grow in a coupled manner, because a non-faceted phase can follow the growth direction of the other phase. Here, the two

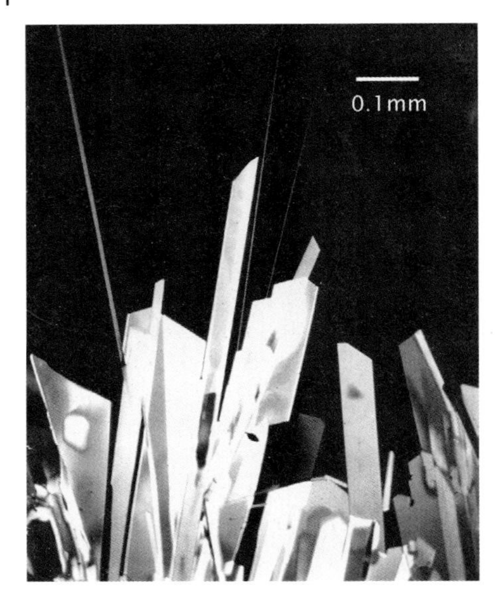

Figure 27.11 Growth of azobenzene–benzil, a Class III eutectic alloy.

phases grow essentially independently. Of course, there is an interaction between the two, because the composition around one phase is enriched in the other component of the eutectic.

Class III eutectics were studied extensively by physical chemists many years ago, because eutectics between many organic compounds are in this class. There was a period where chemists were studying the growth of Class III eutectics, and metallurgists were studying the microstructures of Class I eutectics, and the two are so different that there seems to have been no cross fertilization of ideas between the two groups.

27.3 Analysis of Lamellar Eutectics

At the interface between the liquid and the growing crystal, the total interface undercooling can be written as the sum of three terms, one due to the local composition, one due to the local curvature of the interface, and the third is the kinetic undercooling.

$$\Delta T = \Delta T_C + \Delta T_r + \Delta T_K \quad (27.1)$$

For Class I eutectics, the first two terms are much larger than the kinetic undercooling, ΔT_K. For typical laboratory growth rates, the first two will be a few tenths of a degree, while the kinetic undercooling will be less than a millidegree, so the kinetic undercooling will be ignored in the following analysis of lamellar growth.

27.3.1
Curvature of the Interface

The local curvature of the interface provides the necessary undercooling to incorporate the phase boundaries into the solid. That is, the local curvature of the interface increases the undercooling required for the transformation. The curvature of the interface depends on the shape of the interface as imposed by the angle which the interface makes in the groove at the inter-phase boundary. The average undercooling of the interface created by the curvature is exactly what is needed to incorporate the energy of the inter-phase boundary into the solid. So the very high local energy associated with creating the inter-phase boundary is spread across the interface by the curvature. This can be demonstrated as follows.

The local undercooling due to curvature along the interface between the alpha phase and the liquid is (see Equation 15.3):

$$\frac{L\Delta T}{T_M} = \frac{\sigma_{\alpha L}}{r} \tag{27.2}$$

Where $\sigma_{\alpha L}$ is the surface tension of the interface between the alpha phase and the liquid, and r is the local radius of curvature of the interface. The total free energy difference due to the curvature of, say, the alpha phase is obtained by integrating this along the interface, as illustrated in Figure 27.12.

$$\int_0^{S_\alpha/2} \frac{dx}{r(x)} = \int_0^{S_\alpha/2} \frac{\frac{d^2z}{dx^2}}{\left[1+\left(\frac{dz}{dx}\right)^2\right]^{3/2}} dx = \sin\tan^{-1}\left(\frac{dz}{dx}\right)\Big|_0^{S_\alpha/2} = \sin\theta_\alpha \tag{27.3}$$

where S_α is the width of the alpha phase lamellae, z is the interface height in the growth direction, and x lies in the plane of the interface, perpendicular to the lamellae, as in Figure 27.12.

The integral of the curvature has a simple form, which depends only on the slope of the surface at the two ends of the range of integration. It is independent of the shape of the interface in between the two ends. The average value of the σ/r from the center of a beta lamella to the center of an alpha lamella due to curvature is:

$$\left\langle\frac{\sigma}{r}\right\rangle = \frac{1}{S}\left[\int_{-S_\beta/2}^{0} \frac{\sigma_{\beta L}}{r(x)} dx + \int_0^{S_\alpha/2} \frac{\sigma_{\alpha L}}{r(x)} dx\right] = \frac{1}{S}\left[\sigma_{\alpha L}\sin\theta_\alpha + \sigma_{\beta L}\sin\theta_\beta\right] \tag{27.4}$$

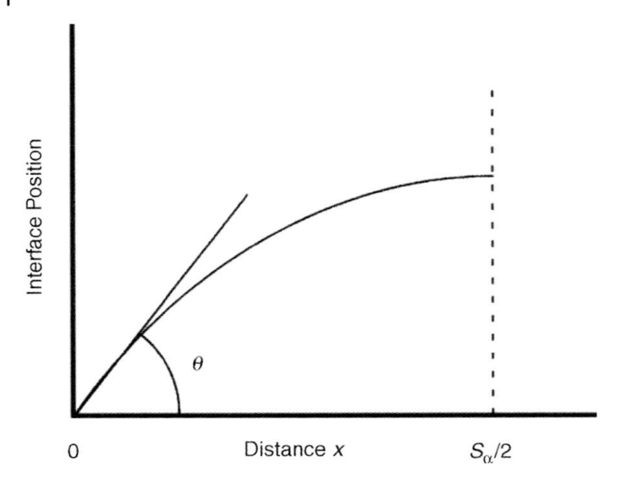

Figure 27.12 Interface shape.

Equilibration of the angles between the boundaries which meet where the phase boundary intersects the interface is illustrated in Figure 27.13.

The balance of forces giving the equilibrium configuration is:

$$\sigma_{\alpha L} \sin \theta_{\alpha} + \sigma_{\beta L} \sin \theta_{\beta} = \sigma_{\alpha\beta} \tag{27.5}$$

Combining Equations 27.4 and 27.5, the average of the local undercooling across the interface due to the curvature of the interface is:

$$\left\langle \frac{\sigma}{r} \right\rangle = \frac{\sigma_{\alpha\beta}}{s} \tag{27.6}$$

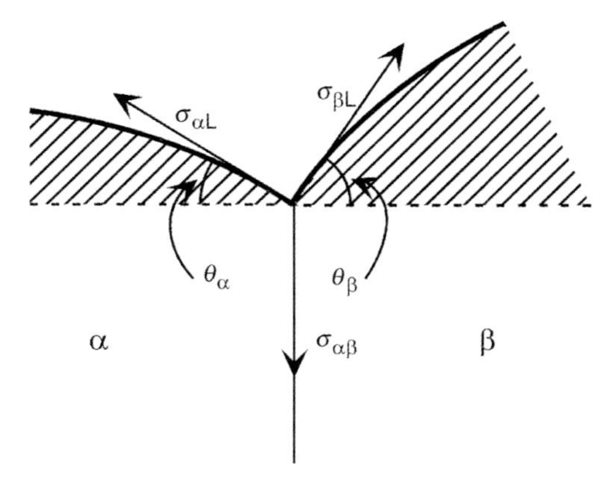

Figure 27.13 Equilibrium boundary angles.

The total undercooling due to curvature of the interface has the simple form:

$$\frac{L\Delta T_r}{T_M} = \frac{\sigma_{\alpha\beta}}{S} \tag{27.7}$$

But this is exactly the total undercooling required to grow the phase boundaries into the solid, which can be demonstrated using Figure 27.14.

In a volume $V = hjl$, where h, j, and l are shown in Figure 27.14, there are n phase boundaries, each with an area hj. The total area of interface per unit volume is thus n/l, which is equal to $1/S$, where S is the spacing between the phase boundaries. So the free energy required to grow phase boundaries with surface tension $\sigma_{\alpha\beta}$ and spaced S apart is given by:

$$\frac{L\Delta T}{T_M} = \frac{\sigma_{\alpha\beta}}{S} \tag{27.8}$$

This is identical to Equation 27.7, which was derived by integrating total undercooling due to curvature along the interface. The local curvature of the interface is required by the geometry of the phase boundaries. The local undercooling at each point along the interface, which is required by this curvature, results in a total free energy which is exactly what is needed to incorporate the phase boundaries into the solid. This is a rather neat result. The depth of the groove, and the curvature of the surface it produces, spreads out the free energy needed to incorporate the grain boundary over some area of the interface. This also works at the groove which forms for the incorporation of grain boundary into solids.

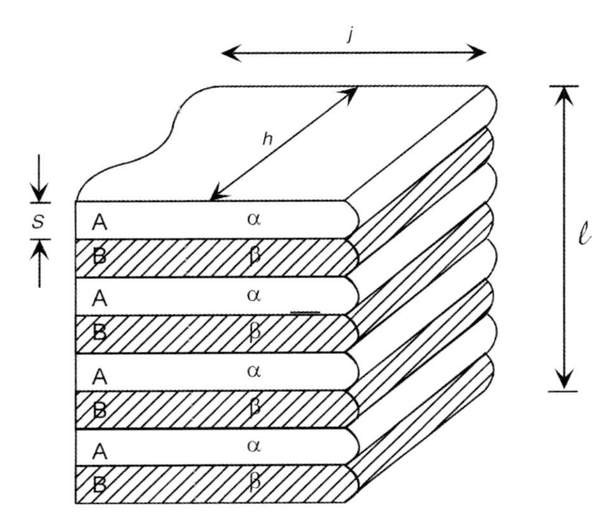

Figure 27.14 Dimensions of the lamella structure.

27.3.2
Diffusion

The alpha phase rejects B, and the beta phase rejects A. The total amount of B rejected by the alpha phase is the rate at which volume is added to the alpha phase, vSh, times the amount of B rejected by the alpha phase per unit volume, which from the phase diagram, see Figure 27.1, is $(C_{eut} - C_\alpha)$. So the total amount of B rejected is:

$$vSh(C_{eut} - C_\alpha) \tag{27.9}$$

This produces a B-rich liquid ahead of the alpha phase, and an A-rich liquid ahead of the beta phase, as illustrated in Figure 27.15.

For steady state growth, A and B interdiffuse in the liquid ahead of the interface. The concentration profile at the interface depends on this interdiffusion process. This interdiffusion process depends on the lateral flux across the plane that is a projection of the phase boundary into the liquid. The total lateral flux across this plane can be written as:

$$D\int_I^\infty \frac{dC}{dx} h\,dz \tag{27.10}$$

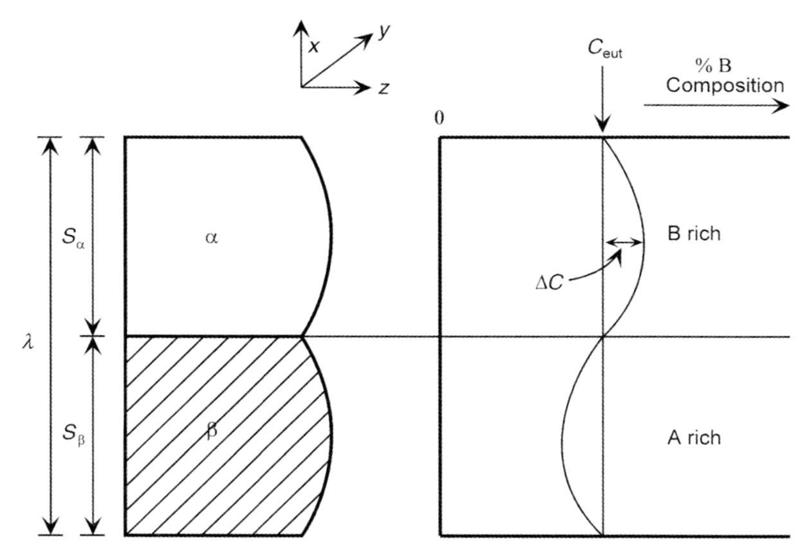

Figure 27.15 (a) The interface profile, and (b) the concentration profile in the liquid just ahead of the interface is shown on the right.

Here D is the diffusion coefficient and h is a length parallel to the lamella, in the y direction, and $\mathrm{d}C/\mathrm{d}x$ is the lateral concentration gradient in the liquid at the projection of the phase boundary. This lateral concentration gradient will be largest near the interface, and will decrease out into the liquid, extending only a distance of the order of S into the liquid. Zener [2] (a diode structure is named after him) who did this analysis for the growth of the lamellar phase (known as pearlite) in iron–carbon alloys, approximated this total flux with an average gradient $\Delta C/(S/2)$, which extended a distance S into the liquid. So he wrote the total lateral flux as:

$$D\int_{I}^{\infty}\frac{\mathrm{d}C}{\mathrm{d}x}h\,\mathrm{d}z \approx \frac{D\Delta C\,h}{S/2}S = 2Dh\Delta C \tag{27.11}$$

The B component rejected by the alpha phase goes across each of the two boundaries of the alpha phase, so half of the rejected amount given by Equation 27.9 goes across each. At steady state, the amount rejected must be equal the amount carried away by the lateral diffusion flux:

$$\frac{1}{2}v\,S\,h(C_{\mathrm{eut}}-C_{\alpha}) = 2D\,h\,\Delta C \tag{27.12}$$

The concentration of B in front of the alpha phase is thus approximately:

$$\Delta C = \frac{vS(C_{\mathrm{eut}}-C_{\alpha})}{4D} \tag{27.13}$$

Multiplying this by m, the slope of the liquidus line on the phase diagram, gives the undercooling, $\Delta T_{\mathrm{C}} = m\Delta C$ at the interface due to composition. So the total interface undercooling is:

$$\Delta T = \Delta T_{\mathrm{C}} + \Delta T_{\mathrm{r}} = \frac{mvS(C_{\mathrm{eut}}-C_{\alpha})}{4D} + \frac{\sigma_{\alpha\beta}}{S}\frac{T_{\mathrm{M}}}{L} \tag{27.14}$$

Here we have the same dilemma as we had with the dendrite analysis. Which spacing, S, goes with which growth rate v? The undercooling can be the same for a large spacing at a slow growth rate or a small spacing at a rapid growth rate. Zener [2] suggested that the transformation would proceed as quickly as possible, so that that relationship between the growth rate and the spacing could be obtained by adjusting the spacing to give the maximum growth rate at a given undercooling. This results in the condition:

$$S^2 v = \frac{4D\sigma T_{\mathrm{M}}}{Lm(C_{\mathrm{eut}}-C_{\alpha})} \tag{27.15}$$

This is usually written in terms of the lamellar spacing, λ, rather that the center-to center distance, S, which we have been using; $\lambda = 2S$.

$$\lambda^2 v = \frac{16\, D\, \sigma\, T_M}{Lm(C_{eut} - C_\alpha)} \tag{27.16}$$

A better analysis of the diffusion field in the liquid replaces the factor of 16 with 15.

Experimentally, it is observed that $\lambda^2 v$ is constant over a wide range of growth rates in many eutectic alloys. However, the numerical value of the right-hand side of Equation 27.16 does not fit very well in systems where the parameters are known. It is also observed that it is possible to change the growth rate by about a factor of two, for the same lamellar spacing. So the overall average spacing which is observed is believed to be dependent on the motion of faults or defects in the lamellar structure, rather than the maximum in the growth rate.

27.3.3 Calculation of Eutectic Interface Shape

The detailed shape of a eutectic interface for a lamellar eutectic can be calculated using the scheme outlined above [3, 4]. A very good approximation to the composition variation along the interface, as illustrated on the right-hand side of Figure 27.15 can be obtained by assuming that the interface is flat, and then solving the diffusion equation using a Fourier expansion. The undercooling is constant locally along at the interface within several lamellar spacings, since thermal diffusion is so much more rapid that compositional diffusion. The interface undercooling is taken up by the two terms, as in Equation 27.14, and so the difference between the total undercooling and undercooling due to compositional changes is due to variations in interface curvature. Starting with the groove angle at the interphase boundary, the curvature can be integrated twice to give the interface shape. The result of such a calculation [3] is shown in Figure 27.16.

27.4 Off-Composition Eutectics

The eutectic microstructure has a unique property. During solidification, the relative volume fraction of each phase can adjust to change the overall composition of the solid. The composition of the solid adjusts so that it is the same as the composition of the starting liquid. A diffusion boundary layer extending a distance D/v into the liquid forms at the interface, so that the interface composition is close to the eutectic composition. The growth proceeds at approximately the eutectic temperature. This is illustrated in Figure 27.17, which is a photograph of a sample in which the composition varies laterally, there is a temperature gradient in the vertical direction. The sample was grown up to where the lamellae stop, and then was allowed to sit in

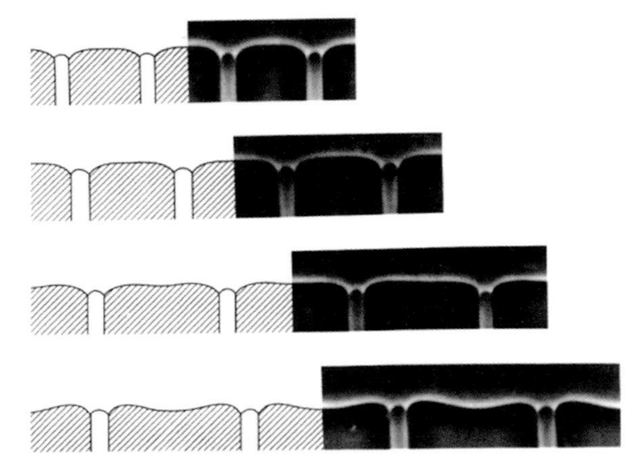

Figure 27.16 Calculated interface shapes compared with experimental interfaces in carbon tetrabromide–hexachloroethane.

the temperature gradient, so that the two primary phases grew up to the local liquidus temperature.

The relative widths of each of the two solid phases in the lamellar region changes across the photograph, which corresponds to the lateral variation in composition. The position of the eutectic growth front, where the lamellae end, varies slightly across the photograph, which indicates that the growth temperature is relatively independent of the composition.

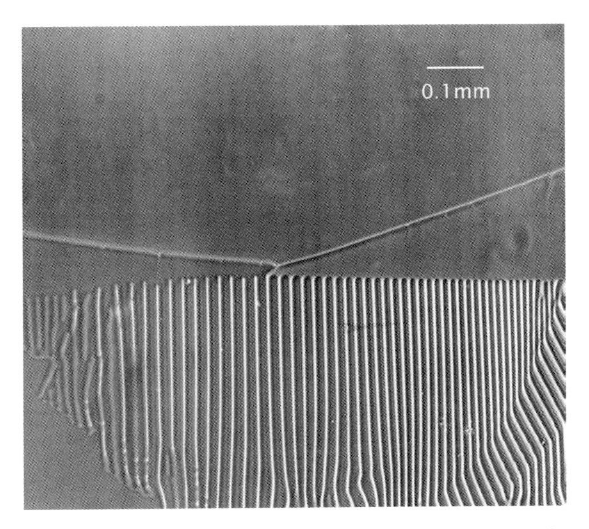

Figure 27.17 Visualization of a eutectic phase diagram in carbon tetrabromide–hexachloroethane.

27.5
Coupled Growth

There is a limited range of compositions and growth rates where the two phases of the eutectic can grow in a coupled manner [5]. Figure 27.18 shows schematically the growth rate for the eutectic, and for dendrites of the alpha and beta phases for a Class I eutectic.

The eutectic grows faster than the dendrites because the distance which the species diffuse is the lamellar spacing, usually a few microns at laboratory growth rates, as compared to a diffusion length of one hundred microns or so for a dendrite growing at the same undercooling. The growth rates into a liquid alloy which is richer in the A component than the eutectic composition are shown in Figure 27.19.

For an A-rich liquid, the melting point for the alpha phase is raised, and the melting point for the beta phase is lowered. The alpha phase can grow at a temperature above the eutectic temperature, where neither the eutectic nor the beta phase can grow, and so there is a region where the dendrites of the alpha phase will grow ahead of the eutectic, as illustrated in Figure 27.20.

In the phase diagram shown in Figure 27.21, the region in which the eutectic grows faster than the dendrites is outlined. In this region, only the eutectic grows. Outside this region, dendrites of one of the primary phases will grow ahead of the eutectic, in which case the microstructure includes regions of the primary phase, intermixed with the lamellar structure.

For a Class II eutectic, one of the phases grows much more slowly than the other, as illustrated in Figure 27.22.

Here it is assumed that the alpha phase has a small entropy of fusion so it grows rapidly, and the beta has a large entropy of fusion, so it grows slowly. Because one

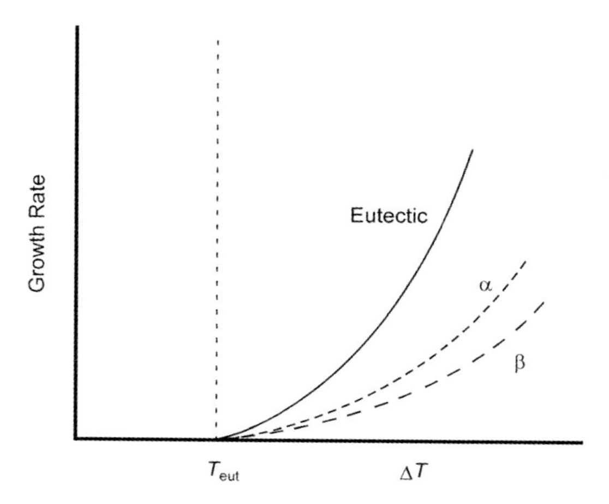

Figure 27.18 Growth rates at the eutectic composition for the eutectic, and for dendrites of the alpha and beta phases.

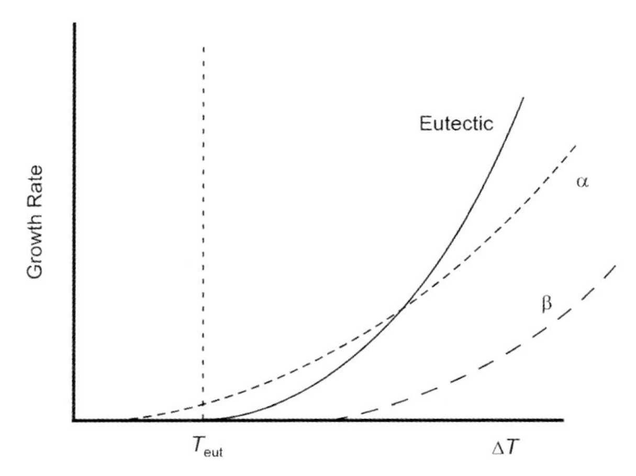

Figure 27.19 Growth rates for the eutectic, and for dendrites of the alpha and beta phases into an A-rich liquid.

phase grows rapidly and the other grows slowly, the eutectic, which involves the growth of both phases, has an intermediate growth rate, which lies between the growth rates of the two primary phases. At the eutectic composition, dendrites of the rapidly growing alpha phase will grow out into the liquid ahead of the coupled eutectic.

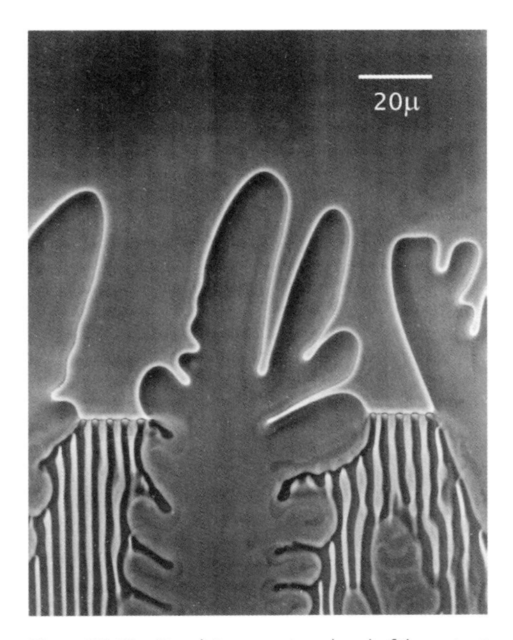

Figure 27.20 Dendrites growing ahead of the eutectic growth front in an off-eutectic-composition carbon tetrabromide–hexachloroethane alloy.

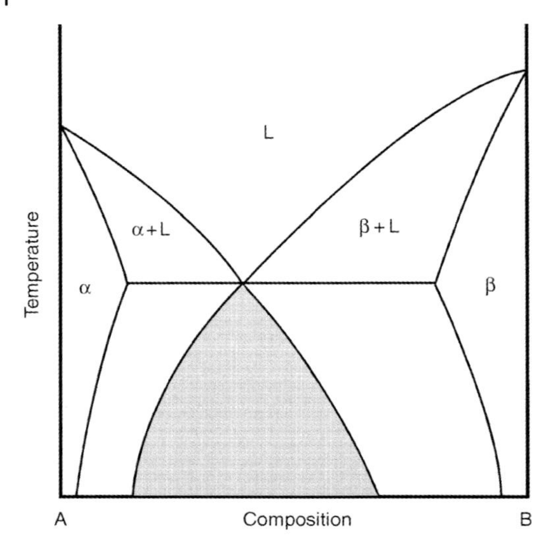

Figure 27.21 The eutectic grows preferentially in the region outlined on the phase diagram.

A coupled eutectic will grow only into B-rich liquid, only in the region where the eutectic grows faster than either the alpha or beta phases, as illustrated in Figure 27.23.

At small undercoolings, primary crystals of the beta phase can grow into the B-rich liquid at temperatures above the eutectic temperature, so there is a temperature interval where it is the only phase that can grow. At larger undercoolings, below the eutectic temperature, the coupled eutectic grows most rapidly, and at even greater

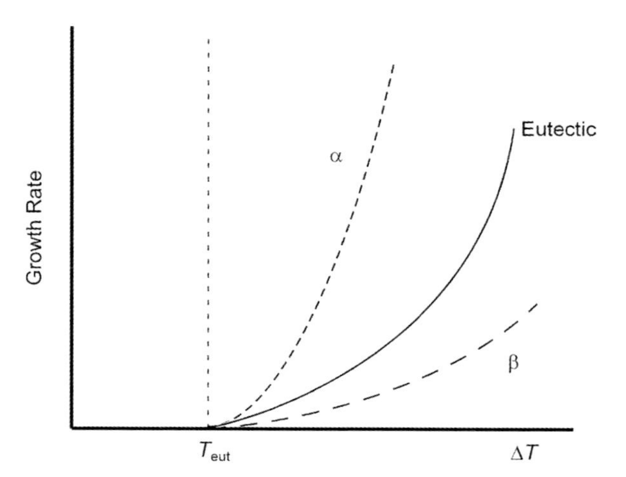

Figure 27.22 Growth rates at the eutectic composition for the coupled eutectic, for dendrites of the alpha and for primary crystals of the beta phase, in a Class II eutectic.

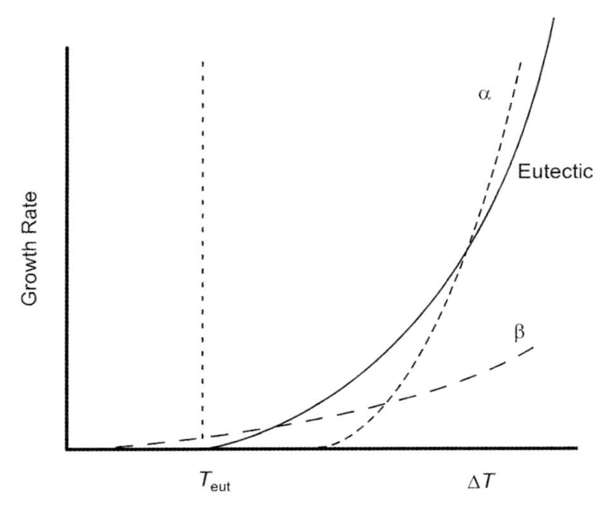

Figure 27.23 Growth rates for the eutectic, and for dendrites of the alpha and beta phases in a B-rich liquid, for a Class II eutectic where beta is the slow growing phase.

undercooling, dendrites of the alpha phase grow fastest. The region of eutectic growth for this case is sketched in Figure 27.24. For finite undercoolings, it does not include the eutectic composition.

Diagrams such as these were first constructed by Tamman and Botschwar [6, 7], and are discussed in detail by Kofler [8] who also devised a scheme for identifying

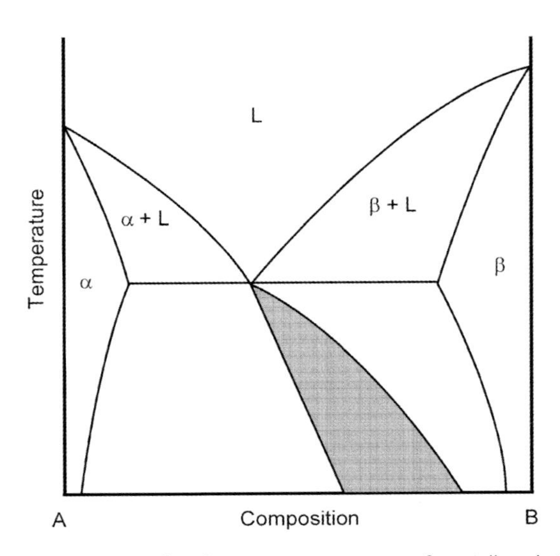

Figure 27.24 The Class II eutectic grows preferentially only in the region outlined on the phase diagram.

organic compounds based on their eutectic temperatures with a few solvents. Tamman and Botschwar measured the crystallization rate of many organic compounds, and also determined the region of coupled growth for the eutectics. The compounds which they studied crystallized with faceted interfaces, and their growth rates were sufficiently slow that they could be readily measured as a function of composition and undercooling. All of these eutectic systems were in class III. Their detailed analysis of the region of coupled growth takes into account the change in growth rate of one phase in the vicinity of the other phase due to the change in composition of the liquid there.

27.6
Third Component Elements

As can be seen in Figure 27.17, the lamellar eutectic can accommodate a range of compositions by adjusting the relative widths of the two phases. This is not true for a third component which is rejected by both primary phases. In this case, the eutectic can grow as a cellular structure superimposed on the lamellar structure, as in Figure 27.25.

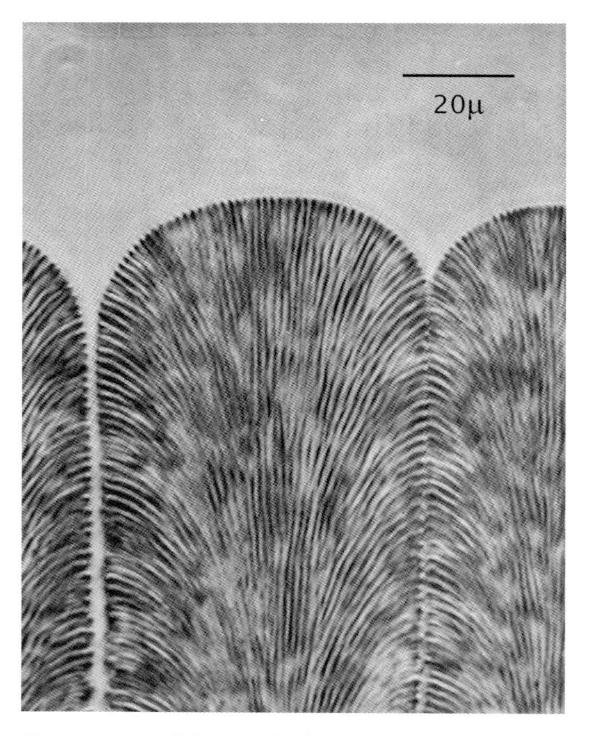

Figure 27.25 Cellular growth of a lamellar camphor–succinonitrile eutectic, with an unknown impurity.

The lamellae tend to grow normal to the growth front, resulting in the lamellar pattern shown. For larger concentrations of a third component, the structure can become dendritic. The resulting regions where the lamellae have grown in a correlated fashion, as in one of the cells in Figure 27.24, are known as eutectic colonies.

References

1 Hunt, J.D. and Jackson, K.A. (1966) *Trans. Met. Soc. AIME*, **236**, 843.
2 Zener, C. (1946) *AIME Trans.*, **167**, 550.
3 Hillert, M. (1960) *Jernkontorets Ann.*, **144**, 520.
4 Jackson, K.A. and Hunt, J.D. (1966) *Trans. Met. Soc. AIME*, **236**, 1129.
5 Hunt, J.D. and Jackson, K.A. (1967) *Trans. Met. Soc. AIME*, **239**, 864.
6 Tammann, G. and Botschwar, A.A. (1926) *Z. Anorg. Chem.*, **157**, 26.
7 Botschwar, A.A. (1934) *Z. Anorg. Chem.*, **220**, 334.
8 Kofler, A. (1950) *Z. Metallk.*, **41**, 221; (1965) *J. Australian Inst. Met.*, **10**, 132.

Problems

27.1. Discuss the conditions under which dendrites grow ahead of a eutectic growth front.

27.2. The diffusion distance at a growth front is given by the diffusivity divided by the growth rate. What is the thermal diffusion distance for a eutectic alloy growing at 1 mm min^{-1}, with a thermal diffusivity of 10^{-5} m^2 s^{-1}. What does this imply about temperature differences over a lamellar spacing at the growth front, which is typically a few microns?

28
Castings

28.1
Grain Structure of Castings

The grain structure of a typical casting [1] is illustrated in Figure 28.1.

There is a region of fine grains at the surface of the mold, which is called the chill zone. This forms when the hot liquid first comes into contact with the cold mold. The extent of the chill zone depends on the temperature of the liquid, the temperature of the mold, and the thermal properties of the two. In some cases, the chill zone can be quite extensive. In other cases, the hot liquid can remelt the first crystals to form, so there is no chill zone in the final structure of the casting.

Dendrites then grow inwards from the chill zone or from the wall of the casting, making grains which are elongated in the direction of heat flow. This is called the columnar region, because of the elongated grains. There is a competition for growing space in this region, and the grains which are oriented so that their dendrites grow straight in from the wall win the competition. So the grains in the columnar region

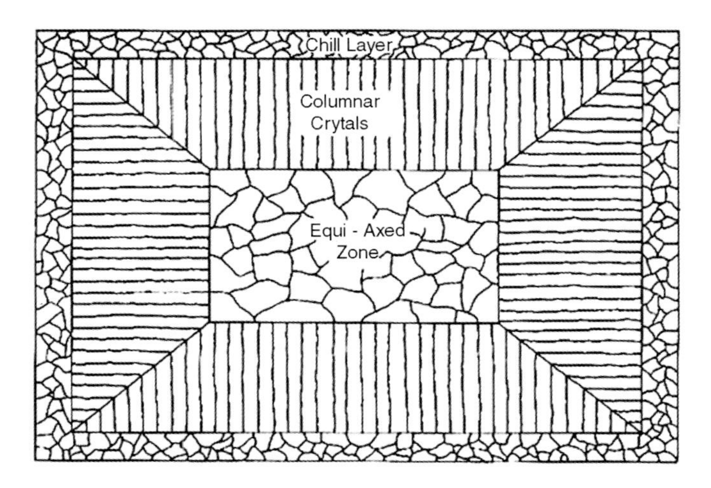

Figure 28.1 Typical grain structure of a casting. (From Hensel [2].)

Kinetic Processes: Crystal Growth, Diffusion, and Phase Transitions in Materials. Kenneth A. Jackson

ISBN: 978-3-527-32736-2

have a preferred orientation texture such that they tend to have the dendrite growth direction along their axes.

The third zone is called the equiaxed zone, because it consists of grains which have roughly the same length in each direction, and they are randomly oriented.

For many years, the origin of the grains in the equiaxed zone was a mystery. The undercooling in the center of the casting is limited to constitutional supercooling, which is not sufficient for homogeneous nucleation, or even for heterogeneous nucleation in the absence of nucleating agents. The equiaxed zone forms because the dendritic structure in an alloy can come apart due to temperature fluctuations [3].

28.2
Dendrite Re-Melting

During dendritic growth of an alloy, the main stem of the dendrite creates a layer around it which is enriched in the second component. The side branches of the dendrite must grow through this enriched layer. As a result, the side branch is richer in the second component close to the main stem, where it grows through this layer. After it grows through the layer, the dendrite grows more rapidly into undisturbed liquid, and it contains less of the second component. As illustrated in Figure 28.2, the dendrites have narrow necks where they join the main stem, and the second component is more concentrated in the neck.

Later in the growth process, the structure coarsens, and these enriched regions can re-melt, so that the arms become detached from the main stem of the dendrite, as illustrated in Figure 28.3.

The arms can also become detached if there is a temperature fluctuation. This is illustrated in Figure 28.4 where a dendrite was first grown slowly, then more

Figure 28.2 Dendrite growing in an alloy. The gray scale in the photograph depends on the concentration of the second component.

Figure 28.3 During coarsening, the arms can become detached from the main stem of the dendrite.

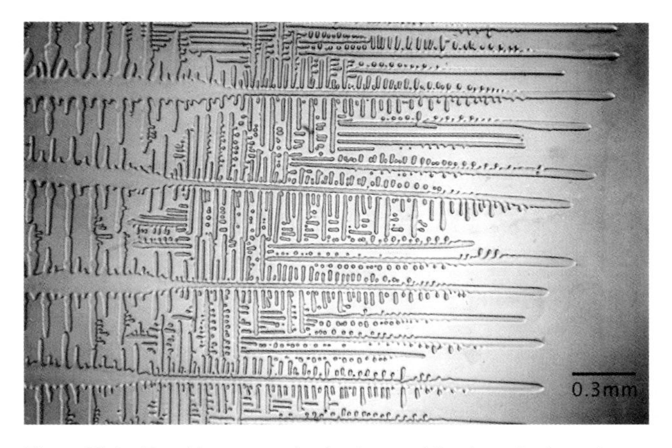

Figure 28.4 Dendrite grown slowly, then rapidly, then slowly again.

quickly by lowering the temperature, and then finally returned to the original slow growth rate.

The dendrite structure was much finer where the growth was more rapid. The fine structure coarsened quickly, and many of the dendrite side branches in this region are detached from the main stem. This process provides many detached crystals, and these crystals are carried by convection currents to the central region of the casting, where they grow into the equiaxed grains.

The process by which this happens is illustrated in Figure 28.5, which shows three stages in the solidification of an ammonium chloride–water casting.

The dark line which moves inwards progressively from the wall is the eutectic growth front. The dendritic region inside this dark line is partly liquid and partly

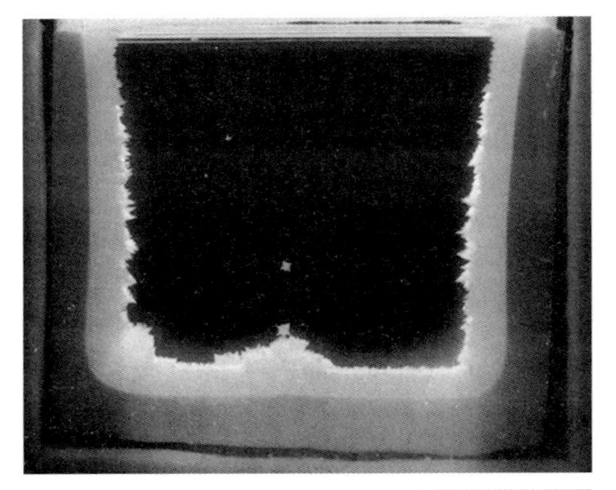

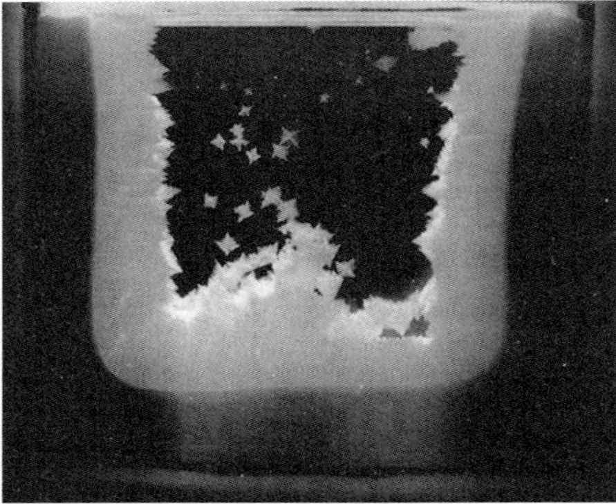

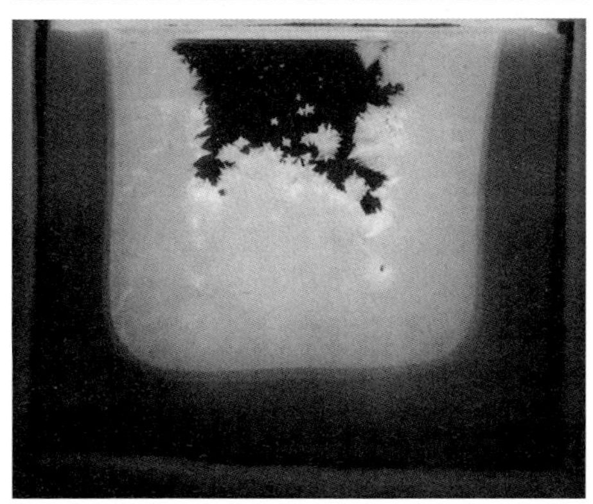

Figure 28.5 Ammonium chloride dendrites growing in a water solution in a 2 in × 2 in × 1/4 in cell. (a) Dendrites grow in from the wall in the columnar region. (b) Detached arms from the dendrites are carried into the center of the casting by convection currents, where they grow dendritically into equiaxed grains. (c) The equiaxed grains fall and collect in the bottom open space in the center of the mold.

solid, and this front is where the remaining liquid freezes as a eutectic, as in Figure 27.20.

In the modeling of castings [4], the regions which are partly liquid and partly solid are called mushy zones, a term which completely ignores the beautifully complex dendrite structure which is there.

The blades for gas turbine engines are made from nickel-based super alloys, which have excellent high temperature creep resistance as well as a reasonable ductility. These turbine blades are made by directional cooling of the casting, so that the whole structure is in the columnar growth region. The mold structure is usually designed so that only one dendrite grows up through the casting. These are termed single crystal blades, even though they contain many small angle boundaries where the dendrite branches meet as the interdendritic liquid finally freezes. The critical point is that there are no lateral grain boundaries, because these provide sites for thermal cracking and corrosion.

The final structure of a casting, including the grain size and distribution, micro-segregation, porosity, and macro-segregation due to the convection depend on the configuration of the casting and how heat is extracted from the mold [5–7]. Specialized computer programs are used to design the mold, the location of vents, and the thermal conditions during solidification.

References

1. Flemings, M.C. (1974) *Solidification Processing*, McGraw-Hill, New York.
2. Hensel, F. (1937) *Trans AIMME*, **124**, 300.
3. Jackson, K.A., Hunt, J.D., Uhlmann, D.R., and Seward, T.P. III (1966) *Trans. Met. Soc. AIME*, **236**, 149.
4. Poirier, D.R. and Heinrich, J.C. (1994) *Mater. Charact.*, **32**, 287.
5. Allen, D.J. and Hunt, J.D. (1979) *Metall. Trans. A*, **10**, 1389.
6. Burden, M.H. and Hunt, J.D. (1975) *Metall. Trans.*, **6**, 240.
7. Burden, M.H. and Hunt, J.D. (1974) *J. Cryst. Growth*, **22**, 99.

Problems

28.1. Prepare a report on “freckles” in castings.

28.2. Prepare a report on macro segregation in castings.

Subject Index

Index by Page

Kinetic Processes: Crystal Growth, Diffusion, and Phase Transitions in Materials. Kenneth A. Jackson

ISBN: 978-3-527-32736-2

e

q

r

Subject Index

Index by Chapter Sections

Kinetic Processes: Crystal Growth, Diffusion, and Phase Transitions in Materials. Kenneth A. Jackson

ISBN: 978-3-527-32736-2